# PEDIATRICS BOA
# MNEMONICS, FR
# REVIEW QUESTIONS & A FIRST-TIME PASS GUARANTEE!

***Efficiency In Studying So That You Can Start Living!***

***Written By Ashish Goyal, MD***

***www.PediatricsBoardReview.com***

## COPYRIGHT INFORMATION



ISBN 978-1-300-45528-8

# MEMBER REGISTRATION

## *WHY REGISTER?*

*** MONEY BACK PASS GUARANTEE:** Submission of this form is required for the money back pass guarantee. It must be submitted within 90 days of your purchase and before you take the exam. This helps PBR release a certain amount of funds for use. As mentioned on the PBR site, this guarantee applies to anyone taking an ABP initial or recertification exam for the first-time. The original PBR purchase must have been made at least 45 days prior to the exam. "Money Back" requests may be made within 180 days of the score release date. For complete details, please visit: http://www.pediatricsboardreview.com/guarantee

*** FREE UPDATES**: I often create last minute e-book bonuses (previous examples include the lab norms section, intrauterine drug exposures, a corrections and errata guide, etc.). These are given to existing members of the PBR community for FREE. If similar tools are created, they'll be sent to you for FREE as well.

*** CORRECTIONS**: Registering will ensure that you are notified of any major corrections.

## *FORM*

PDF download at: www.PEDIATRICSboardreview.com/download/registration.pdf
DOC download at: www.PEDIATRICSboardreview.com/download/registration.doc
ONLINE: www.PEDIATRICSboardreview.wufoo.com/forms/z7x3k1/
PRINT: (*Denotes Required)
*Full Name: ______________________________________________
*Email Address: ___________________________________________
*Product(s) Purchased & Purchase Price

- BUNDLE PACK. PRICE: ______
- HARDCOPY - CORE STUDY GUIDE. PRICE: ______
- ONLINE - CORE STUDY GUIDE. PRICE: ______
- HARDCOPY – PBR Q&A. PRICE: ______
- ONLINE – PBR Q&A. PRICE: ______

*Purchase Date: __________
*I heard about PBR through: ______________________________________
*I purchased PBR to help me study for:

- The ABP Initial Certification Exam
- The ABP Recertification Exam
- Residency training
- Something else (please specify): ______________________________

*I have:

- Never taken this exam before
- Have taken it _____ times
- No plans on taking any standardized exam in the near future.

*I will be taking the exam on __________ (please fill in the month/year, or write "n/a")
*In an ideal world, PBR would create the following resources to help me with my studying:

______________________________ ______________________________
______________________________ ______________________________
______________________________ ______________________________
______________________________ ______________________________

**IF YOU PURCHASED A HARDCOPY PRODUCT THROUGH LULU.COM, AMAZON.COM, ETC., YOU MUST ALSO SUBMIT A COPY OF THE RECEIPT.**

## *SUBMISSION*

*** EMAIL:** Type or scan your info and email it to registration@pediatricsboardreview.com
*** FAX:** FAX THIS FORM TO (775) 854-4637. **Please verify the fax was received.**

# INTRODUCTION TO PBR

## ***ABOUT THE PBR CORE STUDY GUIDE***

Hi. My name is Ashish Goyal and I've written the PBR core study guide to provide you with all of the information you need to pass the American Board of Pediatrics initial certification exam. There has been a very strong effort on my part to use easy to understand language, and to provide you with hundreds of memory aids. This **will** significantly cut down on the amount of time you need to spend on each topic. Your efficiency will allow you to go through the material multiple times and feel confident on the test day that you are well prepared to take the exam.

Before you begin your studying, though, I'd like to share some key pieces of information with you. I took the ABP initial certification exam the year that I graduated residency. I used multiple study guides to prepare and felt "okay" going into it, and felt nauseous coming out of it. I was possible that I had failed a board exam for the first time in my life. I went home and made notes about how I would do things differently if there was a "next time." The passing score at the time was a 410 and I scored a 380. **I failed by 7-9 questions**. The following year I made key strategy changes based on my previous experience. I studied for hundreds of hours as though it was a fulltime job. **I took the test again and felt amazing this time. My score increased by 160 points up to a 540! I passed by 37-39 questions**I was really just happy to pass. **Failing the first time had cost me extra time, money and energy that I would have preferred to spend with my loved ones.**

Prior to writing the PBR core study guide, I didn't really like to share the fact that I had to take the exam twice. The reason I've decided to share this information with you is to hopefully gain your trust in my ability to help you pass this exam based on my own experience. **I'M AN AVERAGE PERSON WHO DID EXTREMELY WELL ON THE EXAM.** I enjoy teaching and I've been able to incorporate my studying techniques and tools into the PBR system. It's easy to follow, focused, efficient, has had GREAT success rates, and is **guaranteed to help you pass**. **All you have to do is FOLLOW THE ROADMAP.**

## *YOUR PBR ROADMAP TO SUCCESS*

If you're one of the many new pediatricians out there wondering if you're going to be in the passing 75%, or the failing 25%, I urge you to follow just a few simple, yet HIGH-YIELD recommendations:

* **PLEASE stick to one primary study guide... this one!** The ***BIGGEST MISTAKE*** you can make is spreading yourself too thin. That's an extremely common trend I've noticed in my discussions with numerous pediatricians looking for advice on how to avoid failing again. Do not spend your time going through other books, DVDs, or live courses. **Go through the PBR core study guide at least 4-5 times.**

- Start with a very fast "run through" in which you get familiar with the content and with the writing style. **Do not spend time cross-referencing** with Nelson's, the Red Book or Harriet Lane. If you have questions, write them in the margins and work on the answer next time.
- The **second time,** answer the questions written in the margins. Do NOT let your curiosity of non-PBR topics distract you (it's a guaranteed waste of precious time).
- The **third, fourth and fifth times** through the PBR, you should be adding more and more information into your long-term memory through repetition and the use of mnemonics. Again, **you must resist that urge** to look up extraneous information.

Your primary goal should be to pass the exam. If you know everything in the core study guide and are a descent standardized test taker, you **will** easily pass. If you try to learn everything in pediatrics, you will get overwhelmed and possibly fail the exam. Also, I spent over well over 300 hours studying for the boards. Try to map out at least **200 hours of studying**.

* **Use PBR's Q&A product(s) to get familiar with very high-yield topics and questions**. The format is short and to the point so that you learn the information you need to without wasting too much time digging through extra information. The questions will help you understand what types of key findings you need to be on the lookout for during your test prep and during the test.

* **Go through about 3 years of PREP® questions.** Please don't go through them all at once. Go through about 25-50 questions every 2-3 days and aim to finish them by the end of September. As you go through the questions, **work on your timing**. If you can average about 1 minute and 15-25 seconds per question, you should be fine for the boards. Do not try to understand why every single incorrect answer is wrong. Just focus on the correct answer, and also on the answer that you chose. **Remember, the AAP writes PREP®, the ABP writes the boards.** PREP® is great, but it's for **CME**. It is not designed to be a study guide for the ABP. So it's a good practice tool, but your highest-yield information will come from this study guide and system.

* **EXTREMELY Important Test Day Tips**: http://www.pediatricsboardreview.com/category/test-day-tips

## *GETTING THE MOST OUT OF THE PBR FORMAT*

* **GRAY HIGHLIGHTING** OR RED FONT: For the PBR hardcopy, gray highlighting is used over a word, phrase or heading that you should KNOW! Those are very high yield topics or have at least been on board exams in the past. In the online PBR, such text will appear in red.
* **PEARLS**: These are pieces of information that help tie key concepts together for you.

- PEARL: There are only a finite number of ways the ABP can test you on a disease process. In some PEARLS, I try to show you ways how you could be presented with information on the exam.

* **MNEMONICS**: These are memory aids designed to increase your efficiency as you go through the study guide. Some of them will work for you, others will not. My mnemonics will give you a feel for how to create memory aids so that you can then create and personalize your own. When creating a mnemonic, use as much action and color as possible. Also, the crazier it is, the more likely it is to stick in your mind.
* (**DOUBLE TAKE**): This means a topic is presented in multiple areas of the book. This is done to limit the number of times you have to bounce around the guide. Most of these topics tend to be very high-yield.
* **NAME ALERTS**: Some disease names are extremely similar, and thus confusing (e.g. Condyloma Lata versus Acuminata, or Shwachman-Diamond Syndrome versus Diamond-Blackfan Anemia). NAME ALERTS serve as reminders to look for these subtle differences.
* ABBREVIATIONS: Some of the disorders are discussed using their abbreviations, other are discussed with their proper name. Please keep that in mind if you are using the online study guide. This is especially important when doing a search on the PBR site.

## ***PBR & AVSAR – THE NONPROFIT CONNECTION***

**WHAT IS AVSAR?** AVSAR is a nonprofit organization I started at the age of 27 to help support other nonprofit organizations that are already doing phenomenal work in the slums of India.

AVSAR stands for the Alliance of Volunteers for Service, Action and Reform. After graduating medical school I spent 1 year volunteering in the slums of Mumbai. The need for help was astounding and conditions were shocking. Six-year-old children worked as child laborers. Their small, agile fingers apparently allow for detailed handwork, such as the creation of stunningly beautiful tapestries. Others spent their days digging through garbage dumps and landfills looking for metal or plastic recyclables that would be sold for about $1 - $2 per day.

During my year of volunteering, I bonded with these children and was compelled to create AVSAR, a 501-c-3 nonprofit organization under the U.S. IRS. AVSAR quickly recruited over 100 volunteers from around the world to do the same thing I did, volunteer in India to work on some amazing projects. We recruited volunteers from all walks of life (college students, dentists, doctors, medical students and even MBA students). My personal success stories included the creation of an efficient Western-style clinic for child laborers in partnership with the Niramaya Foundation, the establishment of an adolescent sex education curriculum and the creation of AVSAR, a nonprofit organization that I believe has the power to effect tremendous change throughout India.

AVSAR has helped thousands of people, and it's likely that the systems created through the organization continue to help thousands more every year. The core volunteer and assistance program was shut down in my last year of residency because of a lack of funding and my competing responsibilities. The creation of PBR has been integral in helping re-launch the organization. **Through PBR I have been able to generate over $50,000, all of which went to the nonprofit organization. To date, I have contributed over $70,000 to AVSAR and hope to officially re-launch the organization in late 2013 or early 2014.** My goal is to make the new AVSAR bigger and better than ever before. While PBR has been transitioned to a regular personal business and is no longer a nonprofit project or company, it's great to know that if AVSAR needs help, PBR and I will be there.

If you'd like to learn more about AVSAR, please contact me by email or through the PBR website. You can learn more about AVSAR at www.avsarindia.org.

If you would like to make a donation, your generosity would be greatly appreciated.

**DONATION OPTIONS**

* **CHECK/MONEY ORDER**: Please mail a check payable to AVSAR, Inc. to the address below.
  AVSAR Donation
  P.O. Box 870835
  Stone Mountain, GA 30087

* **ONLINE DONATIONS: http://www.avsarindia.org/new/Donate%20now.htm**

## *MEMORY AIDS – PEGS AND MNEMONICS*

**MNEMONICS**: Mnemonics are memory aids that assist in helping you recall something. They are used throughout the study guide in an effort to help you study in a more focused and efficient manner.

**PEGS:** Memory "pegs" are typically used to help you remember a list of items. By having 20 pre-memorized pegs that represent the numbers 1-20, you can easily "peg" items to those numbers. For example, if you are memorizing a grocery list of 10 items and the 9$^{th}$ item is a gallon of milk, you could imagine a BLACK CAT drinking from a bowl of milk. In the system below, CAT symbolizes the number 9 since cats are said to have "nine lives." Some of the pegs below will be used in high-yield mnemonics. Go through them a few times to see if you can get the hang of it.

**NOTE:** Some of the peg associations below may not work for you. If you have ideas on how to make them better, please do so for yourself, and consider sharing them with the PBR community.

**TWENTY PEGS**

| # | PEG | EXPLANATION |
|---|---|---|
| 1 | TREE TRUNK | Imagine the number 1 looking like a huge, brown tree trunk with limbs full of green foliage sitting at the top of a lush, green hilltop. |
| 2 | LIGHT SWITCH or LIGHT BULB | A light switch/bulb has 2 positions (ON & OFF) |
| 3 | STOOL | Imagine a dark, cherry wood stool with 3 legs. |
| 4 | CAR | Cars have FOUR doors and FOUR wheels |
| 5 | GLOVE or HAND | A glove has 5 fingers. Consider making MJ's shiny glove your symbol for FIVE |
| 6 | GUN | Another name for a gun is a 6-shooter since they used to only hold 6 bullets. GUNS also kill people and put them "6 feet under" the ground. |
| 7 | DICE or CARDS | Lucky number 7! Think Vegas, think craps or gambling with dice or cards |
| 8 | ICE SKATE | Ice skaters are known for performing a move called the figure 8. Eight also rhymes with skate. |
| 9 | CAT | Ever heard of the phrase, "cats have 9 lives." |
| 10 | BOWLING BALL or PINS | The goal of bowling is to knock down 10 pins. |
| 11 | AMERICAN FOOTBALL or GOAL POST | In American football, a field goal occurs when a football is kicked through two, white, vertical uprights (the goal post). A goal post which look like the number 11. |
| 12 | EGGS | Eggs usually come in a carton that contains "a dozen eggs." |
| 13 | HOCKEY MASK | Unlucky number 13 and the unlucky day/movie "Friday the 13$^{th}$." The main character in the movie, "Friday the 13$^{th}$" is Jason. A hockey mask wearing killer. |
| 14 | ROSE or CHOCOLATE HEART | February 14$^{th}$ is Valentine's Day! So think of a long-stem, red ROSE or perhaps a big CHOCOLATE HEART |
| 15 | PAYCHECK | Every year, your TAXES are due on APRIL 15$^{th}$ and the IRS takes a huge chunk out of your PAYCHECK. Welcome to healthcare. |
| 16 | DRIVER'S LICENSE | Age at which you get a driver's license. Other considerations: CANDLES, CANDY or a BIRTHDAY CAKE for "Sweet SIXTEEN" |
| 17 | MAGAZINE | There is a teen magazine called "SEVENTEEN" |
| 18 | VOTING BOOTH | Age when you become legal in the U.S. and are allowed to vote. |
| 19 | KNIGHTING | Imagine a "KNIGHTING" ceremony |
| 20 | CIGARETTES | A pack of cigarettes has 20 cigarettes in it. |

There are TONS of mnemonics throughout PBR. Many will seem brilliant. Others may not work for you at all. If that happens, please CREATE YOUR OWN. It's initially intimidating, but gets much easier with time.

### OKAY, ARE YOU READY TO GET STARTED WITH THE PBR? LET'S GO!!!

## Table of Contents

# ADOLESCENT MEDICINE

## ***PUBERTY***

**NOTES**: Please note that there is a great deal of overlap and repetition between the puberty section and the Endocrinology section.

* Know conversion from inches to cm. 1 inch is about 2.5 cm!

* Sexual Maturity Ratings (SMR) and Tanner Staging begins with ONE. There's NO ZERO. Tanner/SMR 1 = Prepubertal

* Experts disagree regarding some SMR descriptions. They definitely can't agree on the age at which Delayed Puberty is diagnosed. Don't stress! Questions on the exam should be fairly clear.

### NORMAL PUBERTY TIMELINE

| SMR | Girls<br>Delayed Puberty: 13 – 14 yo Precocious Puberty: 2' signs before 8 yo | Boys<br>Delayed Puberty: 14 – 15 yo<br>Precocious: 2' signs before 9 yo |
|---|---|---|
| 1 | **Basal growth at 5-6 cm/yr**, boyish chest (papilla elevation only), no hair | < 4 ml volume or 2.5 cm diameter of testicle, no hair, baby penis, still **basal rate of 6-8 cm/yr**, no hair |
| 2 | **Accelerated growth at 7-8 cm/yr**, a breast bud is the 1st sign of puberty (palpable, areola enlarges), Hair only along the labia (coarse) | **>4 ml or >2.5 cm** (this is the 1st sign of puberty), hair at base of penis. Penis *may* start to enlarge (usually at SMR 3) |
| 3 | **PEAK ht velocity of 8-10 cm/yr**, elevation of breast contour, areola enlarges, curly hair at pubis, axillary hair begins, acne. This stage is similar to a boy's SMR 3 + 4 combined. "Imagine a girl sitting on a 3-LEGGED STOOL crying because she has hair in her armpit and now has acne!" | **Accelerated vertical (and penile) growth** >12 ml/3.5 cm, Gynecomastia in 50% boys 10-16 yo, resolve in 3 yrs), CURLY hair at pubis. "Think about the 3 Stooges. They all had funny pubertal voices, and the fat one had BOOBS/GYNECOMASTIA and was named CURLY." |
| 4 | **Mound on mound**, enlarged areola. Dense hair, none at the thigh. Menses usually occurs around SMR 3 or 4. | **PEAK height velocity at 10 cm/yr**, no thigh hair, develops **AX**illary hair, acne, and body odor. "Teenage boy with raging hormones is pissed about acne & hair so takes an **AX** to his **4 DOOR CAR** (SMR **4**) which explodes and burns his hair!" |
| 5 | Stop growing at about 16 yo, areola recesses to general contour of breast and the **breasts again look like tanner 3** | >4.5 cm penis, thigh hair, stop growing at ~17 or 18, +facial hair at sides, no more gynecomastia |

## NORMAL PUBERTY PEARLS

Here are some great pearls and shortcuts about normal puberty.

* Girls have adult looking breasts in SMR 3 and 5

* SMR 4 = mound on mound breasts

* SMR 2 to 5 usually lasts about 3 to 5 years in total duration for both sexes

* MENARCHE usually occurs in SMR 3 or SMR 4 (more likely 4), OR within in 2-3 YEARS of the onset of puberty. Amenorrhea does not require workup until 2 years after puberty has ended. Since puberty may take 5 years to complete, it's possible a patient may not need a workup for amenorrhea until 7 years after their breast buds form.

* MENSES/HEIGHT: At the onset of menses, girls are probably within in 1-2 inches (2.5 – 5 cm) of their adult height. Why do I say that? Because they're probably in SMR 4 (which occurs **after** the peak height velocity).

* VAGINAL BLEEDING: Bloody vaginal discharge while in SMR 2 shouldn't happen. Consider a foreign body (e.g. toilet paper) in your differential.

## HEIGHT

For the test, pre-pubertal basal rate for height in both boys and girls is 5-6 cm/year. The peak is 10 cm/yr. Early puberty results in shorter adult height

## GROWTH SPURTS

Elevated alkaline phosphatase can be normal during growth spurts. Hematocrit increases alongside growth spurts

## THELARCHE, ADRENARCHE THEN MENARCHE

(THELARCHE) Breast development → (ADRENARCHE) Hair development → (MENARCHE) Menses

**PEARLS AND MNEMONICS**: "Girls are TAMer than boys." "Boys like to TAP Her!"

- Girls are "TAMer" = Thelarche, then Adrenarche, then Menarche = Breast development → Hair development → Menses. Thelarche ='s first sign of puberty – stage 2. Adrenarche is the same thing as Pubarchy. Breasts: Look most natural at SMR 1, 3, and 5. TAM = "Breasts are higher than Pubic hair which is higher than a Vagina."
- Boys = "TAP Her" = Testicular enlargement, then Adrenarche, then Phallus/Penile enlargement, THEN Height velocity peaks

## (DOUBLE TAKE) AGE RANGE OF NORMAL PUBERTY

Age ranges for puberty = 8 to 13 for girls, and 9 to 14 for boys. If puberty begins at those age ranges, that is **ok**, but **before** that is precocious puberty and **after** that is delayed puberty.

**MNEMONIC**: "Imagine Reece Weatherspoon being pissed because she hasn't hit puberty yet! She puts on a HOCKEY MASK and ICE SKATES and then knocks a huge CHOCOLATE HEART out of the hands of TOMMY CREWS with her hockey stick. He falls backwards and lands on a white CAT! Now they're both upset and crying like little children!" Those are the shortest celebs I could think of to represent pre-pubertal kids. Hockey Mask and Ice Skates = Jason from Friday the 13th and Skates = 8. Those are for the normal age range for puberty in girls. Chocolate heart and Cat = 14 (Valentine's Day) & 9 (nine lives) for the normal age range for puberty in boys.

## ESTROGEN

**Estrogen causes the development of Breasts** + Change in vaginal color + Labial Prominence.

## ANDROGENS

**Androgens cause pubic hair development.** So when you think of pubic hair, think of an androgen related issue.

**PEARL**: If you're presented with an adolescent "girl" with breasts but NO pubic hair, guess what? She's NOT A GIRL! Think ANDROGEN INSENSITIVITY (aka TESTICULAR FEMINIZATION) in someone carrying XY chromosomes.

**PEARL**: If presented with an adolescent female with a history of pubic hair development, but no history of preceding breast development, the patient likely has ANDROGEN EXCESS or LOW ESTROGEN.

**PEARL**: If adolescent presents with isolated PREMATURE ADRENARCHE (pubic hair with no breasts or no increase in testicular size), get bone age films! If the bone age films are within 1 year of the chronologic age, it's OK TO OBSERVE. If not, the patient will need an Endocrinologist to intervene.

**PEARL**: "Girl" with no breasts (or just buds), no hair or only scant hair, no menses → TURNER'S SYNDROME → Get a karyotype.

**PEARL**: Always question whether or not the girl in your question is truly a normal XX girl without any hormonal issues/deficiencies.

## BREAST MASSES – FIBROADENOMAS AND FIBROCYSTIC DISEASE

Breast masses in children are usually benign and due to fibrocystic disease or fibroadenomas. Check the masses at end of menstrual periods. Mammography is needed until patients are much older since their breast tissue is dense. To evaluate, use ULTRASOUND. Excisional biopsy is almost never indicated (only if aspirate is bloody).

* **FIBROCYSTIC DISEASE** is the most common breast disease and is usually bilateral and tender. OCPs may help.

* **FIBROADENOMAS** are unilateral and ESTROGEN dependent (bigger with OCPs/pregnancy). Refer to gynecology if there is bloody aspirate or if it persists beyond after 3 cycles

# ***PUBERTY GONE HAYWIRE***

## (DOUBLE TAKE) AGE RANGE OF NORMAL PUBERTY

Age ranges for puberty = 8 to 13 for girls, and 9 to 14 for boys. If puberty begins at those age ranges, that is **ok**, but **before** that is precocious puberty and **after** that is delayed puberty.

**MNEMONIC**: "Imagine Reece Weatherspoon being pissed because she hasn't hit puberty yet! She puts on a HOCKEY MASK and ICE SKATES and then knocks a huge CHOCOLATE HEART out of the hands of TOMMY CREWS with her hockey stick. He falls backwards and lands on a white CAT! Now they're both upset and crying like little children!" Those are the shortest celebs I could think of to represent pre-pubertal kids. Hockey Mask and Ice Skates = Jason from Friday the 13th and Skates = 8. Those are for the normal age range for puberty in girls. Chocolate heart and Cat = 14 (Valentine's Day) & 9 (nine lives) for the normal age range for puberty in boys.

## PRECOCIOUS PUBERTY

If you feel someone has precocious puberty on the exam → Rule out TESTICULAR OR OVARIAN TUMORS. LOOK at, and FEEL the size/consistency of your patient's gonads. In girls, do a pelvic ultrasound, and in boys, FEEL THE NUTS. Also need to get LH and FSH to look for a central disorder. In boys, isolated LH elevation may be present. For boys, pubarche (adrenarchy/hair growth) + an enlarged phallus without testicular enlargement means there is the presence of excess androgens from outside the normal gonads. Remember, testicular enlargement is the FIRST sign of puberty in boys, so if other signs of puberty exist without testicular enlargement… something is wrong!

* ULTRASOUND is useful to look for adrenal or ovarian masses.
* CENTRAL VS PERIPHERAL: Get LH, FSH and Adrenal Steroid levels to help differentiate

## PSEUDOPRECOCIOUS PUBERTY

If everything is normal except for one abnormality and a good workup has been done, consider the possibility of **PSEUDO**PRECOCIOUS PUBERTY, which means that puberty is occurring **out of order**. Example: Breast development followed by the onset of menstruation, but without pubic hair. This is usually due to sex steroid production by adrenals, ovaries, or testicles.

## ADRENAL ANDROGENS

Adrenal androgens cause body odor, acne and hair development. Etiology of ACNE → androgens. The term **adrenarchy** = hair.

## PREMATURE ADRENARCHE

Premature adrenarche is common in girls. Parents bring them to the office quickly because they are concerned about their hairy/mannish princess. It's usually not a big deal. In boys, it's VERY concerning, but boys are unfortunately NOT brought to the office often enough because parents think boys are supposed to be hairy! If it's an adrenal source, the diagnosis is likely CONGENITAL ADRENAL HYPERPLASIA (CAH) rather than an adrenal tumor.

## CONGENITAL ADRENAL HYPERPLASIA (CAH)

In congenital adrenal hyperplasia (CAH), there is a cortisol and aldosterone manufacturing problem in the adrenal glands. Negative feedback results in high levels of ACTH being released from the pituitary glands → Results in an increase in cortisol precursors → Resulting in more ANDROGENS. (More details elsewhere)

**NORMAL ADRENAL STEROID SYNTHESIS**

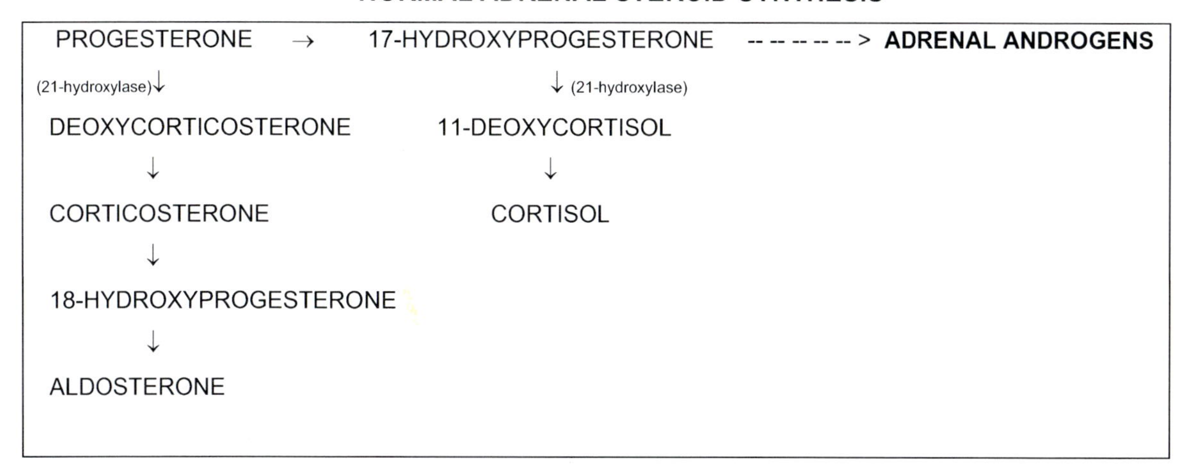


## TROPIC

TROPIC refers to central hormones. Break down big words. So HYPERGONADOTROPIC refers to an excess of central hormones

## PREMATURE THELARCHE

Premature thelarche is defined as thelarche prior to the **age of 4.**

**PEARL**: You might hear a parent say, "my baby had boobies since the day she was born!" This is NOT a big deal unless she's also having menses.

**PEARL**: If 4 years old or older, look out for excess estrogens in the form of exogenous estrogens, an estrogen-secreting tumor, or early activation of the hypothalamic-pituitary axis.

## PREMATURE ADRENARCHE IN GIRLS

Premature adrenarche is defined as having hair development without breast development, or other signs of puberty prior to the **age of 8**. Get bone age films! If they are within 1 year of the chronological age, it's ok to observe at this age!

**PEARL**: If 8 year old or greater, look out for extra androgens in the form of exogenous androgens (oral? topical?), ANDROGEN-SECRETING TUMOR, CAH, or EARLY ADRENAL PUBERTY. ADRENAL glands are typically responsible for the ANDROGENS that result in ADRENArche. Remember that LH and FSH are gonadoTROPS and are from the pituitary, not released from the adrenals or gonads. However, elevation in LH and FSH **can** result in GONADAL androgen production. Ovaries ALSO PRODUCE TESTOSTERONE. If that's difficult to remember, "think of it this way: In CAH, excess progesterone results in excess androgens being formed. So, maybe the same holds true for excess progesterone floating around from ovarian production!"

## PRECOCIOUS PUBERTY IN GIRLS

In girls, precocious puberty is defined as having breasts + vaginal bleeding OR accelerated growth. Get LH, FSH, estrogen and progesterone. Often idiopathic when brain hormones have started early (elevated LH and FSH). If there are elevated gonadal hormones but low brain hormones = BAD = tumor. Look for ovarian tumors with a pelvic ultrasound. May consider getting an MRI to look for a pituitary tumor if neurologic signs are present and/or central hormones are elevated. Treat CENTRAL PRECOCIOUS PUBERTY with a GnRH Analogue (gonadoTROPIN releasing hormone analogue) called Leuprolide or Lupron ®. It's counterintuitive, but it eventually results in the suppression of LH and FSH release and therefore results in suppression of ovarian (or testicular) steroidogenesis.

### PRECOCIOUS PUBERTY IN BOYS

In boys, precocious puberty can be caused by elevated LH alone causing elevated gonadal androgens. HCG can act on the same LH receptors, therefore an HCG-SECRETING TUMOR can also cause it. Look for increased testicular size/volume.

**PEARL/SHORTCUT**: In order to choose an answer that has "PUBERTY" in it (CENTRAL PRECOCIOUS **PUBERTY** or TRUE **PUBERTY**), there **must** be evidence of testicular enlargement **(>4 ml/2.5 cm).** So, if the testes are small (<4 ml/2.5 cm), but there is evidence of extra hair, penile enlargement, and/or a growth spurt… **there is a non-central and non-gonadal** problem → think late onset Congenital Adrenal Hyperplasia or a VIRILIZING TUMOR or EXOGENOUS STEROIDS!

## *DELAYED PUBERTY*

### DELAYED PUBERTY DEFINITION AND PEARLS

In girls, delayed puberty is defined as having no breast buds by the age of 13. It can also be defined as a LACK OF MENSES within 2 years of the presence of both SMR 4 breasts (mound on mound) and pubic hair. So if she has had pubic hair and breasts for over 2 years but hasn't started her period, she's DELAYED!

**MNEMONIC**: No breast buds by 13. A **3 on its side** looks like breast buds! Also, remember the mnemonic about the HOCKEY MASK? And the hockey/ice SKATES for precocious?!?

**PEARL**: Unlike PREMATURE ADRENARCHE, this is **more concerning in girls**, but less commonly diagnosed at the time of onset because "how many parents think it's a bad thing that their little princess hasn't had her period yet? They probably think it's a blessing!"

**PEARL**: If TROPIN (LH and FSH) values are normal, it's likely a gonadal issue.

### PROLACTINOMA

Prolactinoma should be in your differential for a patient with DELAYED PUBERTY.

### CONSTITUTIONAL DELAY OF PUBERTY

Bone age films that estimate an age that is younger than the true chronologic age would be consistent with a constitutional delay of puberty. Patient's height is "normal," but on a lower curve than expected (patient is short but probably not < 3rd percentile. In boys 14-15 year old is considered delayed ("remember the CHOCOLATE HEART?!?, and the CAT he fell on – for precocious puberty?!?"). If bone age films are as mentioned above, it's usually not treated with hormones. But, if asked which one **could** be used, the answer is TESTOSTERONE, NOT GROWTH HORMONE! In girls, 13-14 year old is delayed.

## HYPOGONADOTROPIC OVARIAN FAILURE

For delayed puberty associated with **hypo**gonado**tropic** ovarian failure, consider POOR NUTRITION or an EATING DISORDER as possible etiologies.

## KALLMANN SYNDROME

Kallmann syndrome is defined as HYPOGONADOTROPIC HYPOGONADISM + ANOSMIA. So low TROPS resulting in low GONADAL hormones. Patients can also have hypoplasia of the optic nerve, absence of septum pellucidum (a midline portion of the brain) and a micropenis. "This is the only disease with a smelling issue on the exam, and note that all of the **involved organs are midline** structures! Nose, pituitary, penis, optic nerve (at least the chiasm), septum!"

## HYPERGONADOTROPIC OVARIAN FAILURE

TURNER'S SYNDROME results in **hyper**gonado**tropic** ovarian failure. The TROPINS are elevated because the patient has no real ovaries to provide hormones and negative feedback.

# *SHORT STATURE*

**PEARL**: Note that up to a TWO year delay between bone age & chronologic age is NORMAL

## GENETIC OR FAMILIAL SHORT STATURE

**A child with genetic or familial short stature is** ***born with normal length*** but the height then decelerates over the first 2 years of life to find the new (genetically determined) curve. ***Bone age MATCHES real/chronologic age***. Patient is likely proportional in height and weight. Can also use the midparental height to help guide you. MIDPARENTAL HEIGHT = Average the two parents' heights, then add or subtract 2 inches or 5 cm depending on if it is a boy or a girl.

**PEARL**: If a child's parents were malnourished, mid-parental height goes out the window. The kids will be much taller if they have adequate nutrition. "Ever notice how children of immigrant parents are so much taller than their tiny little parents?"

## CONSTITUTIONAL GROWTH DELAY (& PUBERTAL DELAY)

With a constitutional growth delay, look for normal growth until about 1 year of age, but then the child will be at around the 5th percentile for HT and WT. Probably **not** < 3rd percentile. "Delayed bone age which mirrors height age" So look a short child whose height matches up with what the bone age films show. This is a **DEceleration** phenomenon and may not occur until adolescence, but the key is the **delayed bone age + delayed puberty**. Often a similar family history. No treatment needed. patient will reach normal adult height.

## GROWTH HORMONE DEFICIENCY

Short stature from **growth hormone deficiency** is rare, but findings will include a micropenis or clitoris and hypoglycemia (seizures may be the only clue that a patient has hypoglycemia). There is **DEcelerated growth rate,** but to the point of **crossing lines**. Diagnosed by seeing a lack of GH release following insulin or arginine stimulation

## CONGENITAL GROWTH HORMONE DEFICIENCY

Congenital growth hormone deficiency,will present with a bone age that's approximately 75% of chronologic age + **a decelerated growth rate +** WT percentile that's > HT percentile (unlike constitutional growth delay, in which the percentiles are usually proportional).

**PEARL**: Although this option may be in the list of answers, it is usually NOT a preferred answer choice. Consider other diagnoses such as Crohns Disease, Hypothyroidism, Genetic Short Stature, Constitutional Growth Delay, Turners, Hurler's, Hunters, etc.) because this is often a distracter on the boards! So know what it is, but also what it is NOT.

## ACQUIRED GROWTH HORMONE DEFICIENCY

If a growth hormone deficiency is acquired (probably due to a PITUITARY TUMOR), the patient may have a normal bone age but still sharply **DEcelerated growth rate** (NEEDS Endocrinology evaluation and NEEDS to have an MRI evaluation for CNS tumor).

**PEARL**: When considering one pituitary hormone deficiency, ALWAYS check for other pituitary hormone deficiencies as well (FLATPiG = FSH, LH, ACTH, TSH, Prolactin and Growth Hormone).

## OTHER CONSIDERATIONS FOR SHORT STATURE

* **HYPOTHYROIDISM**: Short + Overweight + Delayed bone age +/- Constipation +/- dry skin
* **CONGENITAL ADRENAL HYPERPLASIA (CAH)**: Hyperandrogenism, premature closure of growth plates. Look for "early puberty" + accelerated growth but a short final height
* **NUTRITION DEFICIENCIES**: Bone age mirrors Chronologic. Patients who acquire nutritional deficiencies will initially have lower WT percentile while preserving the height percentile (become thin while still growing), but eventually HT and WT percentiles will become proportional if the malnutrition continues. If patient has had poor nutrition since birth (3rd world country), HT and WT percentiles will be equal.
* **TURNERS SYNDROME**: Look for amenorrhea, no breasts (or just the presence of breast buds), no/scant hair. It is OK to treat these patients with GROWTH HORMONE (**not** testosterone! Testosterone is for boys with CONSTITUTIONAL DELAY OF PUBERTY, but only in rare cases.)
* **ACHONDROPLASIA**: Disproportionate percentiles + signs of the disease

# *TALL STATURE*

**NOTE**: GENETIC predisposition to being tall should be kept in mind.

## (DOUBLE TAKE) KLINEFELTER SYNDROME (aka KLINEFELTER'S)

Klinefelter Syndrome (aka Klinefelter's) = XXY. Gynecomastia, small testicles/infertile and normal intelligence to MILD MR. May have a mild motor or speech delay. Tall stature with long arms/legs. Low upper to lower segment ratio. Workup = REFER FOR CHROMOSOMAL ANALYSIS!

**MNEMONIC:** Gynecomastia = Kalvin Kline FELT HER BREASTS

**PEARL**: Mild mental retardation may be described as a patient who is "awkward," "below average in school" or even just "shy." Generally, though, these patients have some learning disabilities but their IQ can be normal.

**PEARL**: Someone who is tall with long arms/legs could also be described as having a LOW "upper to lower segment ratio."

## (DOUBLE TAKE) MARFAN'S SYNDROME (aka MARFANS SYNDROME)

Classic features of Marfan's Syndrome (aka Marfans Syndrome) include tall stature with long and thin upper extremities, long fingers, a pectus deformity, joint flexibility/hypermobility and possible cardiac problems. Cardiac problems may include mitral **valve prolapse (MVP), aortic dissection** and mitral or aortic regurgitation. Patients may have a high arched palate and a speech disorder, but do NOT have cognitive deficits. Patient's are also at risk for esophageal perforation.

* **PEARLS:** Patients can have subluxation of the lens, which may also be seen in Ehlers Danlos and homocysteinuria. If they mention SUPERIOR subluxation of lens, pick Marfans. Any patient with Marfans should not be cleared for sports participation until they have had an echocardiogram and an evaluation by a cardiologist. If they mention "arm span greater than height," you're done.

* **IMAGES**: Please do not get distracted by the reading. Look at the images and move on.

- http://trialx.com/curebyte/2011/05/21/marfan-syndrome-photos/
- http://www.hughston.com/hha/b_12_2_4a.jpg

* **MNEMONIC**: http://www.clevelandleader.com/node/6446 Michael Phelps won several gold medals. Isn't that just like winning the Most Valuable Plater (MVP = Mitral Valve Prolapse)?

## HIGH CALORIC INTAKE

Patients with high caloric intake are tall, overweight and can have an **advanced** bone age.

## OBESITY

The high caloric intake in obesity CAN also result in tall stature with advanced bone age. If obesity is due to hormonal/endocrine issue, patient is usually fat but short + delayed bone age. If no bone age is provided in the question, consider Cushing's as the diagnosis.

**PEARL**: Obesity is a risk factor for depression, avascular necrosis of the hip "at the top of the LEGG (LEGG CALVE-PERTHES)

# *GROWTH CHART TRENDS*

**NOTE**: There are MANY questions that present with growth charts. Please be as familiar as possible with the above-mentioned information and the information below. These questions are often difficult and come down a guess between two answers.

## ENDOCRINE DISORDERS

Endocrine disorders often present with the presence, or development, of **short stature**. Patient's weight is often still normal, or possibly even elevated. Considerations include: GH DEFICIENCY, HYPOTHYROIDISM, DIABETES MELLITUS, and CUSHING'S (short/fat... if the patient is tall, it's NOT CUSHING'S)

## CHROMOSOMAL ABNORMALITIES

Microcephaly and dysmorphic features should suggest a chromosomal abnormality.

## INADEQUATE CALORIC INTAKE or MALABSORPTIVE DISORDERS

For a patients with inadequate caloric intake or a malabsorptive disorder, look for an initial decline in the weight curve, THEN a deceleration of the height or length. So if you see that HC and HT are spared while WT falls off, consider CALORIC INSUFFICIENCY.

**PEARL**: If you see a quick drop in weight without significant drop in height yet (low WT for HT), there is a **severe** underlying disorder. Consider a workup for GI (Celiac, malabsorption), Renal or Metabolic disorders.

## SPARING OF HEAD CIRCUMFERENCE

If there is sparing of head circumference while the weight and height are falling on the curves, it's usually due to an ENDOCRINE problem. Sometimes, this can also be seen with GENETIC SHORT STATURE and CONSTITUTIONAL DELAY.

## SMALL HEAD DISORDERS

* **PRIMARY CRANIOSYNOSTOSIS**: Abnormal suture lines, normal brain on imaging

* **PRIMARY MICROCEPHALY**: Brain is **genetically** abnormal

* **SECONDARY MICROCEPHALY**: Normal sutures, abnormal brain on imaging because of some type of disease process or environmental exposure.

# ***AMENORRHEA***

## AMENORRHEA PEARLS

* Always rule out pregnancy, anatomic obstructions and malformations. Remember, menarche is usually during SMR 3 or 4.

* PREGNANCY is a frequent answer.

* LH results in Progesterone production

* FSH results in Estrogen production. Estrogen is responsible for breast development!

## AMENORRHEA WORKUP

Always start an amenorrhea workup with an HCG, if that is negative, then move on to further testing:

*** PROGESTERONE CHALLENGE**

- **If POSITIVE** (patient bled within 2 weeks of administration), that means there is plenty of ESTROGEN but progesterone was missing. Why? Obtain an LH level:
  - High LH level = PCOS
  - Low LH level? → Get PRL (prolactin) and TSH levels. Keep in mind that elevated PRL can be primary disorder or due to hypOthyroidism. If only the PRL level is high, then the diagnosis is likely a PROLACTINOMA
- **If NEGATIVE**, then it means there is NOT enough ESTROGEN. So, obtain an FSH level:

- HIGH FSH: Consider OVARIAN FAILURE (TURNER'S SYNDROME, an Autoimmune disease, chemotherapy, premature menopause). It is doubtful that it is increased due to excess release from the pituitary.
- Low FSH: Means there's a CENTRAL problem = Mass? Prolactinoma? Craniopharyngioma? HypOthyroid?

## PRIMARY AMENORRHEA

* Do a workup for primary amenorrhea if the patient has not had menses in the presence of ANY of the following:

- It's been 2 years since puberty ended. So, if puberty started at age 8 and finished at age 11, then there is no need to do a workup until at least age 13.
- Patient is 14-15 years of age **in the absence** of ANY breast development.
- Patient is 16-17 years of age **in the presence** of breast development.

* **Lab workup** as mentioned in the "amenorrhea workup" section. In other words, after HCG is checked, check LH, FSH, PRL, and/or TSH as per the algorithm. BUT, if there are signs of ANDROGENIZATION, add TESTOSTERONE and DHEA.

* Possible causes of primary amenorrhea include:

- **EXERCISE INDUCED AMENORRHEA**: Look for DELAYED puberty + LOW LH and FSH → Treatment is to reduce exercise and increase calories
- **TURNER'S SYNDROME**: No breasts (or just buds), no/scant hair → Get a Karyotype
- **TESTICULAR FEMINIZATION/ANDROGEN INSENSITIVITY**: Genetically XY = Breasts but no hair!

## SECONDARY AMENORRHEA

Secondary amenorrhea is defined as 6 months without menses after a patient was regular, or 12 months without menses if the patient was previously irregular. The differential includes ASHERMAN'S SYNDROME (intrauterine adhesions), **CYSTIC FIBROSIS,** SARCOIDOSIS, phenothiazines, nutrition, brain tumors, and Tuberculosis.

## ANOREXIA AS A CAUSE OF AMENORRHEA

Suspect anorexia as the cause of amenorrhea if the patient has not had a period for **3 months**. Anorexia = low HR, orthostasis, hypothermia.

## BULIMIA AS A CAUSE OF AMENORRHEA

Bulemia is often associated with **irregular** menses.

## POLYCYSTIC OVARIAN SYNDROME (PCOS) AS A CAUSE OF AMENORRHEA

Polycystic ovarian syndrome is a **clinical** diagnosis. Labs are supportive. Look for an LH:FSH ration > 2 in a patient with acne, irregular menses, excess hair on her body and/or signs of insulin resistance. Can be caused by anything increasing androgens, including Cushing's syndrome and exogenous steroids

## DYSFUNCTIONAL UTERINE BLEEDING (DUB) AS A CAUSE OF AMENORRHEA

Dysfunctional uterine bleeding, or DUB, is usually associated with anemia! Menstrual cycles with < 20 or > 40 days between periods. Usually primary and due to anovulatory cycles → give OCPs. Secondary DUB is often due to other etiologies (including von Willebrand disease) and is rare, but even more likely to be associated with anemia.

## MENORRHAGIA AND AMENORRHEA

During the 1st two years of menstruation, infrequent menstrual cycles, or cycles lasting 2-3 days is OKAY. Menstrual period lasting > 10 days are abnormal and require a menorrhagia workup.

# ***PREMENSTRUAL SYNDROME & DYSMENORRHEA***

## PREMENSTRUAL SYNDROME (PMS)

Premenstrual Syndrome (PMS) is defined as cyclic affective/mood components + somatic components usually occurring within the 5 days prior to menstruation. First line treatment is NSAIDS, second line is Fluoxetine

**PEARL**: Options for first line treatment of PMS or Dysmenorrhea may not include Ibuprofen or NSAIDS in the answer choices. The correct could be disguised as a CYCLOOXYGENASE SYNTHETASE (COX) INHIBITOR!

## PRIMARY DYSMENORRHEA

Primary dysmenorrheal is defined as painful menses. No affective/mood component. It is due to PGE so give NSAIDS. Usual description is CRAMPY pain. The diagnosis is likely PRIMARY DYSMENORRHEA even if the vignette says that the patient didn't respond to NSAID. She probably took the NSAIDS too late! Should be started 2-3 before the usual onset of symptoms. Symptoms can occur before or during menses, and usually do not start until a girl's cycles become ovulatory (6-12 months after menarche).

## SECONDARY DYSMENORRHEA

Secondary dysmenorrheal is due to PELVIC PATHOLOGY, such as endometriosis, pregnancy, ectopic pregnancy, etc.

# ***SOCIAL ISSUES***

## AUTONOMY

Allow kids autonomy within their abilities unless they are a danger to themselves. This means they should be able to make decisions for themselves regarding chronic illness if you deem they are able to understand the implications of their decisions.

## BREAST EXAMS

Breast exams should be taught to teens

## RAPE/PTSD

Rape and PTSD are associated with nightmares, a desire NOT to be alone, poor appetite, poor grades, abdominal pain, vaginal discharge (probably from an STD).

## DEPRESSION

Depression is associated with weight changes, poor concentration. Differential is wide, including hypothyroidism, nutritional deficiencies, infections, rheumatologic diseases and drugs. The younger the patient, the more severe the depression.

## OSTEOPOROSIS

Risk for developing osteoporosis increases with smoking and LOW estrogen exposure (late puberty/menses). "Estrogen is often given to little old ladies (post-menopausal) to protect them from osteoporosis!"

## ALCOHOL AND TOBACCO

Children are now starting to use alcohol and tobacco at the young as 12 years old! Smoking has shown a decline in the last decade.

**MNEMONIC**: "Imagine kids having sleepovers and sneaking out in the middle of night after drinking and smoking cigarettes. They stumble to their neighbor's yard and throw A DOZEN EGGS at their home!"

## MARIJUANA

Marijuana use is now common amongst 14 year olds. Can cause GALACTORRHEA (even in boys!)

## INHALANTS

Inhalants are abused more frequently in middle school than in high school.

## CONDOMS

Always recommend the use of condoms!

## PLAN B

Plan B (aka "the morning after pill") can be taken up to THREE days after unprotected intercourse or condom failure. It prevents ovulation, fertilization and implantation, but does NOT do anything if implantation has already occurred. "Imagine a lazy, lazy, lazy girl who forgot to take her OCPs for a few days. She parties on a FRIDAY and has sex without a condom. The clinics are all closed on Saturday, but lucky for her she can afford to be lazy this time and wait to seek medical attention until THREE days later when the clinics open again."

## PAP SMEARS

If a patient presents with a known or likely STD, it is OK to do a PELVIC EXAM, but you should hold off on doing a PAP SMEAR of the cervix because the inflammatory changes can result in a false positive! Thin vaginal discharge is not a contraindication. Mucopurulent discharge suggesting an STD is a contraindication. If she's on her period, don't do a pap smear.

## ASCUS

In general, recommendations regarding pap smears are much more aggressive now. If a pap reveals ASCUS, a repeat colposcopy is warranted in 6 months.

## HUMAN PAPILLOMA VIRUS (HPV)

Human papilloma virus (HPV) is the most prevalent STD. Yes, it's considered an STD. Only a small percentage of patients actually get warts.

## CHLAMYDIA TRACHOMATIS

Chlamydia trachomatis is the most prevalent BACTERIAL STD. Much more on chlamydia in the ID section.

## MOTOR VEHICLE ACCIDENTS

Motor Vehicle Accidents are the leading cause of adolescent morbidity and mortality. Next is HOMICIDE. "Cars are common to **every** adolescent neighborhood in the country, guns and violence are common only in certain ones."

## GUNS

50% of homes have guns. 75% of teen homicides are due to guns. 90% of suicide attempts are successful. A gun in the home increases the risk of teen suicide by five times. The AAP recommends against having guns in the home. If a gun is present, it should be unloaded and locked out of reach from children in a SEPARATE location from the bullets (which should also be locked away out of reach).

**AAP Resource**: http://www.aap.org/healthtopics/safety.cfm

**PEARL**: Simply using gun LOCKS have **not** been shown to decrease injuries from firearms.

## HOMOSEXUALITY

Homosexual students are MORE likely to abuse substances.

## SELF CONSENT

"Mature minors" may self consent.

## DRUG SCREENING

Drug screening in the office should always be voluntary.

**PEARL**: Parents should be encouraged to have "the talks," with kids around the age of 10. This includes an open discussion about sex and illicit drugs.

## EXOGENOUS ANABOLIC STEROIDS

If taking by IM injection, exogenous anabolic steroids are detectable for 6 or more months. If orally, then 1 month or less. Exogenous **growth hormones** give a similar laboratory and side effect profile as exogenous anabolic steroids, but it's not detectable as a substance of abuse. Side effects of exogenous anabolic steroids include:

* Girls: Hirsutism, deep voice, early closure of growth plates

* Boys: Gynecomastia, high voice, hypogonadism, acne, oligospermia

- **MNEMONIC**: Girls start to look/sound like boys and vice versa

* Both: Liver dysfunction, low HDL, high LDL, emotional lability, HTN.

## *EATING DISORDERS*

**NOTE**: Eating disorders are common at times of transition for overachievers. In general, weight loss should not exceed 3 lbs/week.

### ANOREXIA

In patients with anorexia, look for laxative abuse, diuretic abuse (supportive labs include hypOkalemia, alkalosis, and HIGH FeNa!), emesis, bradycardia, ST depression, prolonged QT, Ventricular Tachycardia (V-tach), hypothermia, orthostasis. History must include a distorted body image, WT **< 15% expected**, fear of weight gain, 3 consecutive missed periods.

**MNEMONICS**

* proLONGed QT and ST Depression = "a LONG-faced anorexic QT/Cutie is a little too **S**kinny/**T**hin... even so, she still can't stop being DEPRESSED because her body just doesn't look right when she looks in the mirror!"

* Bradycardia and Hypothermia = "The super skinny anorexic girl's body is STARVED so her heart beats slower and her body produces LESS HEAT ENERGY (conservation mode)."

### BULIMIA

Bulimia = Overeating + Induced emesis. Patients tend to be a little overweight. Signs and symptoms include enamel erosion (from the acid), enlarged parotids, elevated amylase, irregular menses, hypOkalemic hypOchloremic metabolic ALKalosis and lanugo hair. Any patient with dehydration, an abnormal EKG or suicidal ideation in the setting of bulimia should be hospitalized.

### REFEEDING SYNDROME

Refeeding syndrome can cause hypOmagnesemia, hypOphosphatemia, rhabdomyolysis, delirium due to thiamine deficiency, and seizures.

**PEARLS**:

* Excessive Vomiting, Laxatives, Diuretics = hypOkalemic hypOchloremic metabolic ALKALOSIS

* Refeeding syndrome: "When the body is in starvation mode, AT**P** is not being made. When sources of phosphorus (food) are reintroduced, too much is taken up to restart making AT**P** therefore resulting in dangerously low hypophosphatemia."

### OVERWEIGHT VERSUS OBESE

A BMI that falls in the 85-95% range for sex and age is considered overweight. If the BMI is >95%, the patient is considered obese.

**PEARL**: Do not use the absolute number of 25 as your cutoff for categorizing a patient as overweight. Rather, use the percentage for sex and age to guide you.

# ***SCROTAL MASS***

## TESTICULAR CANCER

Testicular cancer results in a painless scrotal mass. When doing the cancer workup, get HCG and AFP levels. If either is high, get a biopsy, scrotal ultrasound, CT of the chest/abdomen/pelvis for staging. Treatment will include at least an orchiectomy.

**PEARL**: Assume that any PAINLESS scrotal mass is CANCER until you can prove otherwise (by physical exam findings, symptoms, testing)

## HYDROCELE

Hydroceles are painless, change shape with position and valsalva, **superior & anterior to testes**, transilluminates, usually resolves by 1 year old (if not consider surgical intervention)

## SPERMATOCELE

Spermatoceles are painless, **superior and posterior** (basically on top of the epididymis), **no change with position**, nodular, also transilluminates. Contains sperm. No treatment needed.

## VARICOCELE

**V**aricoceles are generally painless. It is a dilatation of the pampiniform **V**enous plexus.

**MNEMONIC**: "Bag **ov**' **V**orms" on the LE**1**<T **SIDE."** Described often as feeling like a bag of worms and is usually on the left side (the F in the mnemonic is made with a **V**.

## INGUINAL HERNIA

Inguinal hernias are often **PAINFUL**. The scrotal mass changes with position and has no transillumination.

# ***TESTICULAR AND PENILE ISSUES***

## TESTICULAR TORSION

With testicular torsion, there is an absence of the CREMASTERIC REFLEX. Stroking the inner thigh normally elicits scrotal elevation. If that reflex is absent then it is highly suggestive of testicular torsion. Another symptom is increased pain when the scrotum is elevated.

## TORSION OF THE APPENDIX TESTES OR EPIDIDYMIS

Torsion of the appendix testes, or of the epididymis, is painful but benign. Look for the Blue Dot sign (more commonly seen with torsion of the appendix testes than the appendix epididymis). Treat with NSAIDS.

**MNEMONIC**: Do you remember the BLUE-DOT SPECIALS at KMART®? Imagine a huge blue-dot special on a big bottle of Advil®! This should remind you of the very benign treatment.

## EPIDIDYMITIS

Patients with epididymitis present with DYSURIA, frequency, fever and NO discharge. Pain is typically unilateral. Etiologies include CHLAMYDIA, GONORRHEA, and E. COLI (especially if there's a history of anal sex).

## ORCHITIS

Orchitis is also known as testicular inflammation. Look for unilateral pain, possible fever, NO discharge, NO dysuria. Etiologies include MUMPS, and it can also be secondary to epididymitis. If testicular atrophy is present there is an increased risk of CANCER. If both sides are involved, there is an increased risk of infertility!

## BALANITIS

Balanitis is an inflammation/infection of the glans.

## PHIMOSIS

Phimosis is defined as a tight foreskin that prevents retraction over glans which persists beyond age 3. For most boys, foreskin is fully retractable by 3 years of age. BALANITIS occurring prior to 3 years of age is associated with phimosis.

## PENILE EPIDERMAL INCLUSION CYSTS

Penile epidermal inclusion cysts are small, white, benign bumps located on the glans

# ENDOCRINOLOGY

## THYROID DISORDERS – KEY TERMINOLOGY

DEFINITIONS of Thyroglobulin, Thyroxine-Binding Globulin, Thyroxine, and Free T4

*** THYROGLOBULIN:** This protein found only in the thyroid itself. It is used to make thyroid hormones.

*** THYROXINE-BINDING GLOBULIN;** TBG is responsible for carrying thyroid hormones in the blood.

*** THYROXINE:** THYROXINE = T4 = Hormone that is **BOUND** to TBG and in the blood and **INACTIVE**

*** FREE T4:** FREE T4 = FT4 = The **ACTIVE** hormone in the blood

## *HYPOTHYROIDISM*

### THYROXINE-BINDING GLOBULIN DEFICIENCY

Thyroxine-binding globulin deficiency is an X-Linked disease in which the newborn screen (NBS) reveals **normal TSH but LOW T4 values**. NBS is measuring T4 (BOUND hormone) which is low because there is a TBG deficiency! Get a FREE T4 as the next step. It should be NORMAL. Another option is to get a TBG level, which should be low. **NO treatment indicated**. So if TSH is normal but T4 is low on NBS → **NO TREATMENT.**

**PEARL**: TSH is the barometer for the thyroid. If it's normal, there's probably no clinical problem.

### HYPOTHYROIDISM & CONGENITAL HYPOTHYROIDISM

Hypothyroidism and congenital hypothyroidism are diagnosed with **elevated TSH and a low Free T4** (get T4 if FT4 not offered on exam). If asked how to screen, choose TSH. In utero, some of mom's thyroxine crosses the placenta so babies may by asymptomatic at birth. **Signs of hypothyroidism** may include puffiness, large tongue, hoarse cry, umbilical hernia, hypotonia, large anterior fontanelle (AF), open posterior fontanelle (PF), constipation, and mottling. **Possible etiologies include**:

* Dysgenesis is the most common reason

* Abnormal thyroid development somewhere between the base of the tongue and the normal position. Mass would be midline but **not cystic**. Removal can result in **worsening** of hypothyroidism. Once a patient is started on Thyroxine, check T4 and TSH in 1 month.

* **Acquired** Hypothyroidism is more common than Congenital Hypothyroidism. If a patient is euthyroid up to age 5 yo, IQ should be fine. If the patient is short, it is OKAY to start the patient on meds that can help with catch up growth.

* HASHIM**O**T**O**S THYROIDITIS = CHRONIC LYMPHOCITIC THYROIDITIS: Hashimoto's Thyroiditis, or chronic lymphocytic thytroidistis, results in HYP**O**THYR**O**IDISM. Labs include +anti-TPO or +anti-thryroglobulins, low T4 with an elevated TSH. Patient is likely to have a **painless, firm GOITER/thyromegaly**.

- "Infiltration/inflammation/destruction? of the gland results in low thyroid hormone (so low T4) which results in and elevation in TSH. Sometimes, T4 or FT4 may be normal because of the extremely high TSH. A rare complication is transient thyrotoxicosis. Radioactive Iodine Uptake in Hashim**O**t**O**s is LOW. "Thyroid is being destroyed so how could it take up any iodine?"

* **PEARL**: If TSH is abnormal on the NBS, start Levothyroxine now (and get labs). You can always ask questions later. Urgency is due to the tremendous cognitive delays that can occur if therapy is not started by 4 weeks. Continue medication until repeat labs are back.

## THYROGLOSSAL DUCT CYST

A thyroglossal duct cyst is a midline lesion on anterior neck. As many as half of all thyroglossal duct cysts can get infected, which then increases the chances of recurrence. Therefore the treatment of choice is surgical excision. For the exam, if they describe a midline cystic lesion, choose this as the diagnosis, and **remove it**!

**PEARL**: Don't get confused with a RANULA: Painless mucous CYSTIC usually near the inner lips or under the tongue. Might be midline. Clear contents. Treated by removal.

## THYROID NODULES

**Thyroid nodules are more likely to be malignant in kids,** so must be worked up. Get an ultrasound to better assess size and location. If ultrasound is not a choice, get radioactive imaging. COLD NODULE = INACTIVE **TISSUE** = **BAD**!

# *HYPERTHYROIDISM*

## GRAVES DISEASE = HYPERthyroidism

Graves disease (aka Grave's Disease) causes hyperthyroidism due to the presence of "thyroid-stimulating immunoglobulin." Signs/symptoms may include an infiltrative ophthalmopathy, emotional lability, weight loss, heat intolerance and possible LID LAG. TSH should be VERY low/absent! Radioactive Iodine Uptake is HIGH in Grave's since it needs lots of iodine to make all of the THYROXINE being released. Treatments options include methimazole, iodine ablation, a beta blocker (propranolol for symptomatic relief) and thyroidectomy. Methimazole and PTU inhibit T4 production (not secretion). GOITERS in patients with hyperthyroidism are cells that are FULL of thyroid hormone so it can take months to become euthyroid.

**PEARL:** Some of the symptoms of hyperthyroidism may be disguised as "hyperactivity, disorganized thinking and trouble sleeping."

**PEARL:** If a patient has a goiter, more information is needed to differentiate between hypothyroidism and hyperthyroidism

**MNEMONIC**: PTU can be quite toxic so it is NOT a first-line agent. P-T-U = Potentially Toxic, UGH!

## NEONATAL THYROTOXICOSIS (aka NEONATAL GRAVES DISEASE)

Neonatal Thyrotoxicosis (aka Neonatal Graves Disease) occurs when **maternal** thyroid stimulating antibodies cross over and causes symptoms in the immediate **newborn period.** Symptoms include tachycardia, SVT, and tremors. Occurs in < 10% of babies born to mom's with Grave's Disease. For a pregnant women with Graves Disease, **P**TU is ok to give during **P**regnancy.

**PEARL**: If a patient has symptoms that are suggestive of both Neonatal Thyrotoxicosis and an Inborn Errors of Metabolism (IEM), look at the age of onset! IEMs do not result symptoms in the immediate newborn period.

## CALCIUM AND VITAMIN D RELATED DISORDERS

### (DOUBLE TAKE) HYPERCALCEMIA

Hypercalcemia is defined as a calcium level > 12. Hypercalcemia shortens the ST segment, therefore it also shortens the QT interval. Polyuria occurs due to an osmotic diuresis. Patient's can also get nausea, vomiting and change in mentation and possible abdominal pain. Initial treatment for hypercalcemia should be IV HYDRATION. Possible etiologies include:

* FAMILIAL **HYPO**CALCI**URIC HYPER**CALC**EMIA**: The disorder is fairly benign. Nothing to do.

* WILLIAMS SYNDROME (aka William's Syndrome)

* VITAMIN D & VITAMIN A ingestion

* THIAZIDE DIURETICS

* SKELETAL DISORDERS: Including dysplasias, skeletal **immobilization and skeletal/body casting**. For hypercalcemia related to immobilization, treat with **IVF and loop diuretics**.

- **PEARL: For hypercalemia due to immobilization, treat with IV fluids and loop diuretics**

* HYPERPARATHYROIDISM

* MALIGNANCY

* **MNEMONIC:** "Bones, stones, abdominal groans and psychiatric moans." This refers to the classic mnemonic used to describe the complications of hyperparathyroidism.

- BONES: Osteoporosis, osteomalacia, pathologic fractures, osteitis fibrosis cystica (aka Von Recklinghausen's disease)
- STONES: Nephrolithiasis, nephrocalcinosis and nephrogenic diabetes insipidus.
- ABDOMINAL GROANS: Nausea, vomiting and constipation.
- PSYCHIATRIC MOANS (or OVERTONES): Coma, delirium, depression, fatigue and psychosis.

### (DOUBLE TAKE) HYPOCALCEMIA

For **hypo**calcemia, look for a calcium level < 8.5 or, or an ionized calcium level < 4.5. **Hypo**calcemia is much less common than hypercalcemia, but has much greater morbidity associated with it and is thus tested more often on the pediatric boards. Can range from minimal symptoms due to mild hypocalcemia from hyperventilation to HORRIBLE outcomes. Symptoms of hypocalcemia include: **Paresthesias, Tetany**, Carpopedal Spasm/Tetany (also known as **Trousseau's sign**), **Chvostek's sign** (abnormal reaction/tetany of the facial nerve when tapped), Seizures that do not respond to benzodiazepines, Laryngospasm (resulting in tachypnea. Can look/sound like Croup), **prolonged QT**.

* **MNEMONIC**: Calcium gluconate is **given** to stabilize the most important **muscle** in the body, the HEART! If you can remember that a low calcium results in numerous "muscle symptoms/signs," that should help you on the exam.

* **MNEMONIC**: Most of us have had an anxious patient in the emergency room who complains of transient numbness and tingling in their hands and/or feet… it's due to transient hypocalcemia from hyperventilation. If you can remember this association, you can remember that all of these tetany-like symptoms are related to hypOcaclemia!

* **MNEMONIC**: The **CH**vostek Sign (aka **CH**vostek's Sign) has to do with tapping the **Cheek** and looking for a twitch. **CH**eeky **CH**vostek sign! If all else failed with an anxious and hyperventilating patient, wouldn't you want to slap their cheek?

* **MNEMONIC**: Carpopedal spasm due to hypocalemia = Trousseau's sign = "TROUSER'S" Sign = Think of it affecting the extremities!

* **EARLY HYPOCALCEMIA**: Refers to within the first 3 days of life. Your differential should include asphyxia, **IDM** (Infant of a Diabetic Mother), **maternal** hyperparathyroidism and IUGR.

* **LATE HYPOCALCEMIA**: Refers to the after 7 days, or during the 2nd week of life or later.

- DIGEORGE/22Q11 DELETION
- VITAMIN D DEFICIENCY
- HYPOPARATHYROIDISM: Causes low calcium and high phosphorus
- PSEUDOHYPOPARATHYROIDISM: Autosomal dominant disorder in which receptors ar resistant to PTH. Serum PTH will be high. Calcium is low, phosphorus is high. Also look for clinodactyly (short/stubby 4th fingers/toes), developmental delay and moon facies.
- HYPERPHOSPHATEMIA
- RENAL FAILURE: Can't hydroxylate Vitamin D to 1,25, so less calcium is absorbed. — Vit D dep rickets
- NEPHROTIC SYNDROME: The low albumin causes the measured calcium to look low. Get an IONIZED calcium to confirm. Or, calculate the corrected calcium by using the shortcut of adding 0.8 to the calcium level for every drop in albumin of 1 g/dL.
- HYPOMAGNESEMIA: Be sure to CORRECT the magnesium!
- ALKALOSIS: Shifts ionized calcium to the protein-bound form, resulting in less of the active form.
    - **HYPERVENTILATION:** Causes a transient respiratory ALKALOSIS, hypOcalcemia and paresthesias!
- RHABDOMYOLYSIS: Initially there can be hypocalcemia. During the later recovery phase, there can actually be hypercalcemia. Calcium deposits into the muscles (let's assume that's the reason for the hypocalcemia).
- ETHYLENE GLYCOL: Look for mention of calcium oxalate crystals in the urine.
    - **MNEMONIC**: Just assume that the calcium lost in the urine as calcium oxalate crystal causes the hypocalcemia.
    - (DOUBLE TAKE) CALCIUM O**X**ALATE CRYSTALS: There are two types of calcium oxalate crystals. One is square and looks like it has an X on it. That one very common. Calcium oxalate stones are the most common ones in humans. and the other is elongated and looks like a rod. The long one is the one usually associated with **ethylene glycol poisoning**.
        - **IMAGE**: http://meded.ucsd.edu/isp/1994/im-quiz/images/caloxal.jpg
            - **MNEMONIC**: calcium oxalate crystals look like a big X on a crystal.
        - **IMAGE**: http://alturl.com/mg3xv (This one is related to ethylene glycol and kind of looks like a long can of beer, or needles).

## VITAMIN D & ITS EVALUATION

The liver sends 25-Vitamin D to the kidneys where it gets hydroxylated to 1,25 Vitamin D (the active form). If looking for a **nutritional** deficiency, obtain a 25-Vitamin D level. **SUPPLEMENT with 1,25 Vitamin D** (the ACTIVE form). Typically, 25-Vit D is the first one you should check (especially if they ask for a screen).

**MNEMONIC**: Where is the Vitamin D produced that carries 2 numbers with it (1 and 25)? TWO organs = TWO kidneys = TWO numbers (1 and 25)!

**MNEMONIC**: Which one is the active form? Think of it this way… if you ingest a calcium containing food in its natural form, it will first go to the gut, then the liver, then the blood, then finally the kidneys! So keeping the above mnemonic in mind, it's the Vitamin D with TWO numbers!

**PEARLS**: Here are some ways Vitamin D deficiency could present

* African-American (AA) breastfed child whose mom is not on Vitamin D supplementation

* African-American (AA) breastfed child whose mom is not getting enough sunlight

* Child with symptoms consistent with malabsorption

* Child with a history of epilepsy who is on anti-seizure medications

## (DOUBLE TAKE) RICKETS

Findings of Rickets may include widening of wrist and ankle physes (growth plates), bowed legs, pain, decreased growth rate, anorexia, enlarged costochondral junctions (rachitic rosary), pigeon chest, delayed suture/fontanelle closure, frontal bossing (thick skull), or bad tooth enamel. There is no singular lab pattern for Rickets. It can be due to Vitamin D deficiency secondary to one of multiple different disorders and you **MUST** learn the lab patterns associated with the different disorders:

*** NORMAL (or LOW) CALCIUM + LOW PHOSPHORUS**

**This pattern represents FAMILIAL HYPOPHOSPHATEMIC RICKETS (aka "VITAMIN D RESISTANT RICKETS"). It is an X-linked DOMINANT renal disorder**. There is a defect of phosphate reabsorption in the proximal tubule AND a defect of the kidney to convert 25-Vitamin D to 1,25 Vitamin D → Treat with oral phosphate supplementation and avoid hypOcalcemia by giving the active/oral form of Vitamin D (1,25). Labs = Normal or low calcium, LOW serum phosphorus, **HIGH ALKALINE PHOSPHATASE, normal Vitamin D 25,** PTH IS NORMAL since calcium is usually normal.

- **PEARL**: For the exam, they will probably keep it simple and avoid giving you a low calcium level

*** NORMAL CALCIUM + LOW PHOSPHORUS**

This represents **INITIAL** VITAMIN D DEPLETION. Low Vitamin D results in low phosphorus reabsorption. There is a compensatory increased PTH that temporarily normalizes calcium.

- **PEARL**: For the test, they probably want you to focus on FAMILIAL Hypophosphatemic Rickets. The differentiating lab would be low Vitamin D level (25) in early Vitamin D depletion, versus normal in Familial Hypophosphatemic Rickets.

*** LOW CALCIUM + LOW PHOSPHORUS**

This represents SEVERE VITAMIN D DEFICIENCY resulting in poor absorption of calcium and phosphorus from the gut. PTH should be high.

*** LOW CALCIUM + NORMAL PHOSPHORUS**

This represents the initial Vitamin D repletion stage of healing Vitamin D Deficiency Rickets

*** LOW CALCIUM + HIGH PHOSPHORUS**

This represents hypoparathyroidism, phosphorus overload, or pseudohypoparathyroidism. All result in perceived (or real) low PTH in the body

*** NORMAL CALCIUM + HIGH PHOSPHORUS**

This represents renal disease, growth hormone excess, or a high phosphorus diet.

## (DOUBLE TAKE) RICKETS OF PREMATURITY

Premature babies have a high Vitamin D, Calcium and phosphorus requirement. Failure to administer these may result in Rickets of Prematurity. Treat with Vitamin D + Calcium + Phosphorus. Have to give all three (Vitamin D alone is insufficient).

## (DOUBLE TAKE) LIVER DYSFUNCTION

Liver dysfunction can result in decreased bile salts in the gut → Vitamin D absorption issues → Vitamin D deficiency.

# ADRENAL DISORDERS

**NORMAL ADRENAL STEROID SYNTHESIS**

| | | |
|---|---|---|
| PROGESTERONE → | 17-HYDROXYPROGESTERONE | -- -- -- -- -- > ADRENAL ANDROGENS |
| (21-hydroxylase)↓ | ↓ (21-hydroxylase) | |
| DEOXYCORTICOSTERONE | 11-DEOXYCORTISOL | |
| ↓ | ↓ | |
| CORTICOSTERONE | CORTISOL | |
| ↓ | | |
| 18-HYDROXYPROGESTERONE | | |
| ↓ | | |
| ALDOSTERONE | | |

## CUSHINGS SYNDROME (aka CUSHING'S SYNDROME)

CUSHINGS **SYNDROME** = **HYPER**CORTISOLISM = POOR growth/slow growth rate, striae, obesity, moon face, buffalo hump and muscle weakness. CUSHING'S **DISEASE** = Hypercortisolism due to a CENTRAL cause

**PEARL**: If the patient is tall, it's NOT Cushing's. If tall, obese and has stretch marks, consider high caloric intake.

## ADDISONS DISEASE (aka ADDISON'S DISEASE)

ADDISONS DISEASE (aka **AD**DISON'S) = **AD**RENAL INSUFFICIENCY = Electrolyte shifts (hyperkalemia or hyponatremia) can result in weakness, myalgias, malaise, nausea and vomiting. HypOglycemia occurs from a lack of cortisol. ACTH levels are **HIGH** and can result in hyperpigmentation. When an ACTH stimulation test (Cosyntropin) is performed, a normal response (a rise in cortisol levels) **does not** occur. Patients may have elevated levels of ADH. This is an appropriate elevation and should NOT be diagnosed as SIADH.

* **MNEMONICS and PEARLS:**

- Keeping track of which eponym refers to cortisol deficiency or excess can be tough. Instead of ADrenal Insufficiency, call it "**AD**renal **D**eficiency." So **AD**dison's **D**isease = "**AD**renal **D**eficiency" = **AD D** and **AD D** = **ADD**ison's!!!
- It also means Aldosterone **D**eficiency. Aldosterone helps with sodium retention and potassium excretion. In "AD D," it is deficient, resulting in **hyperkalemia and hyponatremia**. If you ever see this combination of electrolytes on a chemistry panel, have a HIGH suspicion for some type of aldosterone deficiency.
- If they talk about a patient having a really good "tan," they may be referring to Addison's related hyperpigmentation.
- "Think of the rise in ADH levels as an appropriate effort to retain water due to insufficient mineralocorticoid (aldosterone)!"

* **PRIMARY ADDISONS DISEASE** is the most common reason for adrenal insufficiency in children. Results in slow autoimmune destruction of the adrenal gland.

* **ADRENAL INSUFFICIENCY** can also be due to infection, due to adrenal hemorrhage (results in very abrupt signs of Adrenal Insufficiency) or idiopathic.

* **SECONDARY ADRENAL INSUFFICIENCY** = A pituitary issue = **LOW** ACTH. There is **NO** hyperkalemia or hyponatremia. ACTH stimulation with Cosyntropin **does** result in improved cortisol levels. Patients sometimes have other midline defects. Treatment options include:

- **PEARL**: Sodium and Potassium levels ore ok in secondary adrenal insufficiency because the **Rennin-Angiotensin (R-A) System is fine** (since the adrenal glands are NORMAL and aldosterone **is** being produced).
- **MNEMONIC**: Secondary adrenal insufficiency is sometimes associated with midline defects. "This makes sense since the pituitary is also a midline structure!"

* **TREATMENT**

- Maintenance therapy for Primary Adrenal Insufficiency is Hydrocortisone (to replace cortisol) and Fludrocortisone (to replace aldosterone).
- Maintenance therapy for Secondary Adrenal Insufficiency is **just** hydrocortisone (to replace cortisol). Since R-A System is intact, fludrocortisone is not needed.
- SALINE + GLUCOSE + IV HYDROCORTISONE for **ADRENAL CRISIS**. Signs and symptoms of adrenal crisis include nausease, vomiting, mailaise, hyperkalemia and hyponatremia.

## CONGENITAL ADRENAL HYPERPLASIA (CAH)

CONGENITAL ADRENAL **HYPER**PLASIA (CAH) = LEADS TO A BUILD UP OF ANDROGENS AND A **LACK OF CORTISOL AND ALDOSTERONE**. For the American Board of Pediatrics, CONGENITAL ADRENAL **HYPER**PLASIA usually ='s 21-HYDROXYLASE DEFICIENCY (not 11-Hydroxylacse Deficiency)! **Shocky, septic picture**. The newborn screen usually tests for the most common form of Congenital Adrenal Hyperplasia, which is 21-hydroxylase deficiency. The newborn screen is considered positive if there is a high 17-hydroxyprogesterone level. Can develop adrenal insufficiency and a crisis-like picture with abnormal electrolytes and shock-like symptoms. Boys may have HYPERPIGMENTED scrotums from elevated ACTH being released from pituitary. Girls usually present at birth with ambiguous genitalia. If the newborn screen is abnormal but the physical exam is ok, repeat a 17-hydroxyprogesterone level. In pregnancy, if the baby has a first degree relative with a history of CAH, get gene testing on fetal cells. **TREATMENT** requires both Hydrocortisone (for the lack of cortisol) **and** Fludrocortisone (for the lack of aldosterone), just like primary Adrenal Insufficiency (or Primary **ADD**isons).

**PEARL**: In boys, testing for this on the newborn screen is extremely important. Since 21-Hydroxylase Deficiency results in increased androgens, male genitalia is often NORMAL. So, **boys often do not present at birth**. Girls, however, have ambiguous genitalia due to the increased androgens are much more likely to present at birth, and therefore get timely treatment if they present with a "septic shock" type of picture.

**PEARL**: Late onset congenital adrenal hyperplasia (AKA Nonclassic Congenital Adrenal Hyperplasia) can present with signs of early puberty.

**PEARL/MNEMONIC**: You could be asked about which part of the adrenal gland makes a deficient hormone. Try to remember the layers of the adrenal cortex as **GFR**. Also, use the following as a memory aid: "The deeper you go, the sweeter it is." Yes, it's a sexual mnemonic. The layers go from salty to sweet. G = Glomerulosa = Aldosterone (salty!). F = Fasciculata = Cortisol/Glucocorticoids (sweet!). R = Reticularis = Sex steroids (deep and sexy!)

## 21-HYDROXYLASE DEFICIENCY

21-Hydroxylase Deficiency is the autosomal recessive disorder that's usually responsible for Congenital Adrenal Hyperplasia on the pediatric boards. During pregnancy, if either parent is a known carrier (or has the disorder), oral Dexamethasone should be given to the pregnant mother until the sex of the baby can be determined (checked by amniocentesis or chorionic villus sampling). If the sex is male, treatment may stop (even if the baby has the disorder) because an excess build up of androgens during pregnancy has no major side effects on males. After birth, the baby will need steroid treatment. If the baby is found to be female **and** has the disorder on genetic testing, treatment should continue. This is extremely important and is done to prevent female virilization.

**21-HYDROXYLASE DEFICIENCY RELATED CONGENITAL ADRENAL HYPERPLASIA**

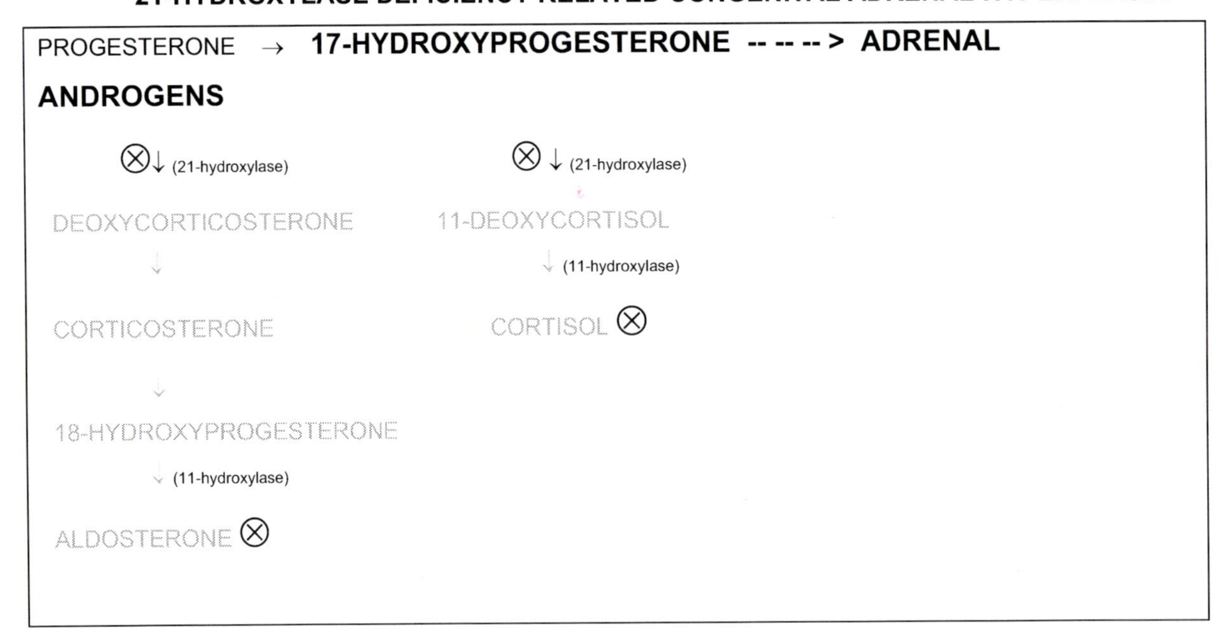


## 11-HYDROXYLASE DEFICIENCY

11-HYDROXYLASE DEFICIENCY can also cause Congenital Adrenal Hyperplasia, but is much less common. This is probably **not** the answer that the ABP is looking for, but be familiar with it. If a patient has symptoms/labs consistent with Congenital Adrenal Hyperplasia **and also has** hypertension, then this is the answer. This is a non-salt wasting form of CAH, so sodium levels are NORMAL.

**MNEMONIC**: 11 = 1-1 = H = HTN. The 11- in 11-Hydroxylase Deficiency can be made to look like the H of HTN, or "1-1TN." It works even better on paper. Try it!

## 17-HYDROXYLASE DEFICIENCY

17-Hydroxylase deficiency is an extremely rare cause of Congenital Adrenal Hyperplasia. There is decreased cortisol and sex steroid synthesis, and increased mineralocorticoid synthesis. This results in Congenital Adrenal Hyperplasia symptoms + HYPERTENSION + hypokalemia. You can also have a metabolic alkalosis.

## PANHYPOPITUITARISM

In panhypopituitarism, low ACTH levels result in decreased production of androgens, so this CAN present with ambiguous genitalia and/or micropenis in males. WILL be accompanied by hypoglycemia because of decreased cortisol production. Can be part of Prader Willi or Kallmann ("nose/midline issues").

# ***AMBIGUOUS GENITALIA & CHROMOSOMAL ABNORMALITIES***

**NOTE**: This section is ripe for questions on the smallest nuances. If you understand a few key points, you'll sail through these questions. Most review books simply give you the name of a disease and the associated findings for this section. Understanding the PEARLS below will be instrumental in helping you understand these conditions.

**PEARLS:**

* The default structure/phenotype of all embryos is the FEMALE phenotype (hence, it's an EASY and PASSIVE process), including the creation of Mullerian Ducts. Those need to REGRESS to give a male INTERNAL phenotype.

* For embryos with a Y chromosome, the PRESENCE OF ANDROGENS and FUNCTIONAL RECEPTORS create the male EXTERNAL genitalia. BOTH must be present. So, if there are no androgens, a penis and scrotum are not created from the default vagina like structure.

* The "Y" chromosome also means the embryo should produce the MULLERIAN INHIBITOR HORMONE/FACTOR/SUBSTANCE (MIH/MIF/MIS), which results in the regression of the default FEMALE **INTERNAL** duct structures.

**MNEMONIC**: Sorry ladies, but for the sake of this mnemonic, just go with the idea that women are passive/easy. "Becoming a woman is PASSIVE/EASY. Becoming a MAN is HARD!" HARD… MAN… I think you get the picture.

## AMBIGUOUS GENITALIA

In patients with ambiguous genitalia, ask yourself: Is the penis malformed? Is the meatus normal looking? Are testes present? Could it be a large clitoris that looks like a penis?

## MICROPENIS

A micropenis is one that **LOOKS NORMAL**, but it is < 2 cm in outstretched length. TESTES are present in the scrotum. Look for any signs that of possible GROWTH HORMONE DEFICIENCY, PRADER-WILLI, KALLMANN SYNDROME or PANHPOPITUITARISM.

## TESTICULAR FEMINIZATION

TESTICULAR **FEMINIZATION** = XY karyotype = X-LINKED RECESSIVE = "te**X**ticular feminization" = Receptor insensitivity to androgens so no male external genitalia develop even though testes are present. Look for a blind ending vagina, lack of uterus and lack of ovaries in what may appear to be a phenotypic female. Testes may be found in the inguinal canal. May even present as PRIMARY AMENORRHEA in a "female" with BREASTS but no pubic hair!

**PEARLS:**

* If the test mentions a family history of MATERNAL "AUNTS" who are STERILE, they are probably XY TE**X**TICULAR FEMINIZED **UNCLES**.

* Patients are XY, so MIH IS PRESENT. Therefore, there are NO **INTERNAL** FEMALE STRUCTURES

* Androgens are also present, but the receptors are insensitive. Therefore, the default programming kicks in and **EXTERNAL** female genitalia (blind vagina) is formed.

* Since patients are not sensitive to androgens, there is NO ADRENARCHE/PUBARCHE. Estrogen receptors work, so the phenotypical female WILL DEVELOP BREASTS. + Breasts, NO hair, NO menses.

## MULLERIAN INHIBITOR HORMONE DEFICIENCY (aka MIH RECEPTOR DEFECT)

A child with Mullerian Inhibitor Hormone deficiency (MIH receptor defect) is born with normal testes/penis (androgens work), but **also** a rudimentrary uterus and fallopian tubes. The uterus and fallopian tubes did not regress because of the lack of MIH, or the lack of working MIH receptors. This child's karyotype will reveal an X and a Y.

## MALE PSEUDOHERMAPHRODISM

**MALE** PSEUDOHERMAPHRODISM = INCOMPLETE MASCULINIZATION: Once again, XY karyotype. Genitalia looks AMBIGUOUS or FEMALE because of inadequate androgen production during critical stages of fetal development.

**PEARL**: The word MALE should key you in on the fact that this patient is XY (in case they present this in a matching section).

## TRUE HERMAPHRODISM

True hermaphrodism refers to a patient with BOTH testicular and ovarian tissue. The condition is rare. If you see this as an answer choice, it's probably a decoy.

## (DOUBLE TAKE) TURNER SYNDROME (aka TURNERS)

Turner Syndrome (aka Turners) has a high association with Aortic Coarctation. Look for a WEBBED NECK (also called CYSTIC HYGROMA). May have a history of having only breast buds, but they do NOT have breasts. Short stature, pedal edema, wide spaced nipples, short 4$^{th}$ and 5$^{th}$ metacarpals and may have either scant or no pubic hair. Primary amenorrhea occurs because of streak ovaries and ovarian failure. They **DO** have a uterus. High FSH always means ovarian failure (top 2 reasons are Turners and autoimmune ovarian failure) = HYPERGONADOTROPIC HYPOGONADISM. Renal anomalies including a possible HORSESHOE kidney.

**IMAGE**: http://bioh.wikispaces.com/file/view/turner.jpg/33628275/427x427/turner.jpg

**IMAGE**: http://www.endocrineonline.org/gif%20box/Ullrich.gif

**IMAGE**: http://newborns.stanford.edu/images/footedemaDAClark.jpg

**MNEMONIC**: Association between XO and coarctation of the Aorta - "XO kind of looks like AO (abbreviation for aorta), doesn't it?"

**MNEMONIC**: Coarctation of the Aorta and Horseshoe Kidney = In case you didn't know, Tina TURNER was married to a man who was reportedly abusive. His name was Ike. Now, imagine Tina TURNER'S hand thrusting through Ike's chest and squeezing his AO... He TURNS and hits her in the head with a HORSESHOE!" KEY: Squeezing his AO = Coarctation of the Aorta. Horseshoe trauma = Horseshoe Kidney.

### (DOUBLE TAKE) KLINEFELTER SYNDROME (aka KLINEFELTERS)

Klinefelter Syndrome (aka Klinefelters) = XXY. Gynecomastia, small testicles/infertile and normal intelligence to MILD MR. May have a mild motor or speech delay. Tall stature with long arms/legs. Low upper to lower segment ratio. Workup = REFER FOR CHROMOSOMAL ANALYSIS!

**MNEMONIC:** Gynecomastia = Kalvin Kline FELT HER BREASTS

**PEARL**: Mild mental retardation may be described as a patient who is "awkward," "below average in school" or even just "shy." Generally, though, these patients have some learning disabilities but their IQ can be normal.

**PEARL**: Someone who is tall with long arms/legs could also be described as having a LOW "upper to lower segment ratio."

## *DIABETES MELLITUS*

### HONEYMOON PERIOD

In diabetes mellitus, the "honeymoon period" usually occurs soon after the diagnosis of **Diabetes Mellitus Type 1**. A lower insulin requirement occurs as the remaining islet cells begin to function again after the glucose toxicity is relieved via exogenous insulin injections. These remaining islet cells begin to produce insulin again and relieve the patient from having such a high insulin requirement.

**MNEMONIC:** After the beginning of any budding relationship comes a short time referred to as the "honeymoon period." Why should it be any different between a patient and their relationship with DM?

### HEMOGLOBIN A1C

A hemoglobin A1C of 8 is OKAY in kids. Tighter control is dangerous. An A1C of 10 or higher is BAD and can result in poor growth and hepatomegaly.

### SOMOGYI EFFECT & DAWN PHENOMENA

Somogyi effect and Dawn phenomena have been tested less in recent years. Both phenomena produce MORNING HYPOGLYCEMIA. Need blood sugars (BS) checked multiple times at night to differentiate between these two.

* **SOMOGYI EFFECT:** The hypoglycemia occurs VERY, VERY early in the morning. Cause is excess insulin administration at night resulting in hypoglycemia around 2 or 3 AM (patients may report diaphoresis or nightmares in the middle of the night). Look for REBOUND HYPERGLYCEMIA at 6 AM. Treatment = DECREASED nighttime insulin

- **MNEMONIC:** Imagine "Mr. Miyagi from that karate movie is having trouble reading his insulin vial. What's the worst thing he could do? He gives himself TOO MUCH insulin every night. Treatment, get him some glasses so he can self administer the right dose!"

* **DAWN PHENOMENON:** Hypoglycemia at 6 AM due to a PHYSIOLOGIC RELEASE OF GROWTH HORMONE. Treat by giving the nighttime insulin dose LATER.

- **MNEMONIC:** Hypoglycemia is at 6 AM... "At the break of DAWN!" Was named after the time of day that the hypoglycemia occurs. Dawn is the time at which daylight first begins.

## HYPOGLYCEMIA

Hypoglycemia is generally considered to be a glucose level < 60. Give 15 grams of fast acting carbohydrates (corn syrup, crackers, juice, sugar, soda) and recheck in 15 min. If still < 60, repeat carbohydrate bolus. If giving dextrose via IV, the general rule is 3 ml/kg of D10.

**MNEMONIC/SHORTCUT:** "Hawaii 5-0 rule" = Divide 50 by the percentage of the solution you have available to give to obtain the **ml/kg** of the dextrose bolus. For example, if you have D10 available and you want to treat a 9 kg hypoglycemic child, you would give 50 ml of D10.

* 50 ÷ d**10** = 5. So you would give 5 ml/kg of D10 to this 9 kg child, which equals 45 ml.

## DIABETIC KETOACIDOSIS or HYPEROSMOLAR NON-KETOTIC HYPERGLYCEMIC STATE

DIABETIC KETOACIDOSIS or HYPEROSMOLAR NON-KETOTIC HYPERGLYCEMIC STATE = D**K**A or HONK = Replace fluid over 36-48 HOURS, NOT over 24 hours because of the risk for cerebral edema related to rapid osmolar shifts. If the patient has a low sodium level, it's probably PSEUDOHYPONATREMIA and does NOT need hypertonic saline. Start with NS boluses. D**K**A = **K**+ deficit. Give up to 60 meq/L of potassium if the K+ is low. Add glucose to IV fluids once blood glucose (BS) is < **300**. DON'T give bicarbonate because there are only rare indications to it (pH < 7.1).

## PSEUDOHYPONATREMIA

PSEUDOHYPONATREMIA results when glucose levels exceed the upper limit of normal by 100. This results in a decrease in Na+ by 1.6 for every increase in glucose by 100.

## ACANTHOSIS NIGRICANS

Acanthosis Nigricans represents insulin resistance. Associated with a low HDL and high TG's. Hyperglycemia is less often seen because as insulin sensitivity decreases, insulin production increases to match the need. Hyperglycemia is seen if a patient finally has enough insulin resistance to become diabetic.

## METABOLIC SYNDROME

"METABOLIC SYNDROME" = Truncal obesity, low HDL, high TG, high BP +/- FBG > 100. Acanthosis nigricans is a NOT part of the definition.

# OB/GYN & SOME STDs

## ***OBSTETRICS***

### ORAL CONTRACEPTIVE PILLS (OCPs)

* Absolute Contraindications to oral contraceptive pills (OCPS): Pregnancy, liver disease, **hyperlipidemia**, breast cancer, breastfeeding at < 6 weeks

* Relative Contraindications: Hypertension

* Anticonvulsants: Interfere absorption so instruct patient to use a backup method of contraception. Consider using an OCP with more ESTROGEN

### CONCEPTION

50% of all conceptions occur within 6 months of a girl/woman losing her virginity. 20% of all pregnancies occur within the 1st month!

### PRENATAL CARE (PNC)

8 prenatal care visits are considered adequate PNC

### GROUP B BETA HEMOLYTIC STREPTOCOCCUS (GBS)

Any child born to a mother who is GBS+ should be watched in the hospital for at least 48 hours.

### GESTATIONAL DIABETES MELLITUS

Unlike regular DM, gestational diabetes mellitus is NOT associated with congenital anomalies.

### SERUM ALPHA-FETOPROTEIN (AFP) SCREEN

A serum alpha-fetoprotein (AFP) screening is offered at 16 weeks gestation. If elevated (more than 2 times the normal limit), there is likely some type of defect in which the baby's **skin is broken**. This is usually a **neural tube defect**.

* If AFP is elevated, get a fetal ultrasound.

* If the ultrasound is negative, obtain an amniocentesis to get amniotic AFP levels.

* Elevated AFP is also seen in gastroschisis (broken skin!), other abdominal wall defects, fetal demise, renal anomalies, wrong dates and wrong number of kids (twins).

*** LOW SERUM AFP is associated with DOWNs syndrome.**

- **MNEMONIC**: "LOW kind of means, DOWN"

## CHORIONIC VILLUS SAMPLING

A chorionic villus sampling is a fetal karyotype obtained from a placental biopsy. This has to do with CHROMOSOMES. This is the test that is most helpful for **early** detection (usually around 12 weeks) of any genetic/cytogenetic abnormalities you're concerned about. For very high risk pregnancies, may even be considered in the 1st trimester.

**MNEMONIC**: "CHROMOSOMIC" Villus Sampling. Also, imagine standing outside of a CVS at 12' O Clock midnight and craving a bag of chips. They're closed, so throw a DOZEN EGGS at the CVS in anger. (MN and eggs representing 12 weeks)

**PEARL**: "CHROMOSOMIC" Villus Sampling has to do with chromosomes, so it SHOULD NOT be performed for suspected neural tube defects (those are usually due to a folate deficiency).

## AMNIOCENTESIS

An amniocentesis screening is offered at 15-18 weeks, but may be performed **at the very beginning of the 2nd trimester/after 12 wks for high suspicion or concern**. Like "CHROMOSOMIC" Villus Sampling, this helps look for identifiable genetic issue. This test **also** can also help identify neural tube defects (via AFP levels leaked into the amniotic fluid from the neural tube defect).

* There's really only one 3rd trimester amniocentesis indication for amniocentesis, and that's to assess fetal lung maturity.

## MATERNAL SERUM TRIPLE SCREEN OR QUADRUPLE SCREEN

The maternal serum triple screen or quadruple screen are other 2nd trimester options and are done with a simple serum sample. They **screen for DOWNS SYNDROME,** but can also help with screening for abdominal wall issues and even some renal anomalies. For the test, they would only expect you to know what pattern of lab values is considered positive for Down Syndrome (Trisomy 21). It is positive for DS if the following pattern is present:

* LOW AFP: **Low = "Down" = Down Syndrome**
* LOW Uncongugated Estrio**L** = "estri**oLo**w"
* **H**IGH **H**CG Subunit
* HIGH Inhibin A: This is part of the quadruple screen. HIgh inHIbin

## FIRST TRIMESTER SCREENING OPTIONS FOR DOWNS SYNDROME

First trimester screening options for downs syndrome include:

* Chorionic Villus Sampling
* HCG subunit: **H**igh. Usually done along with an ultrasound to look for nuchal translucency

## NON STRESS TEST

A non stress test measures fetal heart rate in response to fetal movements. Used as a measure of the "integrity of the fetal nervous system." If a mom says that her baby is not moving, GET A NON STRESS TEST.

* REACTIVE: GOOD

* NONREACTIVE: That's BAD. If this occurs, get a Biophysical Profile.

* **PEARL**: A non stress test has NOTHING to do with maternal contractions.

## BIOPHYSICAL PROFILE (BPP)

A Biophysical Profile refers to an ultrasound that is done to check breathing, tone, movement, amniotic fluid, heart rate and tone.

## STRESS TEST (aka CONTRACTION STRESS TEST)

A contraction stress test evaluates fetal response to **contractions**. Assesses the fetus for **uteroplacental insufficiency and tolerance of labor.**

**MNEMONIC**: Contractions are STRESSFUL!

## FOLIC ACID

**For any woman who is trying to conceive, 4 mg of folic acid per day is recommended to prevent neural tube defects.** For any woman NOT trying to get pregnant, 0.4 mg (400 mcg) of folic acid per day is recommended.

**MNEMONIC**: **FOUR mg of FOUR**lic acid (not 0.4 mg)

## LUNG MATURITY

* PHOSPHATIDYLGLYCEROL is the best predictor of fetal lung maturity. Why? Because it's not made until after the $35^{th}$ week of gestation.

* LECITHIN-SPHINGOMYELIN RATIO: A ratio > 2 (from the amniotic fluid) indicates fetal lung maturity

## MONOZYGOTIC TWINS

In monozygotic twins, one zygote splits into two, so they must be identical twins. There can be ONE or TWO amniotic sacs, and the twins can share ONE placenta, or they can have TWO (one for each).

* MONOAMNIOTIC = 1 Amniotic sac is contained in the uterus and houses both twins
* DIAMNIOTIC = 2 Amniotic sacs are contained in the uterus, each housing one twin
* MONOCHORIONIC = 1 Placenta in the uterus supports both twins
* DICHORIONIC = 2 Placentas are in the uterus, one for each twin
* **PEARL/SHORTCUT:** The one permutation that is NOT POSSIBLE is MONOAMNIOTIC/DICHORIONIC. So, just remember that if both twins are in ONE amniotic sac, they will share ONE placenta. All other combinations are possible.
* **IMAGE**: GREAT ILLUSTRATION - http://commons.wikimedia.org/wiki/File:Placentation.svg
* **PEARL**: Confused? Break down the words and you'll be fine (- Zygotic, - Amniotic, - Chorionic).

## DIZYGOTIC TWINS

Dizygotic twins occur when two separate eggs were fertilized so these are NOT identical. Called "fraternal twins." **Di**chorionic. For the test, these twins always have their own amniotic sacs and their own placenta (Di/Di).

* **PEARL**: Confused? Break down the words and you'll be fine (- Zygotic, - Amniotic, - Chorionic).

## GYNECOLOGY & SOME STDs

### PARENTAL CONSENT

Parental consent is NOT needed to treat a child/adolescent for STDs or for substance abuse.

### (DOUBLE TAKE) CHLAMYDIA TRACHOMATIS

Chlamydia trachomatis can cause urethritis, conjunctivitis and pelvic inflammatory disease (PID). In neonates, it can cause a pneumonia associated with a **staccato** cough. It's the most common STD. PID can lead to ectopic pregnancies and infertility. Eye infections can lead to blindness. Conjunctivitis in a neonate (less than a month old) should raise concern for this as the etiology (vertical transmission). It's an obligate **intracellular** anaerobe. Getting cultures is difficult, so order PCR of CELLS, secretions or urine. Chlamydia can also cause lymphogranuloma venereum (LGV), which is an STD that initially starts with **small nontender** papules or shallow ulcers that resolve. **Then** a **TENDER** UNILATERAL INGUINAL lymph node appears that can rupture, relieve the pain, and then possibly drain for months.

* SEXUALLY TRANSMITTED DISEASE (STD): Treat with **DOXYCYCLINE** for 7 days. For PELVIC INFLAMMATORY DISEASE, treat for 14 days. For STD recurrence, treatment failure or noncompliance concerns, treat with Azithromycin x 1. Other possible medications include erythromycin, levofloxacin or ofloxacin for multiple doses/day x 7 days.

* CONJUNCTIVITIS: Treat with **oral erythromycin** to eradicate nasopharyngeal colonization which can lead to pneumonia.

* (DOUBLE TAKE) LYMPHOGRANULOMA VENEREUM SEROVAR:

Lymphogranuloma Venereum Serovar is an STD caused by Chlamydia trachomatis. Rare in the U.S. More common in tropical areas. Starts as small nontender papules or shallow ulcers which resolve. Eventually, a **TENDER UNILATERAL INGUINAL** lymph node appears. Pain is relieved when it ruptures. The node can **continue to drain** for months. Treat with DOXYCYCLINE or erythromycin.

* **PEARL**: In general, when you think the diagnosis is due to a Chlamydia species, choose doxycycline if the child is > 8 years of age, or choose a macrolide (usually erythromycin). Also, this is an intracellular organism. Look for the phrase "**intracytoplasmic inclusions**."

### NEISSERIA GONORRHEA

Neisseria gonorrhea creates a smelly, greenish discharge. USUALLY asymptomatic, so consider this in the differential for any **adolescent patient with arthritis**. Treat with Ceftriaxone x1 or oral Cefixime by mouth x1.

* **MNEMONIC**: Gonorrhea is also known as "the CLAP." The reason it's called the CLAP is because on Gram stain, the diplococci kind of look like two hands clapping. "C for the Clap, and C for Ceftriaxone or C for Cefixime PO x1." If you are wondering about "GC," it comes from "GonoCoccus."

* DISSEMINATED GONORRHEA: Once the disease has disseminated, the local symptoms are no longer present! Look instead for a rash and joint involvement. Can also cause meningitis and endocarditis. This is an

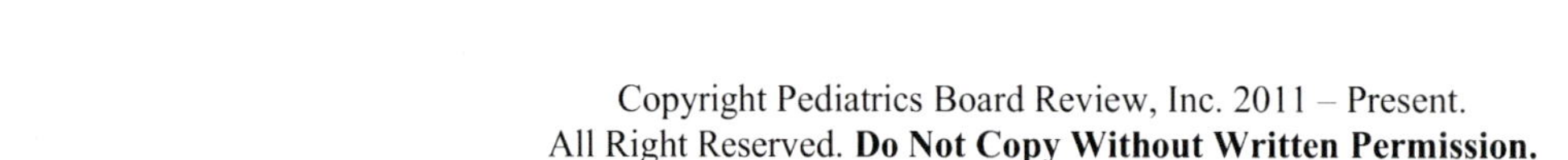

**INTRACELLULAR** diplococci. Fitz-Hugh/PID. When suspecting disseminated Gonorrhea (e.g., single pustule + swollen knee), culture any pustules as well as all orifices.

## NONGONOCOCCAL URETHRITIS

Nongonococcal urethritis is usually associated with scant, mucoid discharge as opposed to the profuse, purulent discharge of gonorrhea. Differential includes Chlamydia, Ureaplasma, Mycoplasma genitalium, HSV, Trichomonas and even adenovirus.

## PELVIC INFLAMMATORY DISEASE (PID)

Pelvic inflammatory disease (PID) can result in **potentially serious outcomes** (sepsis, infertility) so it should be **treated on the basis of your clinical diagnosis.** If there is no improvement on appropriate outpatient therapy and cultures are negative, obtain an ultrasound to evaluate for a tuboovarian abscess. Also obtain an RPR to look for Syphilis and consider getting HSV PCR (not HSV titers since they are nonspecific). Treatment options depend on the setting:

* **OUTPATIENT**: Ceftriaxone (or Cefoxitin) x 1 + (Doxycyline BID x 14 days or AZITHROmycin x1) +/- Metronidazole (optional)

* **INPATIENT**/IV: Cefotetan (or Cefoxitin) + Doxycycline. Alternatives: Clindamycin + Gentamicin, or Augmentin (Ampicillin/Sulbactam) + Doxycycline. Inpatient therapy is reserved for patients who are pregnant, have intractable nausea and vomiting, high fever, or for those who failed outpatient therapy.

## FITZ-HUGH CURTIS SYNDROME (aka PERI-HEPATITIS)

FITZ-HUGH CURTIS SYNDROME (aka PERI-HEPATITIS) due to Gonorrhea or Chlamydia. **LFTs are NORMAL.** Do a pelvic exam and obtain cultures if this is suspected. Do NOT order LFTs as your next step.

## (DOUBLE TAKE) SYPHILIS

Syphilis is caused by TREPONEMA PALLIDUM. Non-treponemal tests (+RPR or VDRL) can be FALSE positives so you need to **do a confirmatory treponemal test (FTA). If a mom was RPR+ but the FTA was not done, obtain the FTA in the baby**. FTA doesn't correlate with disease activity, the non-treponemal tests do, and they may eventually disappear. So once disease presence is confirmed with FTA, look at disease activity with non-treponemal titers to help guide management. If mom was treated and the baby's titers are lower than hers, it's safe to assume those are just the mom's IgGs that crossed placenta. There is NO NEED TO TREAT, just follow the titers. **If mom was treated < 1 month ago, TREAT.** If mom was given Erythromycin, TREAT because it doesn't cross the placenta.

* CONGENITAL SYPHILIS: Baby born with maculopapular rash, HSM, **PEELING SKIN, PERFORATED NASAL SEPTUM** or **HUTCHINSON TEETH**. These teeth are peg-shaped (cone-like) teeth but also have a **central notch** that is extremely specific for congenital syphilis. Treat with PENICILLIN (PCN).

* CONDYLOMA LATA: Refers to SECONDARY SYPHILIS, in which white-gray coalescing papules are seen.

* **PEARL:** If the FTA is positive but VDRL is negative, also consider LYME DISEASE (BORRELIA BURGDORFERI).

* **NAME ALERT/MNEMONIC**: Condyloma LATA (aka "condyloma FLATa," are much more FLAT than Condyloma ACUMINATA (found with HPV infections).

* **NAME ALERT/MNEMONIC**: Peg teeth are also found in patients with Incontinentia Pigmenti (aka "incontinentia PEGmentia")

* **IMAGE**: (PEG-SHAPED TEETH) - http://alturl.com/cysjv

* **IMAGE**: (HUTCHINSON TEETH) - http://alturl.com/jc6nf

## BACTERIAL VAGINOSIS (aka GARDNERELLA)

**Bacterial vaginosis (Gardnerella vaginalis), has a Homogenous**, smelly/fishy odor. Discharge can be white, yellow or **gray.** Associated with anything that changes the usual flora, including IUD's and antibiotic use. Look for CLUE CELLS. The pH is Alkaline and > 4.5.

**MNEMONIC**: "the GARDNER found a missing CLUE in the GARDEN. It was a dead and smelly FISH!"

## (DOUBLE TAKE) TRICHOMONAS VAGINALIS

Trichomonas vaginalis is a protozoa. Look for yellow-green discharge, bubbly, **frothy,** motile flagella and pH > 4.5. Can be intensely pruritic. **Strawberry cervix**. Treat with one dose of metronidazole. Can cause nongonococcal urethritis.

**MNEMONIC**: Imagine 2 girls having this conversation. "Can you believe that dirty old man TRICKED MONA, our favorite BUBBLY stripper, into putting a STRAWBERRY into her va-jay-jay?!? Gross! I heard it still had a long green STEM on it!" Sorry for the gross mnemonic, but hopefully it helps you associate the stem with the motile TAIL/FLAGELLA of Trichomonas.

## (DOUBLE TAKE) HERPES SIMPLEX VIRUS (HSV)

* **HERPES SIMPLEX VIRUS (HSV) AS AN STD:** The initial flare is often very painful. Pain may precede the presentation of lesions. Look for multiple, painful ulcers or vesicles on the labia or penis. Can have lymphadenopathy. The vesicles are CLUSTERED on an ERYTHEMATOUS BASE. Lesions can also be ULCERATIVE. Diagnose by obtaining a viral culture or HSV PCR. The Tzanck smear is not specific for HSV. Treat with ORAL Acyclovir x 7 days (not topical). Treat babies with **IV** Acyclovir.

- **IMAGE**: http://www.coldsoreblisters.com/images/herpes1.jpg
- **PEARL**: HSV can be associated with a very painful infection called a HERPETIC WHITLOW (typically of a thumb or finger).
- **IMAGE**: http://img.medscape.com/pi/emed/ckb/pediatrics_general/960757-964866-1733tn.jpg

* **HERPES SIMPLEX VIRUS ENCEPHALITIS (HSV ENCEPHALITIS):** Look for fever, seizures and mentioning of the **temporal** lobe on CT brain. Treatment is STAT **IV acyclovir, followed by a lumbar puncture** to obtain fluid for PCR testing. EEG might show PLEDs (periodic lateralizing epileptiform discharges).

* **HERPES SIMPLEX VIRUS GINGIVOSTOMATITIS:** Oral and perioral/vermillion border lesions/vesicles. Gingiva is friable and malodorous. There is associated lymphadenopathy.

- **IMAGE**: http://img.medscape.com/pi/emed/ckb/pediatrics_general/960757-964866-1734tn.jpg

## VAGINAL FOREIGN BODY

Malodorous (or bloody) vaginal discharge, think foreign body. Remove the object and prescribe sitz baths.

## ULCERS versus DISCHARGE

Sometimes treating by clinical suspicion alone is fine while the workup is pending. Use some of these high yield facts to help guide your decision.

* **DISCHARGE**: Think Chlamydia, Gonorrhea, Bacterial Vaginosis, Trichomonas or a foreign body. Anytime you treat for suspected Gonorrhea, ALSO treat for Chlamydia (much more common)

* **PAINLESS ULCERS**: Think Syphilis, LYMPHOGRANULOMA VENEREUM/LGV (due to CHLAMYDIA)

* **PAINFUL ULCERS**: HSV (painful, coalesce and can have a herpetic whitlow on the thumb), HEMOPHILUS DUCREY: "DO CRY," BEHCET'S SYNDROME (aphthous & genital ulcers + uveitis + arthritis + GI symptoms).

## VAGINAL DISCHARGE AT BIRTH

In an otherwise well baby, if you note pink or white vaginal discharge in the perinatal period, that is OKAY. It's probably due to mom's ESTROGEN. If, however, you note signs of bleeding or petechiae, consider VITAMIN K DEFICIENCY.

## VAGINAL ADHESIONS (PENILE ADHESIONS for boys)

VAGINAL ADHESIONS (or PENILE ADHESIONS for boys) usually resolve without intervention. If they are present in a girl and she has dysuria, treat using ESTROGEN cream

## BARTHOLIN GLAND CYSTS

In patients with bartholin gland cysts, look for a large and tender vulva, along with a sebaceous cyst. The cyst will look flesh colored and arise from the skin. This is painful, but not associated with discharge, bleeding or STDs.

## SEXUAL ABUSE IN GIRLS

If you diagnose an STD in a child, consider SEXUAL ABUSE.

* **PEARL**: Look for clues in the question that the patient has been abused. Perhaps something about a child who has a babysitter, or even a relative who acts as the caretaker during the day.

* PEDICULOSIS PUBIS = CRABS = LICE

* GONORRHEA

* CONDYLOMA ACUMINATA (HPV)

* CHLAMYDIA

  - **NOTE**: Chlamydia can be due to vertical transmission and be culture positive up to 3 years of age

# ALLERGY & IMMUNOLOGY

## ***<u>HAY FEVER, FOOD ALLERGIES AND ALLERGIC RASHES</u>***

### HAY FEVER

Hay fever due to pollen takes a few years to develop. So, if the patient is a child < 3 years of age with rhinitis in the winter, consider other diagnoses.

* **R**agweed frequently causes **R**espiratory symptoms along with the usual sneezing, allergic conjunctivitis, rhinitis and tearing of hay fever.

### CHRONIC RHINITIS

The differential for chronic rhinitis includes HAY FEVER, SINUSITIS, CYSTIC FIBROSIS, FOREIGN BODY, a nasal POLYP and VASOMOTOR RHINITIS

### VASOMOTOR RHINITIS

Triggers for vasomotor rhinitis may include emotions, cold wind, change in temperature (or humidity) and pollution. Typical environmental allergens are NOT triggers.

### SKIN TESTING

SKIN TESTING = AEROALLERGEN TESTING

* FALSE **negatives** can occur when patients are on antihistamine, or antidepressants!

* NEGATIVE PREDICTIVE VALUE: NPV of skin testing for foods or inhalants is <u>excellent</u>.

* POSITIVE PREDICTIVE VALUE: PPV is good for inhalants but not good for foods.

- **<u>PEARL/SHORTCUT:</u>** All skin-testing results should be considered fairly reliable, regardless of whether they are positive or negative, for food, aeroallergen, or inhalant), The one **EXCEPTION** being a POSITIVE skin-testing result for food allergies.

### IMMUNOTHERAPY

About 50% of patients with hay fever respond to immunotherapy. There's a 0.5% chance (1 in 200) of having a severe reaction during therapy. If it happens, it will usually occur within 30 minutes. Most likely during either peak pollen seasons or during the within the first year of therapy (rapid build up phase).

### RADIOALLERGOSORBENT TESTING (aka RAST)

Since radioallergosorbent testing (RAST) is IN VITRO (blood) testing, it is NOT affected by taking antihistamines.

### FOOD ALLERGIES

Early introduction of solids results in an increased chance of food allergies and may predispose children to obesity later in life. ABDOMINAL PAIN may be the only sign of impending anaphylaxis. If the patient has a history of a food allergy, GIVE IM EPINEPHRINE.

* **PEARL**: After an exposure, symptoms may continue to EVOLVE even up to 2 hours later. It is highly unlikely that the ONSET of an allergic response will start 2 hours after the food exposure.

* **INFANTS/TODDLERS**: Allergies to eggs, milk, soy or wheat are common. Will usually outgrow these by 5 years of age.

* **OLDER CHILDREN**: Allergies to shellfish and peanuts are common.

  - **PEARL**: There is NO contraindication to giving contrast in kids with a shellfish allergy. Just give it. No need to pre-treat.

## PEANUT ALLERGY

Although peanuts are legumes, for patients with a peanut allergy you do not need to recommend against exposure to all legumes. Instead against exposure to TREE NUTS since they are often processed in the same factories as peanuts, and because there is a 25%-50% likelihood that these patients ALSO have a tree nut allergy.

**PEARL**: "Peanut" is a misnomer because it is not a tree nut. Most tree nuts have the word NUT in them (e.g., hazelnut, macadamia nut, etc.). Legumes typically have the word BEAN in them (e.g., black bean, fava bean, lima bean, etc.). Don't get confused on the peds boards.

## FOOD "SENSITIVITIES"

Foods like spicy foods and beans may be associated with abdominal discomfort +/- flatus.

## ECZEMA (aka ATOPIC DERMATITIS)

Breastfeeding x 6 months, or using hypoallergenic formula DOES DELAY THE ONSET of eczema (aka atopic dermatitis), but does not reduce its incidence. The contribution of early food ingestion to the development of atopic dermatitis is controversial. Eggs, fish, milk, peanut, soy, wheat, and strawberries are the foods thought to possibly contribute. Get the patient tested for allergies to them before restricting their diet.

## URTICARIA

If urticaria is due to a bee sting, give EPINEPHRINE **and** refer the patient to an allergist. If there is a life-threatening reaction (urticaria, edema and airway compromise), the patient will need venom immunotherapy.

## CHRONIC URTICARIA (> 6 wks)

Chronic urticaria is usually an autoimmune, IgE mediated phenomenon. It can result in urticaria occurring at any time of the day or night. Often due to a food exposure.

## ARTIFICIAL FOOD COLORING

Artificial food coloring does NOT cause urticaria. Old reports were based on poor studies.

## (DOUBLE TAKE) ANAPHYLAXIS

The following is a review of some common issues to be aware of in case a child presents with anaphylaxis:

* FOOD EXPOSURE: Watch for 4 hours before discharging from the peds ER.

* RESPIRATORY COMPROMISE: Intubate and quickly secure the airway **first**. THEN give EPInephrine (1:1000, "3 zeroes for 3 letters of EPI"), diphenhydramine, IVF and steroids.

* BETA BLOCKERS: They blunt the response to EPInephrine. Give glucagon to reverse the beta blocker effect and then give EPInephrine (again).

* ANAPHYLACT**OID** REACTIONS: Result from mast cell degranulation. These are not true IgE mediated reactions. May be seen with use of NSAIDS, Opiates, Vancomycin and Contrast. Pretreat with steroids and diphenhydramine.

## FIXED DRUG REACTION

A fixed drug reaction presents as a violaceous rash that presents on the same part of the body every time a particular medication is taken

## TRUE MILK PROTEIN ALLERGY

True milk protein **allergy** is a true IgE mediated allergy. Look for GI symptoms (vomiting, diarrhea, GI bleeding) and **extraintestinal** symptoms (eczema, wheezing) to make the diagnosis! Refer to an allergist and prescribe casein **HYDROLYSATE** formula (contains hydrolyzed cow milk proteins). **NOT soy** (there are some reports of cross reactivity). Can also try asking the mother to discontinue ingestion of all milk products.

**MNEMONICS**

* On the peds boards, if "milk protein **ALLERGY**" is an answer choice, it is referring to a true IgE mediated ALLERGY so it should present with more than just GI symptoms.

* **M**P = **M**ilk **P**rotein. The **M** of **M**ilk can be made to look like a **V** for **V**OMITING. The **P** of **P**rotein can be made to look like an **R** for **R**ASH

M P
V R

## (DOUBLE TAKE) PROTEIN INDUCED ENTEROCOLITIS

Protein Induced Enterocolitis is **NOT** IgE mediated. Can cause problems from birth, but often presents as FAILURE TO THRIVE (FTT) after months of already being on **cow milk formula**. This **CAN** occur in breast fed babies, but only if enough whole protein from mom's diet enters the breast milk. Usually presents after switching from breastfeeding to **formula**. 40% also sensitive to soy.

* **MNEMONIC**: It's so much easier if you just call it "FORMULA INDUCED Enterocolitis." Side note: An old term for this is "Allergic Proctocolitis," but hopefully you will not see that on the pediatric exam.

* SYMPTOMS include diarrhea +/- emesis and possibly BLOODY stool or flecks while on formula, but **not on clears or non-whole protein containing formulas**.

* DIAGNOSIS: Best test to establish diagnosis is a flexible sigmoidoscopy with biopsy. NOT allergy testing.

* TREATMENT: If the mom is breastfeeding, try asking mom to stop drinking milk products. If that doesn't work, **change to HYDROLYSATE formula** (formula containing HYDROLYZED cow milk proteins) or to an AMINO ACID DERIVED FORMULA. Protein induced enterocolitis typically **resolves** once the GI tract matures.

## (DOUBLE TAKE) LACTOSE INTOLERANCE (aka LACTASE DEFICIENCY)

It is **not** common for kids < 5 years old to have lactose intolerance (aka lactase deficiency). So for the pediatric boards, if the child is less than 5 years of age, suspect a different diagnosis!

* SYMPTOMS: Diarrhea +/- abdominal pain.

* **HYDROGEN BREATH TEST**: Can be used to help diagnose a lactase deficiency (as well as bacterial overgrowth). Patient takes a carbohydrate load, if unable to digest because of a lack of lactase, the bacteria will digest the carbs and release hydrogen (measured in the breath).

* TREATMENT: SOY milk. It does not contain lactose. It contains sucrose.

* **PEARLS:** Stool is NOT malodorous and does NOT have food particles. These patients do NOT vomit and do NOT have an associated rash.

* **MNEMONICS:** Most of you probably know of at least one friend, relative or patient that had lactose intolerance. Did they ever have a rash or vomiting after eating desert? Never! I'm guessing they had LACTAID in their fridge (my residency roommate certainly did). "**LACT**ose comes from the **LACT**ating breasts of women & cows. NOT soy beans."

# *IMMUNOLOGY*

**KEY PEARL:** There is a TREMENDOUS overlap within the ID, metabolic, immunologic, inborn errors of metabolism and genetic sections. Therefore, the entire immunodeficiency section is highly represented on the exam in the form of answer choices. Knowing the pertinent positive AND negative findings is key to narrowing your choice down to one.

## EPINEPHRINE PEN

Epinephrine pens are for patients with a history of severe allergic reactions or a history of anaphylaxis due to environmental exposures. Prescribe 3 epinephrine pens. Should be kept at room temperature and used on the BARE SKIN at the outer thigh. Replace after expiration (usually 6 months).

## TYPES OF HYPERSENSITIVITY REACTIONS

* Type 1 Hypersensitivity Reaction = IgE = Anaphylactic. Think of PCN allergy. PCN is the only antibiotic that can be skin tested. If a reaction occurs > 24 hours after PCN is administered, it is NOT a PCN allergy and there is NO indication for skin testing. Skin testing done in obscure cases typically results in > 90% of cases being ruled out as PCN allergy.

* Type 2 = Anti**B**ody mediated = **B**-cell = Humoral

* Type 3 = Immune Complex

* Type 4 = Delayed hypersensitivity/T-cell Mediated/Cellular immunity. The classic example is Poison Ivy. This includes CD4 cells and problems that arise from HIV/AIDS.

* **MNEMONICS**:

- "**1** = **A** = **A**naphylactic" or "**1** = **I**gE = **1**gE"
- "**2** = **B** = anti**B**ody" = humoral
- "**3** = **C** = immune **C**omplex"

- "**4** = **D** = **D**elayed." Or, if presented with the classic example of poison "**IV**" = the Roman numeral **IV** = Type **4**.

## (DOUBLE TAKE) ANAPHYLAXIS

The following is a review of some common issues to be aware of in case a child presents with anaphylaxis:

* FOOD EXPOSURE: Watch for 4 hours before discharging from the peds ER.

* RESPIRATORY COMPROMISE: Intubate and quickly secure the airway **first**. THEN give EPInephrine (1:1000, "3 zeroes for 3 letters of EPI"), diphenhydramine, IVF and steroids.

* BETA BLOCKERS: They blunt the response to EPInephrine. Give glucagon to reverse the beta blocker effect and then give EPInephrine (again).

* ANAPHYLACT**OID** REACTIONS: Result from mast cell degranulation. These are not true IgE mediated reactions. May be seen with use of NSAIDS, Opiates, Vancomycin and Contrast. Pretreat with steroids and diphenhydramine.

## DRUG HYPERSENSITIVITY SYNDROME

Drug hypersensitivity syndrome is seen with Sulfa drugs, Dapsone & Aromatic seizure medications. Results in Fever + Lymphadenopathy (LAD) + Rash +/- Visceral involvement.

## ANTICONVULSANT HYPERSENSITIVITY SYNDROME

Anticonvulsant hypersensitivity syndrome refers specifically to DRUG HYPERSENSITIVITY SYNDROME due to SEIZURE MEDICATIONS. Fever + Lymphadenopathy (LAD) + Rash +/- Visceral involvement. Seen with Carbamazepine, Phenobarbital and Phenytoin. Can turn into STEVENS JOHNSONS SYNDROME (SJS) or TOXIC ENDODERMAL NECROLYSIS (TEN).

## IGE MEDIATED MEDICATION HYPERSENSITIVITY

Consider IGE mediated medication hypersensitivity in any patient with a pruritic rash/hives +/- SOB occurring within 1 hour of the second dose of a medication.

## PENICILLIN (PCN) ALLERGY

**Desensitize patients with a penicillin allergy if** +SYPHILIS IN PREGNANCY or if you are presented with a CYSTIC FIBROSIS (CF) patient with resistant Pseudomonas

## SERUM SICKNESS

Serum sickness results in arthritis, nephritis and a rash. Seen with **cefaclor** and minocycline. Treat with anti-venom.

## BEE STINGS

Bee stings cause an IgE mediated reaction. If symptoms resolve with therapy in the peds ER, may discharge home. If not, admit to inpatient.

## POISON IVY, POISON OAK & POISON SUMAC

The term **RHUS** DERMATITIS is used to describe the skin manifestations of poison ivy, oak, and sumac. If the rash is in a sensitive area or is severe, give oral STEROIDS for 2-3 weeks. Shorter courses can result in a re-flare.

**MNEMONIC**: If the boards talk about exposure to a plant with 3 leaves, they are talking about one of the above-mentioned plants. "LEAVES OF THREE, LET THEM BE!" The answer to a matching question might be RHUS DERMATITIS.

## TYPES OF IMMUNITY

Types of immunity include Humoral, Cellular, Phagocytic and Compliment mediated. They are discussed below. **KNOW** what labs should be ordered to test the functionality of each of these different immune systems:

* HUMORAL DEFECTS: Tests include Isohemagglutinins and/or Serum IgG (or Ig subclasses from previous vaccinations such as varicella, rubella or tetanus). Remember, these are antibody tests.

* CELLULAR DEFECTS: Primary test is a Candida **D**elayed-type hypersensitivity intradermal test. >10mm after 48' means cellular immunity is intact and primary T-cell defects are virtually excluded. If negative, can also check lymphocyte counts, T-cell subtyping with flow cytometry, or lymphocyte stimulation testing.

* PHAGOCYTIC DEFECTS: Phagocytes, like neutrophils, are attracted to cytokines. Tests include Nitroblue Tetrazolium dye test for chronic granulomatous disease or the Dihydrorhodamine Fluorescence Test (DHR).

- **MNEMONIC**: "NEUTROblue" tetrazolium dye test

* COMPLIMENT DEFECTS: Test with a total complement assay (aka Complement 50). This checks for early (C1-4) and late (C5-9) deficiencies.

## CD4 CELL

CD4 cells = Helper T cells

**MNEMONIC**: CD4 = Helper = "4elper" T cells

## CD8 CELL

CD8 cells = Suppressor T cells

**MNEMONIC**: CD8 = Suppressor = "8uppressor" T cells

## NEUTROPENIA

Neutropenia is defined as an ANC < 1000 for children less than 1 year of age, and < 1500 for older children. For the boards, there is a STRONG association with MUCOSAL ULCERATIONS. MUCOSAL ULCERATIONS = NEUTROPENIA. The patient may have a genetic immunodeficiency, or possibly a low ANC due to aplasia (is the patient pancytopenic?).

* Mild neutropenia: ANC < 1500

* Moderate neutropenia: ANC < 1000

* Severe neutropenia: ANC < 500.

* **PEARL**: A key point to remember is that an 11 month old with an ANC of 1050 is NOT neutropenic.

## PEARLS/MNEMONICS FOR BRUTON'S, SCID AND HYPER-IGM

* **PEARL:** This table is summarizes a few KEY findings from some of the disorders you're about to see.

| | LAD/TONSILS | PCP PNA | LYMPHOPENIA |
|---|---|---|---|
| **BRUT**ons | **NO** | NO | |
| SCID | **NO** | **YES** | **YES** |
| Hyper-IgM | PRESENT | **YES** | |
| DiGeorge | | | **YES** |

* **MNEMONICS**: The disorders will be discussed in more detail. Memorize the following mnemonic in the meantime - ABSENT NODES/SMALL TONSILS: "SMOOTH neck/face and oropharynx without any bumps." The smooth neck/face represents a LACK OF LYMPHADENOPATHY on exam. The "oropharynx without any bumps" represents a LACK OF TONSILS.

- BRUTON'S: "use BRUT aftershave so you can have a SMOOTH face/neck!" Now imagine the most muscular man you know using BRUT aftershave. He has a SMOOTH face/neck (no LAD or tonsils on exam), but he also cut himself. He's bleeding and now he starts to cry because he's been too macho to establish care with a PCP. HE DOES NOT HAVE A PCP = PCP pneumonia is NOT seen in Brutons Agammaglobulinemia.
- SEVERE COMBINED IMMUNODEFICIENCY = SCID: "You can SCID across SMOOTH surfaces, like a curvy and SMOOTH NECK."

## PNEUMOCYSTIS CARINII PNEUMONIA (PCP)

PNEUMOCYSTIS CARINII PNEUMONIA (PCP) = PNEUMOCYSTIS JAROVECI PNEUMONIA = FOUND IN HyperIgM and SCID (but not BRUTons).

**MNEMONIC**: Imagine a super HYPER naked preschooler running all over the gravel schoolyard. He's being chased by the teacher when he **SCIDS.** He does a running split and destroys his NADS/NODES. His mommy has to come and take him to his established PCP to get checked out...and maybe to rule out adhd ☺. The destruction of the NADS kind of represents a LACK of lymph NODES (goNADS or goNODES or noNODES). This child DOES have a PCP, which means e IS susceptible to getting PCP/PJP.

## PEDIATRIC LYMPHOcyte COUNTS

Pediatric **lympho**cyte counts are **much higher than adults**. They should be **> 40,** so if the value looks like a normal adult LYMPHOcyte count value, it's too LOW and the patient is LYMPHOPENIC.

**PEARL**: LYMPHOcytes do NOT mean total WBC/leukocyte counts!

## *T-CELL DEFICIENCIES AND COMBINED T-CELL/B-CELL DEFICIENCIES*

**PEARLS**:

*** THIS SECTION IS HEAVILY REPRESENTED ON THE EXAM!**

*** Make a strong effort to memorize the probable age of presentation for the various immunodeficiencies. That can be a HUGE clue if offered up by the ABP to help you narrow the answer down from 2 choices to 1.**

* T-CELL MEDIATED IMMUNITY includes activation of antigen-specific cytotoxic T-lymphocytes, macrophages and natural killer (NK) cells. Can get recurrent **viral, fungal** and bacterial infections. For T-CELL or COMBINED T-CELL/B-CELL deficiencies, look especially for chronic or recurrent **Candida** infections (mouth, nail, scalp).

* All patients with a T-cell defect are at an INCREASED RISK OF LYMPHOMA/CANCER as they age (curious DiGeorge, Hyper-IgM, SCID, wiXo**Tt** Ostrich).

* **SKIN** TESTING with CANDIDA, MUMPS, TETANUS and PURIFIED PROTEIN DERIVATIVE (aka PPD, aka TUBERCULIN SKIN TEST): Can all be use to diagnose **D**elayed-type hypersensitivity ANERGY. This checks for Type **IV**, CELL-MEDIATED IMMUNITY associated with T-cell defects. This includes CD4 cells and problems that arise from HIV/AIDS.

- **PEARL IN A PEARL:** If you are presented with an aids or immunocompromised patient who had a negative PPD, consider further skin testing (usually Candida or Tetanus).
- **PEARL IN A PEARL:** SKIN testing for tetanus is completely different than getting titers for tetanus. Titers are checking for antibody production (B-cell function).

## SEVERE COMBINED IMMUNODEFICIENCY (SCID)

Severe combined immunodeficiency (SCID), is a **T AND B CELL DEFICIENCY** (hence the word COMBINED). The deficiency in T and B LYMPHOcytes results in severe **LYMPHopenia. Look for a low Absolute Lymphocyte Count (ALC) that is well below 40**. These patients do NOT have **LYMPH**adenopathy, but they DO have a small thymus. The presence of a thymus means a bone marrow transplant **is** a viable treatment option. These patients get **all kinds of infections** (viral, bacterial, fungal and opportunistic). Usually present in the **first 3-6 months with otitis media (OM), thrush, diarrhea and dermatitis.** There is **complete absence of T-cell function** (on flourometric analysis of T, B, and NK cells). **This condition usually presents closer to 3 months**. In contrast, a pure B-cell deficiency presents around 6 months of age.

* **Possible presentations: PCP**, a viral pneumonitis that doesn't resolve or recurrent candidiasis that seems to be refractory to treatment. For PCP patients, always consider HIV in the differential too.

* LIVE VACCINES: Do not give SCID patients live vaccines!

* CURE: Bone Marrow Transplant (BMT)

## MNEMONICS & PEARL for SCID and WISKOTT-ALDRICH

Read about the disorders first.

* Imagine that a PCP's office has a small, yellow 3 foot slippin' slide for babies. One of the babies loves OSTRICHES. There's a drawing of an OSTRICH at the end of the slide against a wall (**W-A**ll). The mom gives the baby a little push and the baby SCIDs (on his butt) right into the drawing and the baby's head erases the NADS right off of the OSTRICH! Now the OSTRICH and the SCIDing baby boy have **NO NADS/NODES.**

* SCID and OSTRICH should link the 2 diseases in your mind. **They are both combined T and B cell deficiencies**. The "W-All" is another reminder of wi**X**ott aldrich (the X in wi**X**ott refers to the fact that it's X-linked). BOTH disorders **lack** the presence of lymphadenopathy and both have a small thymus (so a BMT helps).

* **PEARL**: **LEUKO**cyte count may be NORMAL because of a higher number of neutrophils. Any time you see a low ALC or the word LYMPHOpenia, you should think about T-CELL or COMBINED disorders because **MOST LYMPHOcytes ARE T-CELLS!**

## (DOUBLE TAKE) WISKOTT-ALDRICH SYNDROME

For Wiskott-Aldrich syndrome, look for a patient with a history consistent with the triad of SMALL **T**hrombocytes, frequent **I**nfections and **E**czema. Platelets can be small in **size and/or number**. This is an X-linked so it will only be found in MALES. Patients develop both T and B cell problems as they ages. IgM is usually LOW, while IgA may be HIGH. A history may include findings such as petechiae, bloody diarrhea, bloody circumcision, eczema and/or otitis media. There is an increased risk of lymphoma in the 3rd decade of life. BMT is curative.

* **MNEMONIC**: Wiskott Aldrich = **wiX**ot**T ostrich** = Imagine a WISe OSTRICH wearing some nerdy black glasses, a tux and a TIE with the word "T-I-E" spelled down the front.

* **MNEMONIC**: (image)

T I E ↓WiXott IgM↓ ↑Aldrich IgA↑

- **KEY: X** Linked, small T = Represents small platelets (low MPV) and Thrombocytopenia, **I** = Infections, **E** = Eczema, Mirrored W = M and arrows show that Ig**M** is low, Duplicate A refers to Ig**A** and arrows show that IgA levels are increased.

* **PEARL**: In Idiopathic Thrombocytopenia Purpura (ITP), platelets are BIGGER

## CHROMOSOME 22 LOCUS Q11 = DIGEORGE SYNDROME or DIGEORGE LOCUS

**DiGeorge Syndrome** is caused by the deletion of a small piece of chromosome 22 (locus Q11). Presentations and phenotypes VARY. Some patients may have VELOCARDIOFACIAL SYNDROME, others may not. Diagnosed with Fluorescence In Situ Hybridization (**FISH**). Look for an absent or hypoplastic THYMUS. This means a patient may not have any T-CELLS = **LYMPHO**penia. If there are no T cells, then antigens are not being presented to B cells, so they are unable to make antibodies. This results in similar infections to SCID because it's essentially like a combined B and T cell defect. Patients may also present with CARDIAC ANOMALIES, neurologic issues, schizophrenia, renal abnormalities or HYPOPARATHYROIDISM due to missing parathyroid glands. Lool for a patient with a murmur and tetany (missing parathyroid glands results in hypOcalcemia). DiGeorge is associated with **TRUNCUS ARTERIOSUS, INTERRUPTED AORTIC ARCH,** PULM ATRESIA WITH VENTRICULAR SEPTAL DEFECT (VSD) and **TETRALOGY OF FALLOT** (TOF). Even if a DiGeorge patient does not have the classic facial features of VELOCARDIAL **FACIAL** SYNDROME (aka Velocardiofacial Syndrome), they are still at risk for the above-mentioned defects (see below for a description of the syndromic features).

* **VELOCARDIAL FACIAL SYNDROME**: **PALATAL** abnormalities (cleft uvula or cleft palate) + CARDIAC ANOMALIES (see above) + CRANIOFACIAL features (unusual palpebral fissures, broad nasal root and narrow upper lip)

* **BMT will not help because there is no thymus** to educate new T cells. A thymic transplant + BMT could potentially help.

* LIVE VACCINES are not appropriate for these kids.

* **MNEMONIC**: Imagine a monkey named "Curious DiGeorge sitting on the proximal part an elephant's **TRUNC** holding the bloody, distal **DISCONNECTED** part of the **TRUNC** in his hand and twirling it round and round! He sees you and throws the bloody TRUNC to you while screaming CATCH!"

- "TRUNC" refers to the possible Truncus deformity and the fact that it is disconnected refers to the possibility of an interrupted aortic arch!
- CATCH: C = Cardiac anomalies, A = Abnormal Facies, T = Thymic aplasia or hypoplacia, C = Cleft palate and H = Hypocalcemia from Hypoparathyroidism

* **PEARL**: Look for a child with lab values or facial findings consistent with DiGeorge who has a sternotomy scar (post-repair of a cardiac defect)

## (DOUBLE TAKE) ATAXIA TELANGIECTASIA

In ataxia telangiectasia, patients can have recurrent pneumonias or sinusitis + High AFP (from a neural tube issue). The telangiectasias are flat, red spots of dilated capillaries that **DO NOT BLANCH**. Ataxia is from problems with the cerebellum and might be seen on the exam in a child right after he learns to walk. Patients have worsening T-cell function later in life, but in childhood the symptoms are mainly neurologic. There is an increased risk of malignancy in the 3rd decade of life.

**PEARL**: An elevated AFP often refers to an issue with broken skin, which is often found in neural tube issue. Levels are used to screen for neural tube issues in pregnant women.

**PEARL**: RECURRENT PNEUMONIAS are seen all over the exam. You will have to pick out other findings to help you narrow your differential.

**MNEMONIC**: Ataxia = Unsteady Gait = Imagine a homeless, drunk, and wheelchair-bound teen (neural tube defect) who frequently comes to your ED for **recurrent pneumonias** or **sinusitis**.

## COMMON VARIABLE IMMUNE DEFICIENCY (CVID)

In Common Variable Immune Deficiency (aka CVID), B-cells fail to transform into plasma cells. This results in a deficiency of all of the Ig subtypes. CVID also has some T-cell wall defects. Symptoms include recurrent upper and lower respiratory tract infections (recurrent pneumonias!). Patient's can have recurrent herpes and recurrent Zoster (VZV) infections. There is a risk of EBV-associated lymphoma. Treat by giving IV Immunoglobulin (IVIG) in an effort to raise the patient's level of antibodies to "normal."

## B-CELL DEFICIENCIES

**PEARLS:**

* A pure B-cell deficiency will usually present with recurrent bacterial infections (aka pyogenic infections) starting around 6 months of age.

* Encapsulated organisms are very high on the differential, so keep in mind the "HNS" organisms (Hemophilus influenzae, Neisseria meningitidis and Streptococcus pneumoniae).

- **(DOUBLE TAKE) MNEMONIC**: A complete list of encapsulated organisms can be recalled by remembering that "Some Nasty Killers Have Some Capsule Protection." Streptococcus pneumoniase, Neisseria meningitidis, Klebsiella pneumoniae, Haemophilus influenzae, Salmonella typhi, Cryptococcus neoformans and Pseudomonas aeruginosa. Bruton's agammaglobulinemia and sickle cell patients are especially susceptible to encapsulated organisms.

* Because of problems with antibody production, you will NOT FIND THE EXPECTED **TITERS** for bacteria we typically immunize against, including Tetanus, Diphtheria and Streptococcus (aka pneumococcus). If you suspect a B-cell deficiency, test for it by obtaining **TITERS** for something the child was already immunized against, such as **TETANUS** (an ABP favorite when testing for knowledge of Agammaglobulinemia). **Do not get confused** with getting SKIN TESTING for tetanus/ Candida/Mumps/PPD which all test for **D**elayed type-hypersensitivity, Type **IV**, T-cell mediated reactions.

* There are NOT as many viral or fungal infections as seen in T-cell deficiencies.

* **MNEMONICS:**

- The age of presentation (6 months) happens to be around the same age when the mother's immunoglobulins/antibodies begin to wane!
- During well child visits, we inject numerous vaccines to help B-cells make immunoglobulins against various organisms. Most of these are against BACTERIA. So try to tie **B**-cell problems with recurrent **B**ACTERIAL infections. Especially the enCAPSulated HNS organisms that have CAPS on them which need to be popped off with antibodies (antibodies look kind of like a bottle opener).

## HYPER-IGM SYNDROME (aka HyperIgM Syndrome)

In hyper-IGM syndrome (aka HyperIgM Syndrome), the IgM to IgG class switch does not occur due to a missing signal to B-cells from T-cells. There is LYMPHO**cytosis and NEUTROPENIA.** Patients have frequent otitis media and sinopulmonary infections. They also get diarrhea and opportunistic infections, including PCP. Usually starts around 6 months of age. This is a **T-cell abnormality**, so there is an increased risk of lymphoma/cancer. For treatment, give IVI**G** to make up for the missing immunoglobulins and Trimethoprim-Sulfamethoxazole (aka Bactrim). Also give PCP prophylaxis

**MNEMONIC**:

**B**actrim is used for **PCP** proplhylaxis

**P C P**
**b C b Bactrim**

**MNEMONIC**: Think of the LYMPHOcytosis as a compensatory mechanism in which there are high numbers of LYMPHOcytes circulating and releasing elevated levels of IgM to "make up" for the lack of IgG, IgE and IgA. Also, think of the neutropenia as being a relative "lack of neutrophils" on the WBC resulting from an excess of lymphocytes.

**PEARL**: Infections are similar to agammaglobulinemia, but the CBC looks different. Also, these patients can get PCP so suspect this diagnosis in any **HIV negative patient diagnosed with PCP**.

## AGAMMAGLOBULINEMIA (aka X-LINKED AGAMMAGLOBULINEMIA, aka BRUTON'S AGAMMAGLOBULINEMIA)

Agammaglobulinemia (aka X-Linked agammaglobulinemia or Bruton's agammaglobulinemia) is X-linked so it is seen in **B**oys. There is a total absence of **B** cells, which means there are **NO IMMUNOGLOBULINS. NO Ig's!** May have **high T-cell counts**. Patients have **tiny or absent tonsils and no palpable lymph nodes**. It results in recurrent **B**acterial infections and presents around 6 months of age. You might be presented with a child who has a history of "many antibiotic courses." Recurrent infections with enCAPSulated organisms, especially Pseudomonas, Streptococcus pneumoniae and Haemophilus influenza. Look for sepsis, meningitis and recurrent pneumonia. Pneumocystis pneumonia (aka PCP) does NOT occur in this disorder. If you see PCP, think Hyper-IgM or SCID! TREAT with IVIG for life and give prophylactic antibiotics. BMT is curative.
**MNEMONIC**: The age of presentation (6 months) happens to be around the same age when the mother's immunoglobulins/antibodies begin to wane!

## TRANSIENT HYPOGAMMAGLOBULINEMIA OF INFANCY

In transient hypogammaglobulinemia of infancy, B-cells are not deficient, but there are low levels of IgG +/- IgA. **Infections similar to Bruton's** (encapsulated organisms) start around 3-6 months of age. Since the child will start developing his/her own immunoglobulins, NO TREATMENT is necessary. The hypogammaglobulinemia can last for years, but children usually outgrow the problem by about 3-6 years of age. Titers to immunized organisms are normal.

## IGA DEFICIENCY

IGA deficiency is extremely common, but 80% of those who have it are asymptomatic. The ones who do have problems usually develop recurrent sinus and pulmonary infections.

## HYPER-IGE SYNDROME

Hyper-IGE syndrome is an Autosomal Dominant disorder associated with impaired neutrophil chemotaxis and extremely high IgE levels. Look for **E**osinophilia + **E**czema + **E**levated levels of Ig**E**. Eczema often starting in the FIRST WEEK OF LIFE. Recurrent sinus and pulmonary infections (recurrent pneumonias!). Pneumatoceles (often seen on chest X-ray after a pneumonia). **Recurrent "cold" STAPHYLOCOCCUS infections**/abscesses/boils ("cold" because there is no surrounding erythema due to neutrophil chemotaxis impairment). Can also have lung abscesses. Patients can also have **delayed shedding of primary dentition** (tEEth) and skeletal/bone problems (e.g., coarse facial features, scoliosis and fractures).
**MNEMONIC**: Lots of "E's." Look at the bold letters above. Also, for delayed dentition, there are 2 more E's in the word t**EE**th, and if you put the E's on their side, they kind of look like teeth!

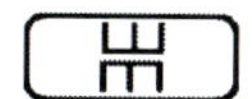

## COMPLIMENT DEFICIENCIES

**GENERAL PEARLS**

* Compliment can help clear pathogens in multiple ways, including by binding to Ag-Ab complexes and also by binding directly onto some types of bacteria and thus promoting phagocytosis.

* NEISSERIA MENINGITIDIS (aka meningococcus or meningococcal disease): Look for recurrent and overwhelming infections due to this bug. Can also result in severe N GONORRHEA infections.

* CH50: If this is normal, then ALL complement pathways are ok (C1-9).

## C1-4 COMPLIMENT DEFICIENCY

C1-C4 compliment deficiency may look similar to Agammaglobulinemia (aka **B**ruton's = **B**oys = anti**B**odies = **B**acterial infections) because this part of the compliment system works closely with anti**B**odies.

## C5-9 COMPLIMENT DEFICIENCY

C5-9 compliment deficiency has an unusual predilection towards **NEISSERIA** infections (**both** MENINGITIDIS and GONORRHEA). Prevent by early meningococcal vaccination.

## C1 ESTERASE DEFICIENCY (HEREDITARY ANGIOEDEMA)

C1 Esterase Deficiency results in Hereditary Angioedema in which there are recurrent episodes of swelling/edema and abdominal pain.

**PEARL**: These patients do NOT have pruritis

**MNEMONIC**: This mnemonic is meant to help you remember which diseases are associated with LOW COMPLIMENTS on lab testing. Keep in mind that it is not specific enough to differentiate between low C3 versus low C3 and C4, etc. Details on these specific diseases are found elsewhere.

* Imagine you're at a circus and they ask for a personal item. You hand them a compliment card you plan to mail to your favorite restaurant. It's in an envelope with a STRIPPED POSTAGE STAMP on it. In a magician's hand, it suddenly starts BURNING UP and flames are shooting off of it. The magician then throws the burning compliment card through a LOOP OF FIRE at a CRYING GREEN GOBLIN that has two NIPPLE-RINGS SHAPED LIKE C'S. The GOBLIN gets angry and becomes monsterously huge. He approaches the audience and stomps on a group of visiting MEMBERS OF PARLIAMENT wearing gray wigs who were quietly watching the show. The only way to appease the green goblin is to keep giving him green M&M's until his CHEEKS/PAROTIDS are so full that they are ready to pop!

- STRIPPED POSTAGE STAMP = POSTSTREPTOCCCAL GLOMERULONEPHRITIS (PSGN)
- BURNING UP = Reminder that compliment is DECREASED in these diseases
- LOOP OF FIRE = SYSTEMIC LUPUS ERYTHEMATOSUS (SLE)
- CRYING GREEN GOBLIN = CRYOGLOBULINEMIA
- NIPPLE-RINGS SHAPED LIKE C'S = HEPATITIS C (because it actually can cause cryoglobulinemia)
- MEMBERS OF PARLIAMENT = MEMBRANOPROLIFERATIVE GLOMERULONEPHRITIS (MPGN)
- M&M's = MULTIPLE MYELOMA
- FULL CHEEKS/PAROTIDS = SJOGRENS SYNDROME

**PEARLS:**

* LEUKEMIA AND LYMPHOMA are two other causes of DECREASED compliment levels

* **NORMAL** COMPLEMENT levels are found in HENOCH-SCHONLEIN PURPURA (HSP), IGA NEPHROPATHY, RAPIDLY PROGRESSIVE GLOMERULONEPHRITIS (**RP**GN, not MPGN) and HEMOLYTIC UREMIC SYNDROME.

# NEUTROPHIL DISORDERS/PHAGOCYTIC ISSUES

## NEUTROPENIA DEFINITIONS

Neutropenia is defined as follows:

* **0-12** MONTHS: ANC of < 1000/microL

* **After 12 months**: ANC of < 1500/microL. An ANC of < 1500 = Mild, < 1000 = Moderate, < 500 = Severe.

* **PEARL**: For the peds boards, there is a STRONG association between NEUTROPENIA and **MUCOSAL ULCERATIONS**. Could be due to an immunodeficiency patient or during a time of aplasia resulting in pancytopenia.

## CHRONIC BENIGN NEUTROPENIA

In chronic benign neutropenia, the ANC is chronically low. Usually noted on the routine 1 year well child visit when a CBC is done. Typically resolves by 2 years of age, it's not really chronic, but it is benign.

## TRANSIENT NEUTROPENIA

Transient neutropenia is common **after viral infections**. No treatment required.

## CYCLIC NEUTROPENIA

Cyclic neutropenia is an autosomal dominant (AD) disorder that is inherited and results in neutropenia lasting for about 1 week every month. Frequent CBCs are required to document the cyclic nature of the neutropenia. This is usually self-limited and doesn't require treatment, but if the patient has infections associated with it, GCSF may be given at the nadir of the cycle. Patients are prone to Clostridium Perfringens infections and ORAL lesions, such as recurrent periodontitis or oral ulcerations.

**MNEMONIC**: "Kind of like a MENSTRUAL CYCLE for your NEUTROPHILS." This is to help you remember that the cycle lasts for about 1 week, and it occurs every month.

## SEVERE CONGENITAL NEUTROPENIA (aka KOSTMANN SYNDROME)

In severe congenital neutropenia (aka Kostmann syndrome), the ANC does not normalize and remains low. Rare condition but associated with severe infections if present.

## (DOUBLE TAKE) CHRONIC GRANULOMATOUS DISEASE (CGD) = SERRATIA

Chronic Granulomatous Disease (CGD) is an x-linked recessive disorder. If you see a boy on the boards diagnosed with Serratia, he's probably got CGD. CGD is the result of an enzyme deficiency within neutrophils that is needed to destroy cell walls and aid in phagocytosis. Chemotaxis is intact. This is not a cell line deficiency so there is **NO LYMPHOpenia and NO NEUTROpenia. This is a functional/oxidative burst issue.** The immune system is unable to destroy certain bacteria and fungi (**Aspergillus, Candida, E. coli, Serratia and Staphylococcus**) because they are all catalase positive organisms. The organisms/infections are usually considered to be **low-grade**. Patients may have **liver abscesses, osteomyelitis, recurrent lymphadenitis and recurrent skin infections**. Can also have GI infections and **granulomas of the skin, GI or GU tract**.

* DIAGNOSIS: **Nitroblue Tetrazolium (NBT) dye test or Dihydrorhodamine (DHR) Fluorescence**

* TREATMENT: Aggressive antibiotics at times of infections

* **PEARL**: Like the name says, it's a "CHRONIC" condition. Children usually present by **5 years of age.**

* **MNEMONICS**:
  - CGD = This is X-linked disorder, ChroniX GranulomatuX Disease. Put an X where it makes sense for you to remember this key fact.
  - NEUTROphil disorder = "**NITROphil** disorder = **NITROBLUE** tetrazolium test which tests for the NITROglycerin oxidative BURST." ChroniX GranulomatuX Disease seems to be all about colors!... BLUE, SERRATIA (RED)!!!

## LEUKOCYTE ADHESION DEFICIENCY (aka LEUKOCYTE ADHESION DEFECT)

In leukocyte adhesion deficiency (aka leukocyte adhesion defect), the adhesion defect results in neutrophils being unable to leave the vasculature to get to infected areas. Look for a black, infected looking cord stump (or other wound) **without any pus/inflammation**, DELAYED separation of cord and delayed wound healing. Specific infections might include periodontitis, perirectal abscesses, skin infections/abscesses and omphalitis.

* LABS will reveal **LEUKOcytosis and NeutroPHILIA**. The WBC will be HIGH even without infections (~20), and may be as high as 100 with infections.

* DIAGNOSTIC tests include a Rebuck Skin Window or CD11/CD18 flow cytometry.

* BMT is curative, and without it patients will die by the age of 1. So if the ABP presents an older child in the vignette, it is probably not LAD.

## CHEDIAK-HIGASHI SYNDROME

In Chediak-Higashi Syndrome, there is NEUTROPENIA, poor neutrophil chemotaxis and platelet dysfunction. Neutrophils containing giant lysosomal granules are seen on microscopy. Patients have OCULOCUTANEOUS ALBINISM and frequent SKIN and LUNG INFECTIONS with Staphylococcus and Streptococcus (can have recurrent pneumonias).

**PEARL**: Don't get confused with Hyper-IgM, which has **Neutropenia + LYMPHOCYTOSIS**.

**MNEMONIC**: Imagine a pair of All American WHITE BOYS. They're brothers and their names are "CHAD & ZACH, the ASHY white boys" with BLONDE HAIR, PINK EYES, and WHITE SKIN that's so ASHY that they get INFECTIONS all over. They like to play with NEUTROglycerine sticks shaped like MAGNETS and blow things up in their backyard (like jars filled with PENNIES).

* **KEY**: WHITE BOYS = Cutaneous Albinism, CHAD & ZACH the ASHY white boys = Sounds like the name of the disease, BLONDE HAIR = Blonde hair, PINK EYES = Ocular albinism, ASHY = Can also refer to white skin, INFECTIONS = Recurrent infections, NEUTROglycerine = A reference to neutrophils, NEUTRO + MAGNETS = Neutrophils that have trouble with chemotaxis, NEUTRO + PENNIES = NEUTROPENIA.

## (DOUBLE TAKE) SHWACHMAN-DIAMOND SYNDROME

SHWACHMAN-DIAMOND SYNDROME is a rare, autosomal recessive (AR) disorder associated with bone marrow failure **(pancytopenia/neutropenia), pancreatic insufficiency**, **bone/skeletal abnormalities** and possible **neutrophil defects** (migration and chemotaxis). Recurrent bacterial infections are common, including **recurrent pneumonias**, skin infections, otitis media and sinusitis. Patients may also have aphthous stomatitis.

* **PEARL:** Due to some overlapping symptoms with Cystic Fibrosis (CF), such as pancreatic insufficiency, you may have to choose between this disorder and CF. Remember, this one has **cell line issues** and the electrolytes and lungs are NORMAL.

* **MNEMONIC**: Imagine a DIAMOND in a SHWACH WATCH being used to make a CREASE in a black PAN with a white handle made of HOLLOW BONE."

- **KEY**: PAN/HOLLOW = **PAN**CYTOPENIA/NEUTROPENIA!!! CREASE = panCREASE, PAN = PANcytopenia, HOLLOW BONE = Aplasia resulting in PANcytopenia and neutropenia.

* **NAME ALERT**: The name alert is for **DIAMOND**-BLACKFAN ANEMIA.

**(DOUBLE TAKE) DIAMOND-BLACKFAN ANEMIA**

DIAMOND-BLACKFAN ANEMIA is an aplastic anemia due to **pure red cell aplasia**. That means that unlike other aplastic anemias, ONLY THE RED CELL LINE IS AFFECTED. Red cells are MACROCYTIC, anemia is typically severe, patients present around 3 months of age and they may have TRIPHALANGEAL THUMBS and craniofacial anomalies. Look for low reticulocyte counts. If asked about treatment, use steroids chronically and transfuse for severe anemia.

* **MNEMONIC**: Imagine looking at a standing BLACK FAN with THREE GIANT PLASTIC BLADES. The blades are BLACK, but are SHAPED LIKE RED CELLS.

- **KEY**: BLACK FAN = Name, THREE = TRIphalangeal thumbs and 3 months old at presentation, GIANT = MACROCYTOSIS, PLASTIC = aPLASTIC anemia, SHAPE = Reminder that this is a pure RED CELL aplasia.

* **PEARL**: If you see a patient with anemia and icterus or hyperbilirubinemia suggesting hemolysis, it is NOT Diamond-Blackfan Anemia. Since there is a red cell aplasia, the anemia is from a lack of production, NOT from hemolysis.

* **NAME ALERT**: The name alert is for SHWACHMAN-**DIAMOND** SYNDROME.

## ***IMMUNOLOGY TESTS, A RECAP***

### SKIN TESTING

**Skin testing** WITH PURIFIED PROTEIN DERIVATIVE (aka PPD, aka TUBERCULIN SKIN TEST), CANDIDA, MUMPS and TETANUS can all be use to diagnose **D**elayed-type hypersensitivity ANERGY. This checks for Type **IV**, CELL-MEDIATED IMMUNITY. T-cell mediated immunity includes activation of antigen-specific cytotoxic T-lymphocytes, macrophages and natural killer (NK) cells. This includes CD4 cells and problems that arise from HIV/AIDS.

## TITERS

If you suspect a **B**-cell (aka Humoral) deficiency, test for it by obtaining anti**B**ody **TITERS** for something the child was already immunized against, such as **TETANUS** (testing for tetanus titers in patients with Agammaglobulinemia, a B-cell deficiency, is a an ABP favorite way of testing your knowledge). Could also test for titers against Diphtheria and Streptococcus/Pneumococcus. **Do not get confused** with getting SKIN TESTING for tetanus/Candida/Mumps/PPD which all test for T-cell mediated (aka cellular) immunity.

**PEARLS:**

* Infections related to B-cell deficiencies rarely occur before 6 months of age because of presence of maternal antibodies.

* ANTIBODY SUBCLASSES: If IgG levels are in the normal range but you still have a high suspicion for a B-cell/Humoral defect, consider checking for the presence of Ig subclasses for Tetanus, Pneumococcus, etc.

## CH50

**CH50** is a great screen to test the entire complement system. Order if a patient has repeated, serious pyogenic/**bacterial** infections.

## NITRO**blue** TETRAZOLIUM (NBT)

NITROblue Tetrazolium (NBT) test is to test for Chronic Granulomatous Disease (CGD). Dihydrorhodamine (DHR) Fluorescence can also be used.

**MNEMONIC**: "NEUTROblue" referring to "ChroniX GranulomatuX Disease" being a NEUTROphil disorder. NITROblue is a reminder of the "NITROglycerine" oxidative burst defect. BLUE is a reminder that this disease has an association with colors (RED = the color that SERRATIA makes)

# CARDIOLOGY

## *EKG FINDINGS*

### RIGHT ATRIAL ENLARGEMENT (RAE)

In right atrial enlargement, the P wave is shaped like an "m" in V1, or it is > 3 mm

### NEGATIVE T WAVE

The T wave is negative in newborns in lead V1. The T wave becomes positive/upright after the first week of life.

### PREMATURE ATRIAL CONTRACTIONS (PACs)

Common in neonates and mostly benign. If a patient is on Digoxin, this could potentially be a concerning finding.

### PREMATURE VENTRICULAR CONTRACTIONS (PVCs)

If PREMATURE VENTRICULAR CONTRACTIONS (PVCs) are described/shown as **monomorphic** (of the same morphology) and they diminish with exercise, these are BENIGN.

### EKG CHANGES DUE TO ELECTROLYTE DISTURBANCES

EKG changes related to hypokalemia, hyperkalemia, hypocalcemia and hypomagnesemia are electrolyte disturbances that are fairly "testable" for the boards. Potassium changes would be most likely tested.

* HypOkalemia: Prolonged QT, flattened T-waves, ST segment depression, U-waves and PVCs

* HypERkalemia: Peaked T waves, absent or flat P waves, widened QRS and Ventricular Tachycardia. Associated with electromechanical dissociation. Treat with potassium lowering drugs (sodium polystyrene, insulin + glucose infusion) and give CALCIUM GLUCONATE.

- **MNEMONIC**: If you imagine the PQRST of a single beat having a little P, a wide QRS and a large T, it's almost as though the extra height in the peaked T waves is causing the QRS complex and the P to be stretched out!"
- **PEARL/SIDE NOTE:** hypERmagnesemia is also treated with CALCIUM GLUCONATE infusions. MAG sulfate is used as a tocolytic and can cause prolonged hypOcalcemia in babies.

* HypOcalcemia: Prolonged QT interval. Can also result in muscle weakness and tetany. There are no changes with hypercalcemia.

* HypOmagnesemia: Prolonged QT interval and prolonged PR interval

* SODIUM CHANGES: There are NO EKG changes with hyponatremia or hypernatremia.

* **MNEMONICS:**

- "**MELT PC**'s because they take too **LONG**! Buy an apple instead." Problems with low **M**agnesium,
- **E**rytheromycin, **L**evofloxacin, **T**CAs, low **P**otassium and low **C**alcium can result in a **LONG** QT.
- LOW electrolytes result in a sLOW interval

## *NORMAL HEART RATES*

### NORMAL HEART RATES

Normal heart rates in children are not not well defined, so use these values as a guideline:

* NEONATES: 70 – 160 is considered normal. Less or more means bradycardia or tachycardia

* OLDER CHILDREN: 50 – 120 can be considered normal.

### SINOATRIAL NODE (SA node)

The sinoatrial node Intrinsically fires at about 80 beats/minute

**MNEMONIC**: The "S" of SA node looks like the "8" of 80 beats/minute. "8A node"

### ATRIOVENTRICULAR NODE (AV NODE)

The atrioventricular node (AV node) intrinsically fires at 60 beats/minute

### VENTRICLE

The ventricle intrinsically fires at 40 beats/minute (so you could see a rate of 40 in a patient with 3rd degree AV Block)

**MNEMONIC**: From higher up to lower (SA to AV to Ventricle), the intrinsic rates decrease by 20 from 80 to 60 to 40 beats/min. Keep "8A" Node beating at 80 beats/min in mind and you should be fine.

## *ARRHYTHMIAS*

**PEARL**: Studying arrhythmias in too much detail has been low yield in recent years. This is likely due to the importance placed on congenital heart disease, murmurs and septal defects. Even though the information presented in this section is fairly complete, consider reading it for a "high level" understanding only.

### BRUGADA SYNDROME

Look for a right bundle branch block (RBBB) + ST segment elevation. This example shows normal ST segments on the left and ST segment elevation on the right.

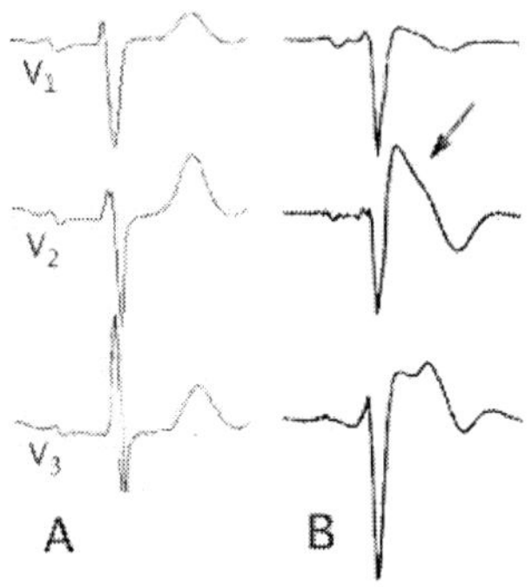

V1
V2
V3
A
B

### SUPRAVENTRICULAR TACHYCARDIA (SVT)

In supraventricular tachycardia (SVT), the heart rate is usually **> 240** and is often seen in patients 3-6 months of age. May be noted incidentally during breastfeeding. Vagal maneuvers may be tried for 20 seconds, including water or ice to the face and valsalva if the patient can cooperate. Carotid massage and orbital pressure is not universally recommended due to concerns for soft tissue injury, especially in children less than 1 year old. If vagal maneuvers do not work, try **ADENOSINE** (may repeat once at a higher dose). If the patient is unstable or the patient is less than 12 months old, use synchronized DC electrical cardioversion. May also try DILTIAZEM or VERAPAMIL, but these are usually reserved for children older than 12 months. After the patient is back into sinus rhythm, repeat an EKG to look for WPW. Treat children with SVT with a chronic beta blocker (**propranalol**) for 1 year to prevent recurrence. If the patient is resistant to treatment, use radiofrequency ablation.

**PEARLS:**

* SVT can cause heart failure due to ineffective filling times. In babies, a finding of HEPATOMEGALY is more common than crackles.

* For the boards, if you see a regular, narrow complex tachycardia that's REALLY fast, pick adenosine if possible.

## WOLFF-PARKINSON-WHITE SYNDROME (WPW)

In Wolff-Parkinson-White Syndrome, pre-excitation results a wave of depolarization that bypasses the AV node. The result is a short PR (less than 0.12s) and a widened QRS (greater than 0.12s). A DELTA WAVE is present due to fusion of the QRS and pre-excitation wave. Tachycardia is treated the same as any SVT, but if a WPW patient has ATRIAL FIBRILLATION OR FLUTTER, AV NODAL AGENTS (BETA BLOCKERS, DIGOXIN and VERAPAMIL) are CONTRAINDICATED because they increase the refractory period of the AV node, thereby allowing faster passage of the electrical signal down the accessory pathway. This can deteriorate into ventricular fibrillation and death. For atrial fibrillation or flutter, use PROCAINAMIDE.

## VAGAL MANEUVERS/ADENOSINE

Vagal maneuvers increase the refractory period, or temporarily block, the AV node (AVN).

## AV NODE REENTRANT TACHYCARDIA (AVNRT)

In AV **node** reentrant tachycardia, the tachycardia originates at the node itself. P waves are buried because of the simultaneous signal going towards the atria **and** the ventricles from the AVN. AVNRT **will** respond to vagal maneuvers and adenosine.

## AV REENTRANT TACHYCARDIA (AVRT)

In AV reentrant tachycardia, there is an extra pathway (like WPW). A circuit exists which includes the atrium AND the ventricle but bypasses the AVN. Adenosine and Verapamil are contraindicated because blocking the AVN could make this extra circuit go even FASTER.

## ATRIAL TACHYCARDIAS

Atrial tachycardias do NOT respond to adenosine or vagal maneuvers. This includes atrial fibrillation and atrial flutter.

**MNEMONIC**: If signals from the atria are firing away on their own, they will continue to do so even if a signal that is further down the pathway (AVN) is blocked.

## ATRIAL FIBRILLATION & ATRIAL FLUTTER

Atrial fibrillation and atrial flutter are **not** usually seen in normal children, so look for a cause of atrial dilation or tachycardia (e.g., hyperthyroidism). May use adenosine to slow rate for diagnostic purposes, but not for treatment. Treatment options include beta blockade, calcium channel blockers, digoxin, cardioversion and ablation. These do not slow the atrial rate, but they do slow the ventricular rate by AVN blockade.

## VENTRICULAR TACHYCARDIA (VT or VTACH)

Etiologies of ventricular tachycardia include digoxin **toxicity**, hyperkalemia and prolonged QT. Treat with Amiodarone. May also use lidocaine or cardioversion.

## PROLONGED QT

Prolonged QT can lead to TORSADES DE POINTES. It can be caused by hyp**O**calcemia, hyp**O**magnesemia (which can be caused by hypocalcemia), hyp**O**kalemia, tricyclic antidepressants (TCAs), Levofloxacin and Erythromycin.

* **PEARL:** Think about Prolonged QT in ANY patient that presents for **loss of consciousness (LOC), drowning, syncope and even seizure**. Patient may have had a prolonged QT and gone into Torsades!

* FAMILIAL PROLONGED QT SYNDROME: Prescribe a beta-blocker.

* TORSADES: May try to **prevent** with a beta blocker or a defibrillator (AICD). **Treat** with MAGNESIUM SULFATE. Can also consider overdrive pacing for treatment.

* **MNEMONICS:**

- "**MELT PC**'s because they take too **LONG**! Buy an apple instead." Problems with low **M**agnesium, **E**rytheromycin, **L**evofloxacin, **T**CAs, low **P**otassium and low **C**alcium can result in a **LONG** QT.
- LOW electrolytes result in a sLOW interval

# *HEART BLOCKS (AV Blocks or AVB)*

## FIRST DEGREE AV BLOCK

In 1st degree AV block (AVB), everything looks normal except for a prolonged PR interval. Diagnostic if greater than 0.20s (5 small blocks, 1 large block). Benign.

## SECOND DEGREE AV BLOCK

There are two types of 2nd degree AV block (AVB):

* TYPE 1 = MOBITZ TYPE I = WENCHEBACH: PR interval widens until there is a dropped QRS complex. It is **benign**.

- **MNEMONIC**: "Won (one) = Widening = Wenchebach

* TYPE 2 = MOBITZ TYPE II: Normal looking EKG except for a nonconducted P wave (dropped QRS). Can progress rapidly to complete heart block, so a **PACER is indicated**.

## THIRD DEGREE AV BLOCK = COMPLETE HEART BLOCK

In third degree AV block (AVB), P waves and QRS complexes are completely independent of each other. A **PACER is USUALLY indicated**. Here are the specific indications:

* Persists > 7 days after surgery
* Congenital 3rd degree AVB with symptoms
* Congenital 3rd degree AVB with CONGENITAL HEART DISEASE (CHD) and HR < 75
* Congenital 3rd degree AVB without CHD and HR < 55
* **PEARL**: For 3rd degree AVB, look for child who is bradycardic except for when s/he is crying

### BUNDLE BRANCH BLOCKS

Make the diagnosis of a left or right bundle branch block by locating the RsR' (aka "rabbit ears")
* RBBB: RsR' is where the leads on the RIGHT side of the chest are placed (especially V1 & V2).
* LBBB: RsR' is where the leads on the LEFT side of the chest are placed (especially V5 & V6).

## *SEPTAL DEFECTS*

### CARDIAC SHUNT PEARLS & MNEMONICS

**PEARLS**: When it comes to isolated septal defects, a RIGHT to LEFT shunt is generally much more serious than a LEFT to RIGHT shunt. In a LEFT to RIGHT shunt the blood is simply getting oxygenated multiple times, but in a RIGHT **to LEFT** shunt, oxygenation is being bypassed resulting in possible hypoxemia, cyanosis and death!

* NOTE: In cyanotic heart disease, this pearl may not apply because sometimes septal defects are a corrective measure and the blood on the RIGHT side has some oxygenated blood that needs to reach the periphery via a RIGHT to LEFT shunt.

**MNEMONICS**: Imagine the circulatory system as a ONE WAY CIRCUIT (or a ONE WAY RACETRACK). "At any point in the cardiopulmonary circuit/racetrack, GOING TO THE LEFT (or SHUNTING TO THE LEFT) is BAD because you will leave the vascular system (or crash and burn).

### ATRIAL SEPTAL DEFECTS (ASD)

Most atrial septal defects (ASDs) are asymptomatic and do not require repair. Findings on exam may include a **FIXED SPLIT S2**, a parasternal impulse/heave or a soft LUSB mid-systolic murmur ("takes time to get enough blood in atrium to make a sound").

* A **SECUNDUM ASD** is the most common type of ASD. Repair is required if the LEFT TO RIGHT shunt progresses to a ratio of > 2:1 (technically if > 1.7:1). Ratio refers to the volume of blood going into the pulmonary circulation compared the volume of blood going out of the LV.

* **OSTIUM PRIMUM ASD**: Much less common, but more serious so these should be repaired EARLY. Having a primum ASD is not a contraindication to getting pregnant as long as the LEFT TO RIGHT shunt ratio is less than 2:1. If the patient develops a RIGHT TO LEFT shunt, pregnancy is contraindicated and this is now called EISENMENGER'S SYNDROME (may present with CYANOTIC MUCOUS MEMBRANES).

- **LOW YIELD NOTE** (for understanding only): Primum ASD's start off as a shunt to the RIGHT from the high pressure left side. Due to excess flow into the pulmonary circulation, pulmonary hypertension develops and eventually causes a shunt to the LEFT. Once that develops, you have Eisenmenger's Syndrome.

## VENTRICULAR SEPTAL DEFECTS (VSDs)

A ventricular septal defect (VSD) is a **left to right shunt**. May close on their own, but a MEMBRANOUS VSD is less likely to close than MUSCULAR. VSDs are associated with CRI-DU-CHAT and the major **TRI**SOMIES (**13, 18** & 21). Trisomy 21 = DOWN SYNDROME. ASDs are more common in Trisomy 13 and 18 than 21). Physical exam might reveal a **LLSB thrill/murmur that is HOLOSYSTOLIC, a "hyperdynamic precordium,"** or possibly NO murmur and a single S2 if the VSD is very large. Chest X-Ray (CXR) may show increased pulmonary vasculature and biventricular hypertrophy. Treat with diuretics and repair if the patient develops PULMONARY HYPERTENSION, FAILURE TO THRIVE or a shunt that progresses to > 2:1.

**MNEMONIC**: A VSD is a SEPTAL defect, so imagine looking into the home of a three year old where you see a roll of SEPTATED (perforated) stamps with pictures **TRI**cycles on them being licked by a CAT.

## AV CANAL DEFECT

In AV canal defect, there is a left to right shunt. The condition refers to the presence of both an **ASD and a VSD**. Strong association with DOWN SYNDROME (Trisomy 21) and DIGEORGE SYNDROME. Physical exam findings include a LOUD S1 AND S2 and possibly a HOLOSYSTOLIC APICAL MURMUR (due to either the associated mitral regurgitation or due to regurgitation at the combined AV valve). Increased pulmonary flow results in BIVENTRICULAR HYPERTROPHY (BVH) . EKG shows **BVH + LAD**, and CXR shows the increased pulmonary vasculature and BVH (or RVH + "cardiomegaly").

## AV CANAL DEFECT & VSD

Both AV canal defects and VSDs are left to right shunts and can present with symptoms of CONGESTIVE HEART FAILURE (CHF) after a few weeks or months.

# ***MURMURS & SPLITS***

## PATHOLOGIC MURMURS

The following terms are usually used to refer to a high grade, or pathologic murmur. If any of these are included in an ABP vignette, a "wait and watch" approach is probably not the answer:

* Click
* Harsh
* Grade III, or "loud"
* Diastolic
* Pansystolic
* Late systolic
* S4 or 4$^{th}$ heart sound
* S3 or 3$^{rd}$ heart sound that remains after changing positions from supine to sitting.

## MURMUR TERMINOLOGY

* RUSB = Right 2$^{nd}$ intercostal space = 2$^{nd}$ RICS
* LUSB = Left 2$^{nd}$ intercostal space = 2nd LICS

* LLSB = Left 4th intercostal space = 4th LICS

* APICAL: Left 4th intercostal space **near the apex** = 4th LICS **near the apex**

* **PEARLS/MNEMONICS**:

- "STENOSIS" murmurs generally start in the **middle** of a cycle (mid-systolic or mid-diastolic). Stenotic means something is only "half" way open!
- "REGURGITATION" murmurs generally start at the **beginning** of cycle (early systolic or early diastolic). Aortic regurgitation is an exception. Mitral valve prolapse is also an exception (different from MV regurgitation).

## PULMONARY STENOSIS (PS)

Pulmonary stenosis rarely progresses. Much more benign than aortic stenosis. Exam reveals a harsh LUSB murmur, a single S1, a widely split S2 (since when A2-P2 happens, the P2 is delayed even more) and possibly a thrill or a click. There is a low risk of sudden death.

## MITRAL STENOSIS (MS)

Mitral stenosis is a mid-DIASTOLIC murmur heard at the APEX. Associated with RHEUMATIC FEVER.

## TRICUSPID STENOSIS (TS)

Tricuspid stenosis is a mid-DIASTOLIC murmur heard at the LLSB. Also associated with RHEUMATIC FEVER.

## AORTIC STENOSIS (AS)

Aortic stenosis is a progressive condition that is associated with a significant **risk of sudden death**, **exertional** chest pain and syncope. Physical exam findings may include a mid-systolic murmur at the RUSB (right 2nd intercostal space), a **thrill at the suprasternal notch**, a click or a paradoxical split (S2 is single on inspiration and splits on expiration). EKG shows LEFT VENTRICULAR HYPERTROPHY (LVH). Patients should NOT be allowed to play sports. Treat with balloon dilation.

## MITRAL REGURGITATION (MR)

Mitral regurgitation is a holosystolic, blowing apical murmur that may radiate to the axilla.

* Associated with MARFAN SYNDROME and EHLERS-DANLOS SYNDROME

## MITRAL VALVE PROLAPSE (MVP)

In mitral valve prolapse, a mid-systolic click followed by the mid-late systolic murmur. Note that although this is kind of a regurgitant type of murmur, it is not a holosystolic murmur like the others.

* **PEARLS:** The murmurs of MVP and HYPERTROPHIC CARDIOMYOPATHY (HCM or HOCM) are both murmurs that **increase with standing or Valsalva, and decrease with squatting**. There are two ways to differentiate these murmurs.

- Only MVP murmurs are preceded by a CLICK
- HAND GRIP MANEUVER: It **increases** the MVP murmur and diminishes the HOCM murmur.

* **MNEMONIC**: "Baseball's MVP (most valuable player) is a pitcher. Whenever he's STANDING on the mound and HOCing loogies/spit, the crowd gets extremely LOUD to see him in action! If he gets hurt and has to SQUAT, the noise from the crowd suddenly DIMINISHES." This should help you remember the key exam findings for MVP and HOCM.

- **KEY**: GRIP = Increased noise for MVP only: "When the MVP gets his trophy, he stands at a podium, GRIPS it tightly over his head and the stadium erupts in LOUD cheer."

## AORTIC REGURGITATION/INSUFFICIENCY (AR or AI)

Aortic regurgitation (aka aortic insufficiency) results in a "relative mitral stenosis" from the jet of blood going backwards and hitting the mitral valve. This causes an **APICAL, mid-diastolic murmur** (called an Austin Flint Diastolic Murmur). Physical exam reveals **BOUNDING PULSES** (like a PATENT DUCTUS ARTERIOUS/PDA) and the apical mid-diastolic murmur. Aortic regurgitation is associated with **MARFAN SYNDROME**.

## RIGHT UPPER STERNAL BOARDER (RUSB) MURMURS

Aortic Stenosis is a right upper sternal boarder (RUSB) murmur.

## LEFT UPPER STERNAL BOARDER (LUSB) MURMURS

The following are left upper sternal boarder (LUSB) murmurs:

* Atrial Septal Defect (ASD): parasternal heave, mid-systolic

* Pulmonary Stenosis (PS): harsh, single S1, ejection click, wide S2 split

* Patent Ductus Arteriosus (PDA): continuous

* +/- Ventricular Septal Defect: holosystolic

## LEFT LOWER STERNAL BORDER (LLSB) MURMURS

LEFT LOWER STERNAL BORDER (LLSB) murmurs include:

* Ventricular Septal Defects (VSD) (holosystolic)

* Still's Murmur: buzzing/**musical**, **mid-**systolic, non-radiating, up to III/VI, benign, possibly at apex, **diminished with standing**

* Hypertrophic Cardiomyopathy (aka HOCM): Laterally displaced PMI, mid-systolic, increases with Valsalva

* Tricuspid Stenosis: Mid-diastolic, associated with Rheumatic Fever

## APICAL MURMURS

Apical murmurs include:

* Mitral stenosis: Mid-diastolic, associated with Rheumatic Fever

* Mitral regurgitation: Holosystolic and associated with Marfan Syndrome and Ehlers Danlos

* Aortic regurgitation (aka aortic insufficiency): Mid-diastolic murmur resembling MS, BOUNDING pulses

## HOLOSYSTOLIC MURMURS

Holosystolic murmurs include Ventricular Septal Defects (VSDs) and Mitral Regurgitation (MR)

## CONTINUOUS MURMURS

Continuous murmurs include:

* **VENOUS HUM**: This is a **continuous**, vibratory, **right subclavicular murmur**. It is present when sitting and **resolves when supine or the head is turned**. Benign murmur so no interventions are needed.

* PATENT DUCTUS ARTERIOSUS (PDA): A continuous "machine-like" or "rumbling" murmur heard best below the left clavicle.

## BOUNDING PULSE

A bounding pulse can be associated with aortic regurgitation (aka aortic insufficiency), patent ductus arteriosus (PDA), LEFT to RIGHT shunts (including large arteriovenous malformations/AVMs)

## WIDE PULSE PRESSURE

A wide pulse pressure can be noted in:

* Aortic regurgitation (aka aortic insufficiency)

* Patent ductus arteriosus (PDA).

* **MNEMONIC**: Since these are "relatively valve-less" disorders, there is nothing to prevent backflow of the blood during diastole. This results in a very low diastolic blood pressure and wide pulse pressure!

## CRANIAL BRUITS

Cranial bruits can be due to intracranial arteriovenous malformations. Watch out for signs of high output congestive heart failure (CHF).

## CAROTID BRUITS

Carotid bruits are a benign in children.

## FIXED SPLIT S2

A fixed split S2 is almost always due to an Atrial Septal Defect (ASD). Occurs due to persistent L to R shunting which results in persistently higher volumes across the pulmonary valve, and thus delayed closure.

## WIDELY SPLIT S2

A widely split S2 can be found with a RBBB or Pulmonic Stenosis (PS). The normal A2-----P2 changes to A2----------P2 due to delayed closure of the pulmonic valve.

## PARADOXICAL SPLIT OF S2

The paradoxical split of s2 occurs with a LBBB and Aortic Stenosis. S2 splits during EXPIRATION instead of inspiration. A2 is so delayed that it overlaps P2, resulting in a single S2 on inspiration. The S2 then splits on expiration.

# *FETAL CIRCULATION*

## NORMAL CIRCULATION

This is a pictorial representation of normal circulation.

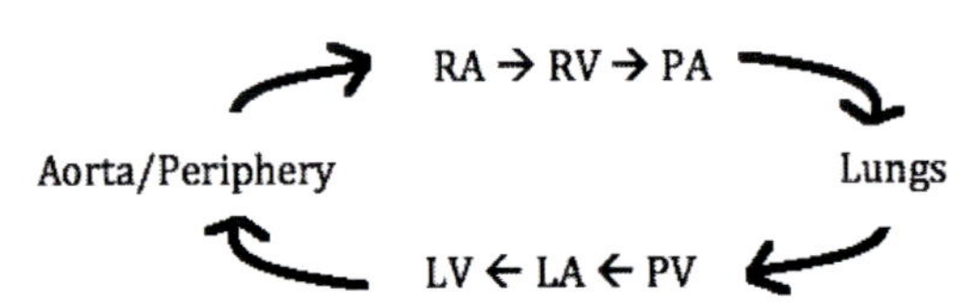


## FETAL CIRCULATION

This is a pictorial representation of fetal circulation.

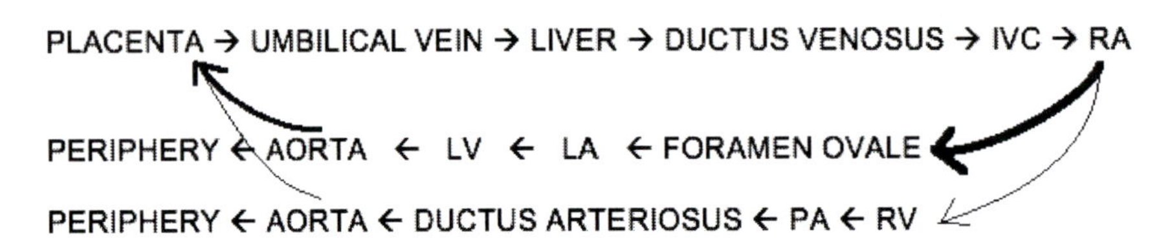


* UMBILICAL VEINS: Take OXYGENATED blood from the placenta to the fetal heart

- From the RA, most of the blood reaches the periphery via the foramen ovale. The rest of the blood goes to the RV to the PA to the Ductus Arteriosus (which connects the PULMONARY ARTERY to the AORTA)

* UMBILICAL ARTERIES: Come off of the descending fetal aorta and take blood back to the placenta

## RIGHT VENTRICLE (RV)

The right ventricle is the main pumping chamber in utero.

# *CYANOTIC CONGENITAL HEART DISEASES (CCHD)*

## PEARL (re: SHUNTS)

Knowing the direction of a shunt can be very difficult. For the exam, it would be surprising if you were asked to name the direction of the shunt in a complex cardiac condition. Information on the direction of shunts is presented in this guide to help you better understand the cardiac disorders. **If it is too confusing, skip over the shunt information**. Other topics are much more high yield.

## CYANOTIC CONGENITAL HEART DISEASES MNEMONIC

Remember the cyanotic heart diseases by using your hand as a guide! Look at your hands as you review this

* 1 = Index finger = TRUNCus arteriosus

* 2 = Index & middle fingers crossed/transposed: TRANSPOSITION OF THE GREAT ARTERIES (TOGA or * TGA)
* 3 = TRIcuspid Atresia
* 4 = TETRology of Fallot (TOF)
* 5 = 5 letters of TOTAL ANOMALOUS PULMONARY VENOUS RETURN (TAPVR)
* 6 = Rub the back of your <u>H</u>and = **H**YPOPLASTIC LEFT HEART
* 7 = Rub the PALM/"PULM" of your hand = PULMONARY ATRESIA
* 8 = **P + P = B** = **P**ERSISTENT **P**ULMONARY HYPERTENSION (PPHTN) = PERSISTENT FETAL CIRCULATION: The "8" kind of looks like two "Ps" from PPHTN stacked on each other

## CYANOSIS ALGORITHM AND PEARL

For cyanosis, give O2. If the child improves quickly, it is likely a PULMONARY problem. If not, give **PGE FIRST** and then start investigations to look for the 8 CCHDs.

## PROSTAGLANDIN (PGE) = PROSTACYCLIN

Prostaglandin (also known as PGE or prostacyclin) is the medication given to keep the ductus arteriosus open. This may be needed when faced with a left-sided obstructive lesion (e.g., coarctation of the aorta, critical aortic stenosis, hypoplastic left heart) or a right-sided obstructive lesion (e.g., tricuspid atresia, pulmonary atresia, TOF with pulmonary stenosis).

## PATENT DUCTUS ARTERIOSUS (PDA)

When a ductus arteriosus fails to close within days to weeks after birth, it's referred to as a patent ductus arteriosus (aka PDA). In utero, the Ductus Arteriosus allows for oxygen rich blood to bypass the lungs and reach the periphery. Blood is shunted from the pulmonary artery to the Aorta via the ductus arteriosus. After birth, the ductus arteriosus should close within days to weeks. If it does not, blood is shunted in the opposite direction (Aorta to pulmonary artery) due to the high pressure in the aorta. A continuous "machine-like" or "rumbling" murmur is best heard below the left clavicle. Treat with INDOMETHACIN or by using a cardiac catheter and coil.

**<u>PEARL</u>**: Do not give NSAIDS, cox inhibitors or any "prostaglandin/prostacyclin synthetase inhibitors" to a patient who has a ductal dependent lesion.

## COARCTATION OF THE AORTA

(This is **NOT A CYANOTIC DEFECT**, but its discussion may help you with the upcoming disorders) Coarctation of the aorta can result in congestive heart failure (CHF) symptoms including a gallop, weak pulses from cardiogenic shock and a "pale" appearance (not cyanotic). Treat with PROSTAGLANDIN (PGE) to keep the ductus arteriosus patent. Since the RV is the main pumping chamber in utero, a child born with Coarctation will have **RVH** on the EKG, not LVH.

**<u>PEARLS</u>:** Due to the coarctation, there is a blood pressure differential between the RIGHT ARM and RIGHT LEG. There is a lower blood pressure distal to the coarctation, thus the RIGHT LEG's pulse is weaker than the RIGHT ARM's.

## PREDUCTAL & POSTDUCTAL SATURATION

The topic of preductal saturation versus postductal saturation can get confusing. Basically, O2 saturations in the RIGHT HAND are considered PREductal. O2 saturations measured in ANY OTHER EXTREMITY is considered a POSTductal saturation.

* PULMONARY HYPERTENSION: There is a RIGHT to LEFT shunt at the PDA as deoxygenated blood is shunted from the PA to Aorta. PREductal O2 sats will be HIGHER than the POSTductal sats because of the shunted deoxygenated blood.

## TRUNCUS ARTERIOSUS (TA)

Instead of two "great arteries," a single TRUNK comes off of the ventricular chambers to create the truncus arteriosus. There is BIVENTRICULAR hypertrophy on EKG and cardiomegaly on chest x-ray. The shunt is bidirectional and there is a strong association with DiGeorge Syndrome (so obtain a **FISH!**). Results in severe CHF and death within months to 1 year. Cyanosis is only **mild**.

## TRANSPOSITION OF THE GREAT ARTERIES (TGA/TOGA)

The "great arteries" are the AORTA and the PULMONARY ARTERY. In Transposition Of The Great Arteries, the LV leads to the PA and the RV leads to the Aorta. Most common cardiac cause for cyanosis on **DOL 1**, and usually presents **within hours**. EKG shows RVH. The two circuits do not connect and are "running in parallel" (see image). Mixing needs to occur in order to support life. Often a VSD is present, but if not, then a septal "defect" needs to be created. To treat, **create an ASD** to allow mixing. Mixing at the PDA also helps (though not as much) so create the presence of BOTH (**ASD and the PDA)** by also giving **PGE**. The ASD (or existing VSD) allows a RIGHT to LEFT shunt (deoxygenated circuit to oxygenated circuit) to be created. CXR shows an **EGG SHAPED and vascular congestion** (due to blood flow from the LV to the PA). There is no associated murmur. In the image below, note the circuits running in parallel. Treatment is an ASD (represented by the crossed arrows).

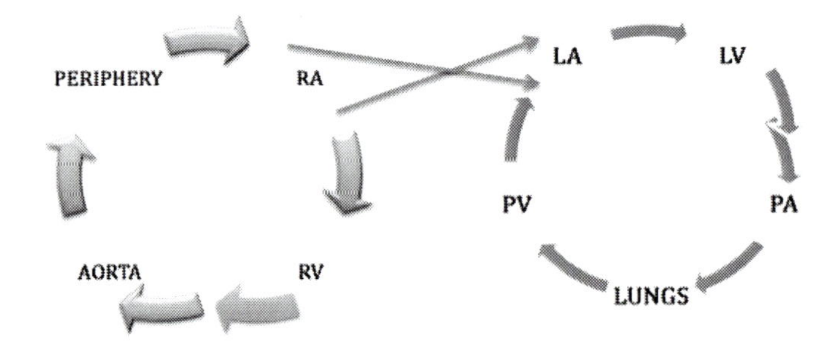


**PEARL**: If you suspect a cardiac cause for cyanosis on DOL 1, this is probably your answer!

## TRICUSPID ATRESIA

In tricuspid atresia, there is essentially NO functional RV since the tricuspid valve did not form. Blood goes from the RA to the LA via a PFO (a R to L shunt). Pulmonary circulation must be maintained through a PDA (with blood shunting from the aorta/periphery to the pulmonary artery, L to R shunt). EKG will show **LVH with**

**LAD (Left Axis Deviation)**. Cyanosis occurs early, usually within hours to days. Treat with PROSTAGLANDIN to maintain the PDA or the baby will eventually die.

## TETRALOGY OF FALLOT (TOF)

Tetralogy of Fallot is the most common cardiac cause for cyanosis in children (of any age). The four primary findings include PULMONARY STENOSIS (PS), RVH, OVERRIDING AORTA and a VSD. EKG shows RAD consistent with RVH. Exam reveals a **PS murmur** (harsh LUSB murmur, a widely split s2, a single S1 and possibly a click). There is a RIGHT **to LEFT** shunt at the VSD because blood cannot flow to the lungs due to the PS. Therefore there is NO pulmonary hypertension or evidence of congestion on the CXR (look for a **clear CXR**). That also means giving **PGE** will result in a PDA shunting blood from the LEFT to the RIGHT (from the Aorta to the PA, with no concerns for directional change from pulmonary hypertension). **CXR** findings include a lack of vascular congestion, BOOT shaped heart and RVH.

* **MNEMONIC**: "you have to **PROV**e it is a TET!" = **P**S, **R**VH, **O**verriding aorta, **V**SD!

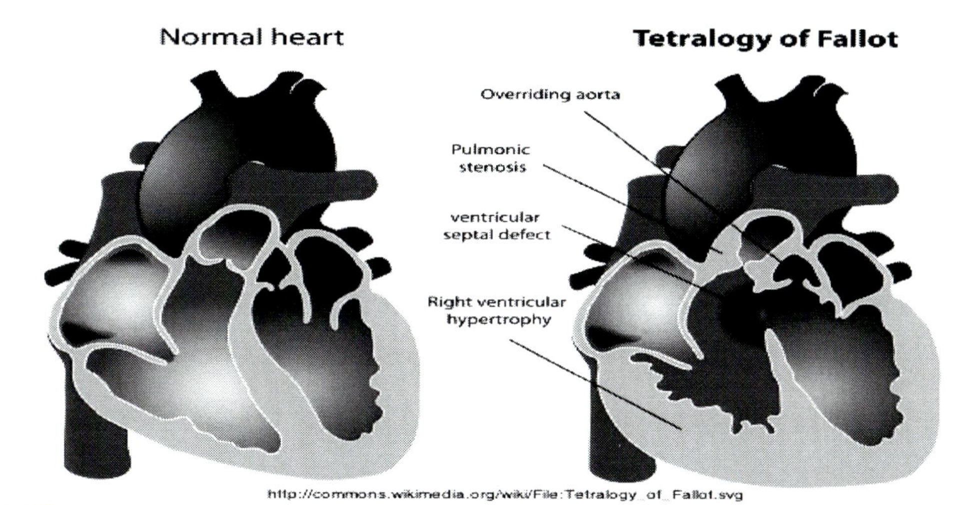

http://commons.wikimedia.org/wiki/File:Tetralogy_of_Fallot.svg

* **TET SPELLS:** Episodes of cyanosis (precise mechanism is uncertain) possibly occurring at times of relative anemia or increased pulmonary vascular resistance. An acute increase in R **to LEFT** shunting occurs which results in cyanosis, possible syncope or even death. Older children squat, younger children's knees are forced to the chest, in order to increase systemic vascular resistance and decrease/reverse the flow of the shunt. Other possible treatments include **morphine**, phenylephrine, propranolol and volume.

## TOTAL ANOMALOUS PULMONARY VENOUS RETURN (TAPVR)

There are four variants (types are not important) of Total Anomalous Pulmonary Venous Return (TAPVR) in which the PULMONARY **VEIN** leads to an anomalous **VENOUS** location. This OXYGENATED blood subsequently returns to the RA. Therefore, the oxygenated blood never enters the systemic circulation. RIGHT side. Since essentially all of the blood is going to the RIGHT side of the heart, there is significant **PULMONARY CONGESTION on CXR.** This is **NOT COMPATIBLE WITH LIFE UNLESS A PFO/ASD IS PRESENT OR CREATED** to allow blood to flow to the LA, LV and then Aorta. CXR shows a heart that is classically described as a **"Snowman"** (some say it looks more like a balloon). TREATMENT is to create or reopen the PFO/ASD.

**MNEMONIC**: "This one is #5 on the list of CCHDs, so imagine your entire hand reaching inside the chest of a SNOWMAN and ripping open an ASD."

## HYPOPLASTIC LEFT HEART

In hypoplastic left heart, everything on the LEFT side of the heart is small. Small left ventricle, small mitral valve, small aortic valve and small aorta. Because of the limited pumping ability, this results in CHF and a L to RIGHT shunt across a PATENT FORAMEN OVALE (PFO). Blood reaches the periphery via a PDA (from PA to Aorta, a R to LEFT shunt). If the PDA is allowed to close, CHF worsens dramatically.

## PULMONARY ATRESIA

Pulmonary atresia is very similar to tricuspid atresia. Since the pulmonary valve did not form, there is essentially no RV and blood goes from the RA to the RV via a PFO. Pulmonary circulation must be maintained through a PDA. EKG will show **LVH**. Cyanosis occurs early. Treat with prostaglandin to maintain the PDA or the baby will eventually die.

**PEARL**: It is unlikely that you would be asked to differentiate between tricuspid and pulmonary atresia.

## PERSISTENT PULMONARY HYPERTENSION = PERSISTENCE OF FETAL CIRCULATION

in persistent pulmonary hypertension (aka persistence of fetal circulation), underlying pulmonary disease causes higher pulmonary vascular resistance. This results in a PDA with shunting from the PA to the Aorta (R to LEFT) and resultant cyanosis. Nitric oxide can be given in an effort to allow greater blood flow into the lungs and thus reduce pulmonary HTN. The PDA will send deoxygenated blood to the **descending aorta**, so look for lower O2 sats in the RIGHT LEG compared to RIGHT ARM (which is getting oxygenated blood that made its way to the lungs and then LV and ascending aorta.. There is NO MURMUR but a possible right, parasternal heave from RVH.

**PEARLS**

* **BLUE/CYANOTIC BABY + CLEAR/BLACK CXR**: CXR is clear due to a lack of pulmonary blood flow. Give PGE immediately to maintain a PDA and a LEFT to RIGHT shunt from the AORTA to the PA.

* FETAL HEMOGLOBIN: Because of the high carrying capacity of fetal hemoglobin, desaturations and cyanosis is REALLY BAD. So, if a neonate has a CCHD and mild anemia with O2 sats in the high 80's, the baby may look okay but this could mean impending doom.

* DEVELOPMENTAL DELAY: MOST children with Congenital Heart Disease (CHD) have at least mild developmental delay and learning problems, but children with Cyanotic CHD have more problems that children with acyanotic CHD.

* **LEFT AXIS DEVIATION: The differential for LAD on EKG in the setting of CHD includes either AV Canal Defect or TRICUSPID atresia.** In tricuspid atresia, this is due to increased blood flow via a R to LEFT shunt at a PFO/ASD (possibly created as treatment) resulting in LVH/LAD.

| CHD | Murmur | Shunt | Notes |
|---|---|---|---|
| TOF | Pulmonary stenosis | R → L at VSD | Shunt at VSD, more as PS worsens. Single S2 because PV component absent. No increase in pulm vasculature on X-ray. |
| TGA/TOGA | NONE | R → L at ASD or VSD. | Cyanosis within hours. Single S2, Egg shaped. Pulm congestion from LV → PA |
| Tricuspid Atresia | Possible systolic. Possible PS sounding murmur | R → L at ASD L → R at a VSD and PDA | Blood goes from RA to LA to LV to RV to PA (with a possible PS murmur) to lungs to LA/LV to Aorta to the Periphery |
| Truncus | Possible systolic | R → L | PA & Ao are joined. Connected to RV & LV |
| TAPVR | | L → R at anomalous vein R → L at ASD | Snowman (to me looks like a balloon) |

## ***RHEUMATIC FEVER & RHEUMATIC HEART DISEASE***

**PEARL**: Know this section cold. Rheumatic Fever (RF) is a VERY testable topic. The only way to exclude it as an answer is to know what it is… and know what it isn't!

### RHEUMATIC FEVER

Rheumatic fever is caused by Group A Streptococcus (GAS, aka Strep PYOGENES) infection.

### JONES CRITERIA for RHEUMATIC FEVER

The two ways to diagnose Rheumatic Fever using the Jones criteria are listed below:

* Patient meets **2 of the MAJOR** Jones criteria. If 2 major criteria are present, **NO evidence of Group A Streptococcus is needed.**

* Patient meets **1 major AND 2 minor AND HAS EVIDENCE OF STREPTOCOCCAL INFECTION** (so a total of 3 criteria + evidence of recent GAS pharyngitis).

**PEARL**: DO NOT order a Streptococcal screen or culture. Since these can positive in **carriers**, order **ASO titers or Streptozyme**, which correlate better with recent infection. Also, negative Streptococcal screen or culture does NOT rule out a recently resolved pharyngitis.

### MAJOR JONES CRITERIA FOR ACUTE RHEUMATIC FEVER

Major Jones Criteria for acute rheumatic fever include:

* An asymmetric, migratory, polyarthritis of the large joints (ankles, knees, wrists)

* Signs of carditis: Valves, myocardium and pericardium can be affected so look for new murmurs, CHF, cardiomegaly and pericarditis.

* Painless, firm subcutaneous nodules

* (DOUBLE TAKE) The transient rash of ERYTHEMA **MARGIN**ATUM (a transient, erythematous, macular and light colored described as "serpentiginous," in which the **MARGIN**s progress as the center clears.

- **IMAGE**: http://www.dermaamin.com/site/atlas-of-dermatology/5-e/470-erythema-marginatum-of-rheumatic-fever-.html

* Sydenham's Chorea: Movements of the face and/or extremities without purpose. Some describe it as "purposeless dancing."

## MINOR JONES CRITERIA FOR ACUTE RHEUMATIC FEVER

Minor Jones criteria for acute rheumatic fever include:

* ARTHRALGIAS: Refers to PAIN without inflammation. Note that this is a MINOR criterion.

* Elevated ESR or CRP

* Fever

* Previous history of Acute Rheumatic Fever

* Prolonged PR

**MNEMONIC for the MAJOR J O N E S CRITERIA:**

* **J**oints: asymmetric, migratory, polyarthritis of the large joints (ankles, knees, wrists)

* **O** looks like a HEART (♥): Carditis = new murmurs, CHF, cardiomegaly and pericarditis.

* **N**odules: Painless and firm subcutaneous nodules

* **E**rythema **MARGIN**atum: The name **MARGIN**atum should remind you of the interesting facts regarding the **MARGIN**s of the rash.

* **S**ydenham's Chorea: Movements of the face and/or extremities without purpose and sometimes described as "purposeless dancing."

**PEARLS:**

* Rheumatoid fever is only caused by Streptococcal **PHARYNGITIS**! It is **not** associated with skin infections.

* **EXCEPTION** to JONES Criteria: If a patient has CHOREA or CARDITIS and also has EVIDENCE or SIGNS of Streptococcal pharyngitis, the patient can be diagnosed with Rheumatic Fever. The rest of JONES criteria do not apply.

## RHEUMATIC FEVER TREATMENT

For treatment of rheumatic fever, use Penicillin VK (oral) x 10 days or Benzathine Penicillin (IM/IV) x 1. If the patient has evidence of carditis, also give Aspirin (ASA) and prednisone

* RECURRENCE: Since there is a high chance of recurrence, administer monthly injections of Penicillin to **prevent** Group A Streptococcal pharyngitis. May discontinue after 21 years of age.

## RHEUMATIC FEVER ASSOCIATIONS

* ASCHOFF BODIES are nodules no the heart or in the aorta. They are **pathognomonic for Acute Rheumatic Fever**. Biopsy reveals inflammatory cells arranged in a "ROSETTE" pattern around a fibrinoid core.

* Mitral, tricuspid and aortic valves can be affected. Lesions can be stenotic or due to regurgitation. Mitral stenosis and tricuspid stenosis are frequently due to Rheumatic Fever. Mitral regurgitation is the most

common murmur of Rheumatic Fever. Possible descriptions might include the **harsh, systolic, apical murmur of mitral regurgitation**, or the **right-sided diastolic murmur of aortic regurgitation (look for evidence of CHF).**

* Endocarditis occurs due to **destruction** of valves, NOT infection.

* Emotional lability is another associated finding in RF.

## *KAWASAKI DISEASE, aka MUCOCUTANEOUS LYMPH NODE SYNDROME*

### DIAGNOSTIC CRITERIA FOR KAWASAKI DISEASE

Diagnostic criteria for Kawasaki Disease include 5 days of fever PLUS 4 out of 5 other findings, including:

* Erythema of the hands/feet

* A non-exudative conjunctivitis

* A non-vesicular rash

* Mucosal involvement (lips, oral mucosa or "strawberry tongue")

* Lymphadenopathy

**MNEMONIC** = "**THERMA/L**, **THERM**a**L** or **THERMA**l":

* **T**emp for 5 days

* **H**and findings (erythema of hands/feet)

* **E**ye findings (**non-exudative** conjunctivitis)

* **R**ash (non-vesicular)

* **M**ucosal involvement (oral mucosa, lips, strawberry tongue)

* **A**denopathy or **L**ymphadenopathy

### SUPPORTIVE DATA

Anemia, elevated ESR, leukocytosis and thrombocytosis support the diagnosis of Kawasaki Disease, but cannot be used to diagnose it.

**PEARLS:**

* Patients may have a "Fever of Unknown Origin" type of presentation

* ANA is negative

### COMPLICATIONS OF KAWASAKI DISEASE

Complications of Kawasaki Disease include **coronary aneurysms** (not during acute phase), coronary arteritis, myocardial infarction, sterile pyuria and liver dysfunction. The cardiac disease causes the greatest morbidity so get an echocardiogram NOW, and then repeat at 2 weeks and 6 weeks. A cardiologist should follow these patients.

### TREATMENT OF KAWASAKI DISEASE

Treatment of of Kawasaki Disease includes **HIGH DOSE ASPIRIN (80 mg/kg) + IVIG**. Cut ASA dose once patient is afebrile for at least 48 hours. IVIG is given as a onetime dose but may be repeated x 1. Continue ASA for a total of 6 weeks.

## ENDOCARDITIS

### ENDOCARDITIS DEFINITION

Endocarditis is defined as an infected heart valve. In general, **STREP VIRIDANS** is the most common etiology in children, followed by Staph aureus and then Staph epidermidis (especially if the valve is prosthetic). Findings may include fever (possibly up to 2 weeks), new murmurs (especially mitral or aortic regurgitation), petechiae, splenomegaly (from CHF), **tender** nodules on fingers/toes (Oslers nodes) and **nontender nodules** on the palms/soles (Janeway Lesions).

### ACUTE BACTERIAL ENDOCARDITIS

Patients with acute bacterial endocarditis present acutely toxic with high fever and possibly a septic picture. The most common etiology is **STAPH AUREUS**.

### SUBACUTE BACTERIAL ENDOCARDITIS

Patient's with subacute bacterial endocarditis present with vague symptoms lasting for weeks, rather than acutely toxic. The most common organism is **STREP VIRIDANS**. Aortic valves are especially vulnerable.

**<u>MNEMONIC</u>**: "VERDE sounds like VIRIDANS and means GREEN in Spanish. Imagine VERDE DECAY in teeth (a common hideout for viridans) entering the blood stream during flossing and settling on heart valves."

### DIAGNOSING ENDOCARDITIS

Negative blood cultures make diagnosing a bacterial endocarditis extremely unlikely. Even if a vegetation is seen on echocardiogram, the likelihood of a bacterial endocarditis in the presence of negative blood cultures is very low.

### TREATMENT OF ENDOCARDITIS

The treatment of endocarditis **VARIES** based on the **type of valve** affected (native vs. prosthetic).

### NATIVE VALVE ENDOCARDITIS

* **<u>MNEMONIC</u>**: NATIVE = "NATIVITY" or natural, like a BABY. If you don't have evidence of a Staphylococcal infection, treat native valve endocarditis the same way you would empirically treat a septic BABY, with Ampicillin and Gentamicin.

* **PENICILLIN-SENSITIVE STREP**: **May use a single antibiotic agent** (Penicillin, Ampicillin, Ceftriaxone or Cefazolin for 4 wks). May also use Ampicillin + Gentamicin for 2 weeks. Again, a single agent is fine if you have sensitivities. Strep viridans is still highly susceptible to PCN.

- **NOTE**: If a patient is allergic to PCN, may use Vancomycin (although it is not as effective as the beta lactams against Strep).

* **PCN-INSENSITIVE STREP SPECIES** (likely Enterococcus): Ampicillin + Gentamicin

- **PEARL**: Although Enterococcus is no longer susceptible to PCN, Ampicillin and Amoxicillin do still work.

* **STAPH AUREUS** (non-MRSA): Any anti-Staph drug is fine (Cefazolin, Oxacillin, Nafcillin or Vancomycin)

* **METHICILLIN RESISTANT STAPH AUREUS** (MRSA): Vancomycin + Gentamicin +/- Rifampin

## PROSTHETIC VALVE ENDOCARDITIS

Treat prosthetic valve endocarditis with IV antibiotics for 6 weeks

* **ACUTE**: Considered to be **within 2 months** of the valve being placed. Treat with emergent **surgery**!

* **SUBACUTE**: > 2 months since the valve was placed. The most likely etiology is Staph Epidermidis, although Staph aureus is also in the differential.

- If known MRSA or MRS**E**, give THREE antibiotics (Vancomycin + Gent + Rifampin). The extra medications are for synergy.
- If known to be regular Staph aureus, use TWO agents: Gentamicin + Cefazolin, Oxacillin or Nafcilin

## PROPHYLAXIS FOR SUBACUTE BACTERIAL ENDOCARDITIS (SBE)

The summary of prophylaxis for subacute bacterial endocarditis is based on the 2007 AHA guidelines.

* **PREVIOUS ENDOCARDITIS:** LIFELONG prophylaxis

* Prosthetic **VALVE or prosthetic material used to repair a VALVE**: LIFELONG prophylaxis

* **UNREPAIRED CYANOTIC** CHD: LIFELONG prophylaxis, or until 6 months after the CCHD has finally been repaired

* **REPAIRED CYANOTIC** CHD with prosthetic material: Prophylaxis for 6 months after surgery. If **prosthetic VALVE** material was used, then lifelong prophylaxis.

* GI/GU PROCEDURES: The new guidelines NO LONGER recommend prophylaxis for GI/GU procedures.

* MEDICATION: Prophylax with Amoxicillin 1 hour before high-risk procedures including most dental procedures (including cleanings), respiratory procedures (bronchoscopy) and procedures involving infected skin.

* **PEARLS/SHORTCUTS**:

- LIFELONG prophylaxis is provided for:
  - Prior endocarditis
  - Any prosthetic valve/prosthetic material used for **valve** repair
  - Unrepaired **CYANOTIC** CHD (not to be confused with Acyanotic CHD like simple ASD's, VSD's, etc.).
- * 6 MONTHS OF PROPHYLAXIS:
  - After ANY other prosthetic material placed in the heart for any reason including repair of cyanotic CHD. Do not confuse prosthetic VALVES (lifelong) with prosthetic MATERIAL used for repair of a CHD (6 mo).

# *MISCELLANEOUS CARDIOLOGY*

## PULSUS PARADOXUS

Pulsus paradoxus is defined as a drop in the systolic blood pressure by > 10 mm during inspiration (4-5 is normal). Seen in:

* Severe Asthma attack

* Myocarditis

* Pericardial Effusion with **TAMPONADE**.

## PERICARDITIS

In simple pericarditis, there are **no signs of CHF** since there is no effusion/tamponade. Findings may include a friction rub, a low-grade fever, muffled heart sounds and chest pain that is worse when supine, better when sitting up and leaning forward. CXR may show an enlarged heart. The likely causes include a recent viral URI or possibly a rheumatologic issue. Obtain a CXR to look for cardiomegaly and then consider doing an echocardiogram and obtaining fluid for cultures and serologies. Treat with NSAIDS.

## PERICARDIAL EFFUSIONS

In pericardial effusions, the heart can look like a "bag of water" heart on chest x-ray. Can also result in Kussmals sign (Increased JVD with inspiration). If you suspect this condition, the "best first test" to order is probably a cheap/quick CXR to look for the classic shape or cardiomegaly. Treat with a PERICARDIOCENTESIS.

## MYOCARDITIS

For myocarditis, look for fever, sudden heart failure symptoms (gallop, hepatomegaly, splenomegaly), pulsus paradoxus or a combination of these. May result in a new holosystolic, apical (mitral) murmur due to distortion of the LV, but there is usually no associated murmur. Differential includes **COCKSACKIE B** (obtain titers), Kawasaki's Disease, Rheumatic fever and a simple viral syndrome.

## EARLY CONGESTIVE HEART FAILURE

If you note signs of CHF within the first week of life, consider obstructive lesions on the left side of the heart as well as a GREAT VEIN OF GALEN MALFORMATION (a type of intracranial AVM that results in a machine-like murmur in the SKULL). Treat with Dobutamine for LV failure and use diuretics if there is pulmonary edema. Provide oxygen, correct an acidosis if present and give **PGE** to maintain a PDA. For children with high fevers and a history of CHF, use antipyretics to prevent high output cardiac failure.

## HYPERTROPHIC CARDIOMYOPATHY = HYPERTROPHIC OBSTRUCTIVE CARDIOMYOPATHY (HCM, HOCM)

Hypertrophic cardiomyopathy (HOCM) is usually an **AUTO DOMINANT** disorder resulting in **septal hypertrophy and pulmonary stenosis**. Patient's are at risk for sudden collapse and death. It is a major cause of sudden death in young athletes. For any child with an affected first-degree relative, a cardiologist should follow him/her for early detection of HOCM as they get older. The murmur of HOCM is unique in that it **increases with standing or Valsalva, and decreases with squatting**. HOCM is more likely in INFANTS OF DIABETIC MOTHERS (IDM) and in children with NOONAN SYNDROME.

**PEARLS:** The murmurs of MVP and HYPERTROPHIC CARDIOMYOPATHY (HCM or HOCM) are both murmurs that **increase with standing or Valsalva, and decrease with squatting**. There are two ways to differentiate these murmurs.

* CLICK: Only MVP murmurs are preceded by a CLICK

* HAND GRIP MANEUVER: **Increases** the MVP murmur and **diminishes** the HOCM murmur.

**MNEMONICS:**

* (DOUBLE TAKE) "Baseball's MVP (most valuable player) is a pitcher. Whenever he's STANDING on the mound and HOCing/spitting loogies, the crowd gets extremely LOUD to see him in action! If he gets hurt and has to SQUAT, the noise from the crowd suddenly DIMINISHES." This should help you remember the key exam findings for MVP and HOCM.

- GRIP = Increased noise for MVP only: "When the MVP gets his trophy, he stands at a podium, GRIPS it tightly over his head and the stadium erupts in LOUD cheer."

* "An astronaut-like MOONMAN stands on the moon. He HOCs/spits a loogie into his RIGHT PALM." MOONMAN = NOONAN, HOCs a loogie = HOCM, RIGHT = Right side of the heart, and PALM = "PALMonary stenosis"

* IDM association: The "D" of Diabetes looks like a hypertrophic septum.

## CARDIOMEGALY and HYPERTROPHY

Cardiomegaly means a "big" heart, where as hypertrophy means a "thick" heart (which can result in cardiomegaly). For the purposes of the boards, these terms are likely synonymous. Cardiomegaly is often a NORMAL finding in athletic kids. Septal hypertrophy/HOCM is not.

## CHEST PAIN

Chest pain in children is common, but it is rarely due to cardiac issues. The cause is often unknown. Refer to a cardiologist only if the chest pain is noted exertion. If a patient has left sided chest pain at rest that increases with inspiration and resolves without intervention within seconds to minutes, it's probably PRECORDIAL CATCH SYNDROME. Consider pericarditis if the history, physical exam or EKG findings are suggestive.

## SVC SYNDROME

SVC syndrome is most commonly due to some type of cancer, especially lymphoma. Symptoms include facial swelling, grayish/dusky skin color, weight loss and night sweats. There is no pulmonary issue and the arterial circuit is unaffected, so the patient's O2 saturations should be NORMAL. Treat with steroids, radiation and cancer-specific therapy.

## MEDIALLY DISPLACED PMI

A medially displaced PMI can be found in:

* DEXTROCARDIA: If presented with a patient whose PMI is on the **RIGHT**, think DEXTROCARDIA.

* PRIMARY CILIARY DYSKINESIA = KARTAGENER SYNDROME: Autosomal recessive disorder. Patients with this disorder have dextrocardia (heart on the right side) and also SITUS INVERSUS (major visceral organs in a mirrored position). In addition to dysfunctional cilia in the respiratory tract, they also have nasal polyps and immotile sperm. If presented with a patient who has nasal polyps, a PMI on the RIGHT side of the chest and a history of chronic sinus or pulmonary infections, think PRIMARY CILIARY DYSKINESIA. Diagnose by obtaining electron microscopy of nasal scrapings.

- **PEARL**: Patients' medical histories can sound very similar to cystic fibrosis patients, so focus on subtleties such as the PMI or an X-ray showing an inversed location of bowel loops.

## PEDIATRIC BLOOD PRESSURE GUIDELINES

Pediatric blood pressure checks should begin around 3 years of age.

* PREHYPERTENSION: A blood pressure that is between the 90th and 95th percentile for height.

* HYPERTENSION: HTN is considered a blood pressure that is > 95th percentile for height on **THREE** separate occasions (with each BP preferably measured 1 month apart). First line therapy is lifestyle modifications (diet/exercise). Differential diagnosis includes: Wrong cuff size, Cushing's, 17-hydroxylase deficiency, Pheochromocytoma, Hyperaldosteronism, Lupus, Renal artery stenosis, Polycystic Kidney Disease (PCKD), Vesicoureteral Reflux (VUR) and Coarctation. Treat essential hypertension with WEIGHT LOSS first, or admit the patient for IV nitrates if extremely high.

- PHEOCHROMOCYTOMA: Labs include elevated serum metanephrines and elevated urine vanillylmandelic acid (VMA, which can also be elevated in neuroblastoma). Treat with alpha blockade **prior** to resection. Because of downregulation of catecholamine receptors, patients are highly susceptible to **orthostatic hypotension**. Children may also have **labile** BP.

## CHOLESTEROL SCREENING = HYPERLIPIDEMIA SCREENING

Risk factors for hyperlipidemia include anorexia, anticonvulsants, a sedentary lifestyle, steroid use and CIGARETTES (biggest risk factor).

* If a patient is obese, has DM or HTN, obtain a **NON**FASTING TOTAL CHOLESTEROL screen. If that level is abnormal (>170), obtain a FASTING LDL.

* If a **parent** has a TOTAL CHOLESTEROL that is > 240, obtain a **nonfasting** TOTAL CHOLESTEROL screen in the child. If that is abnormal (>170), obtain a **FASTING LDL**.

* If the child's parent, aunt/uncle or grandparent has a history of EARLY CORONARY ARTERY DISEASE (CAD diagnosed before 55 years of age), obtain a FASTING **LDL**. If the level is > 110, try diet and exercise alone. If the level is > 160, try diet modifications and then escalate to medications if there is no improvement within 6 months.

- For the exam, the above should be more than sufficient. Guidelines actually state that if the LDL is > 160 and the patient has 2 risk factors, or it is > 160 and there is a strong family history, or the LDL is > 190, you may immediately start cholesterol lowering medications. For the test, choosing lifestyle modifications for any child with an LDL < 190 is probably a safe choice.

## FAMILIAL HYPERCHOLESTEROLEMIA

Familial hypercholeAutosomal Dominant disorder due to a deficiency in LDL receptors. Can result in Xanthomas of the skin.

# DERMATOLOGY

## *GENERAL DERMATOLOGY*

### CONTACT DERMATITIS, A DIAPER RASH

Contact dermatitis, a diaper rash that **spares** the inguinal folds. Treat with more frequent diaper changes and a topical barrier, like zinc oxide.

### (DOUBLE TAKE) CUTANEOUS CANDIDIASIS, A DIAPER DERMATITIS

Cutaneous candidiasis, a diaper dermatitis, can occur secondary to a contact dermatitis or recent antibiotic use. Beefy red rash with papular satellite lesions. This rash goes **into the inguinal folds**. Use a KOH prep to confirm diagnosis and treat with a topical antifungal, such as nystatin or clotrimazole.

**IMAGE** (includes satellite lesions): http://www.pediatrics.wisc.edu/education/derm/tutc/80.html

### ATOPIC DERMATITIS (ECZEMA)

In babies, atopic dermatitis (eczema), **SPARES** the diaper folds/flexural surfaces (but not in older kids). It is **PRURITIC** and LICHENIFIED. Food allergies CAN exacerbate eczema and avoidance of certain foods MAY the delay onset (but not prevent). Breastfeeding also DELAYS the onset. Watch for superinfection if not improving.

**IMAGE**: http://skininfo.org/wp-content/uploads/2010/06/atopic_dermatitis.jpg

### NUMMULAR ECZEMA

Nummular eczema is a coin shaped eczematous lesions usually on the **extensor** surfaces of extremities. Lesions are **uniform**, without any central clearing. Lesions may ooze, crust or have a scaling pattern. Treat with steroids.

**IMAGE**: http://suite101.com/article/nummular-eczema-a98681

**MNEMONIC**: Imagine that you are standing with your arms in abduction and you are balancing silver COINS that are **UNIFORM in color** (without central clearing) balancing on the BACK of both of your arms (**extensor surface**).

### (DOUBLE TAKE) ECZEMA HERPETICUM

Eczema herpeticum is a potentially life threatening disseminated herpes (HSV) infection occurring at sites of skin damage, including sites of eczema. Look for HSV Vesicles + Crusted Lesions. Even if a description is not given of a vesicular rash, have a high index of suspicion for a rash "**not improving with steroids and/or antibiotics**." Diagnose with viral culture for HSV, but do not delay treatment. A Tzanck smear can support the diagnosis. Treat by **STOPPING** topical steroids and/or immunosuppressants and starting **Acyclovir**.

**IMAGE**: http://bit.ly/nkSGQU

## SEBORRHEIC DERMATITIS (aka CRADLE CAP)

Seborrheic dermatitis (aka cradle cap), is a **NONpruritic**, inflammatory, flaky rash with white to yellow scales that usually forms in oily areas (e.g., scalp). Often seen in first two months of life. After that, not very common until adolescence. May treat with topical antifungal agents or mild steroids. Skin may be left with hypopigmented areas, especially in the folds. If asked to name the hypopigmented areas, choose PITYRIASIS ALBA.

**IMAGE**: http://hardinmd.lib.uiowa.edu/dermnet/seborrheicdermatitis18.html

## PSORIASIS

Psoriasis is a **very well defined,** red, flaky rash covered with **silver**-white patches. Can also be described as thick and scaly (like seborrheic dermatitis). Sometimes results in punctate bleeding when scales are removed (called the Auspitz sign). Can occasionally be limited to the diaper area, in which case it goes **into the inguinal folds**.

## GUTTATE PSORIASIS

The "gttate " in guttate psoriasis means "drop like," and describes the shape of these discrete psoriatic lesions. Can be preceded by a **Group A Strep** (Pyogenes) infection.

**IMAGES**: http://www.healthy-skin-guide.com/psoriasis-photos.html

## (DOUBLE TAKE) LANGERHANS CELL HISTIOCYTOSIS (LCH) = HISTIOCYTOSIS X

Langerhans Cell Histiocytosis (LCH), aka Histiocytosis X, is a **PAPULAR** rash that is sometimes associated with petechiae. The rash is located **in the folds** (inguinal folds, supra-pubic folds, perianal area). It can resemble eczema, but the petechiae or PAPULES should guide you towards this diagnosis. LCH is a type of **cancer**. You may be shown lytic bone lesion (possibly of the skull). Diagnose by skin biopsy. Can also be associated with DIABETES INSIPIDUS. Treat by removing the lesion and giving steroids, +/- chemotherapy.

**PEARLS:** Do not confuse this with Wiskott-Aldrich (WiXotT-Aldrich, X-linked, low IgM, high IgA, TIE = Thrombocytopenia, small platelets, Infections and Eczema). Also, if they describe an eczema or seborrheic dermatitis type of rash in a patient with high urine output, LCH is your diagnosis.

**IMAGE**: http://bit.ly/pJIQSN

**IMAGE**: http://bit.ly/pc5dEG

## RASHES THAT SPARE THE INGUINAL FOLDS

Eczema and Contact Dermatitis should be high on your differential for rashes that spare the inguinal folds.

## PRURITIC RASHES

Consider atopic dermatitis/eczema, HSV, scabies, tinea or Varicella (VZV) in your differential of any pruritic rashes.

## KERATOSIS PILARIS

Keratosis pilaris forms due to overgrowth of the horny skin. Can look similar to eczema. May have a mild erythematous background. No treatment is needed.

**IMAGE**: http://bit.ly/rcjX3X

## LICHEN SCLEROSUS

Lichen sclerosus is a chronic, inflammatory, dry, white and somewhat scaly rash that is usually found in the genital area. There is no thickening or sclerosis. There are usually now symptoms, although a small percentage of patients have pruritus. Look for a picture of labia with a rash.

**IMAGE**: http://www.medscape.com/viewarticle/712767_5

## LICHEN STRIATUS

Lichen striatus is a rash that looks like eczema, but is linear or papular and can follow the Lines of Blaschko

**IMAGE**: http://emedicine.medscape.com/article/1111723-overview

**IMAGE**: http://bit.ly/n7GU4p

## ALLERGIC CONTACT DERMATITIS, A TYPE IV HYPERSENSITIVITY SKIN RASH

Allergic contact dermatitis is a Type IV hypersensitivity skin rash requires a **prior exposure**, and tends to be pruritic. See if the location of the rash is in an area where a nickel-containing belt buckle, earring, necklace or other jewelry could have been. Rash may present even after **years** of wearing the irritant. The rash from a nickel exposure is more erythematous and can become lichenified. The classic example of Type IV reactions is the rash of **poison ivy**, or other "leaves of 3" (including poison oak or poison sumac). Regarding a contact dermatitis from these plants, it will not spread once the affected area is washed with soap and water. The fluid from within the vesicles **cannot** spread the rash. This reaction is a Type IV Cell Mediated Hypersensitivity Reaction, and is called a **Rhus** reaction. The rash is vesicular and may be in a linear configuration (where the leaves rubbed across the skin).

* **PEARL**: First exposure may take 1 week to develop the rash as helper T cells proliferate and "remember" the agent. After that, the rash may develop within **hours** of exposure. "No wonder I had to go through the 2-step PPD before starting as an attending!"

* **PEARL**: REMINDERS: A PPD and the skin testing of Candida, Mumps and Tetanus are all Type IV reactions.

* **MNEMONICS:**

- "LEAVES OF THREE, LET THEM BE!"
- Type IV reaction: I + V = the Roman numeral IV = 4, and the 4th letter in the alphabet is **D** = **D**ELAYED. I + V also should you remind you of poison **Iv**y.

* **IMAGE**: http://bit.ly/nWLxen

* **IMAGE**: http://www.lib.uiowa.edu:8080/hardin/md/cdc/4484.html

## (DOUBLE TAKE) BIOTIN/BIOTINIDASE DEFICIENCY

Biotin, or Biotinidase, deficiency may present with a RASH + ALOPECIA + **NEUROLOGIC SIGNS** (ataxia, coma, etc.). Patients may also have lactic acidosis. Treat with Biotin.

**MNEMONIC**: Imagine the TIN MAN from the Wiz of Oz walking with an ATAXIC gait as he SCRATCHES his arm. Notice that he has NO HAIR!

## PAPULAR URTICARIA

Papular urticaria is a rash due to hypersensitivities to the insect bites of bedbugs, fleas and mosquitoes that results in edema, erythema and pruritus. It presents in **RECURRENT CROPS**. Tends to come and go, wax and wane every few weeks or months. Some lesions may be umbilicated. Treat by removing the offending agent (fleas, lice, bedbugs or outside insects).

**PEARL**: You may not be given the history of a specific insect or exposure

MNEMONIC. "CROPular Urticaria." Where do you find insects? In CROPS of course!

**IMAGE**: http://www.skincareguide.com/gl/p/papular_urticaria.html

## VITILIGO

Vitiligo results in hypopigmented macules. Look for a "salt and pepper" type of pattern of re-pigmentation. Often associated with HALO NEVI.

**IMAGE**: http://en.wikipedia.org/wiki/File:Vitiligo2.JPG

## (NAME ALERT) ICHTHYOSIS VULGARIS

Ichthyosis vulgaris is a rash that resembles FISH SCALES. Often seen in atopic dermatitis patients. May attempt treatment with ammonium lactate or alpha hydroxyacid containing agents. The name alert is for lamellar ichthyosis and harlequin ichthyosis.

**IMAGES**: http://www.skinsight.com/adult/ichthyosisVulgaris.htm

## (NAME ALERT) LAMELLAR ICHTHYOSIS

Lamellar ichthyosis is noted at the time of birth in newborns. A thin, transparent film is noted on the body. Eyelashes are missing. Eyelids seem inverted. The name alert is for harlequin ichthyosis and ichthyosis vulgaris.

**IMAGE**: http://bit.ly/okfaoh

**IMAGE**: http://bit.ly/q3unlU

## (NAME ALERT) HARLEQUIN ICHTHYOSIS

Harlequin ichthyosis presents with a newborn that looks much more abnormal than lamellar icthyosis. The covering is hard ("armor-like") and horny. Movement is restricted. Prognosis is poor comparatively. The name alert is for lamellar ichthyosis and ichthyosis vulgaris.

**IMAGE**: http://bit.ly/nhplsa

## PYODERMA GANGRENOSUM

The etiology of pyoderma gangrenosum is unknown, but it is known to e associated with other systemic diseases such as Crohns. Lesions are described as deep, bluish, necrotic and boggy-looking ulcers.

**IMAGE**: http://bit.ly/TrDrCf

**IMAGE**: http://bit.ly/116Vxes

## (DOUBLE TAKE) ECTHYMA GANGRENOSUM

Ecthyma gangrenosum is usually a sign of a **PSEUDOMONAS** infection and possibly sepsis in an immunocompromised patient, especially **LEUKEMIA**! Look for a **neutropenic** patient with black, necrotic, ulcerative lesions with surrounding erythema and edema. Often located in the groin/diaper area.

**IMAGE**: http://www.nlm.nih.gov/medlineplus/ency/imagepages/2647.htm

## GRANULOMA ANNULARE

Granuloma annulare is a chronic skin condition with an annular lesion (circular). It may be slightly pruritic. There are **no scales**.

**PEARL**: Looks kind of like ringworm, but there is **NO SCALING**! Keep this in mind any time you see Tinea as an answer choice.

**IMAGES**: http://anagen.ucdavis.edu/1412/case_presentations/ga/wollina.html

## PITTED KERATOLYSIS

Pitted keratolysis is a condition in which there is **pitted** skin in areas of pressure. There will probably be a history of **strong foot odor**.

**IMAGES**: http://emedicine.medscape.com/article/1053078-media

## DERMATOMYOSITIS

Dermatomyositis results in a a heliotropic, violaceous rash in malar area. Gottron's Papules (erythematous, shiny, pruritic papules over the matacarpals) may be present. Patients will have proximal weakness and possible telangiectasias near the nail folds. Diagnose with a **BIOPSY. CK LEVEL WILL BE HIGH**. These patients can also get calcinosis cutis/calciphylaxis.

**PEARL/REMINDER:** Duchenne Muscular Dystrophy also has elevated CK levels

**IMAGE**: http://emedicine.medscape.com/article/332783-media

**IMAGE**: http://bit.ly/rrzrsY

## STEVENS-JOHNSON SYNDROME (SJS) and TOXIC ENDODERMAL NECROLYSIS (TEN)

The terminology for Stevens-Johnson Syndrome (SJS) and Toxic Endodermal Necrolysis (TEN) varies. Many now view these disorders on a spectrum. SJS and TEN are the same, but TEN is the diagnosis if > 30% of body surface area involved. Look for bullae or erosions followed by hemorrhagic crusting. There may be severe blistering with the Nikolsky sign when pressure is applied. It is a full thickness rash similar to a burn. Skin lesions may look like a BULLSEYE or TARGET lesion, with the center described as DARK, DUSKY or VIOLACEOUS. The target CAN be a blister or vesicle. At least 2 mucous membranes must be involved (most commonly the lips and eyes). If the eyes are involved, this is an ocular emergency!

MEDICATION ASSOCIATIONS: Aromatic seizure medications, penicillins, NSADS and **sulfa** drugs. The rash usually occurs within 2 months of starting the medication.

## ERYTHEMA MULTIFORME

Erythema multiforme is also a confusing topic. It may now also be considered on in the SJS/TEN spectrum, especially if mucous membranes are involved. Distinguishing erythema multiforme minor from erythema multiforme major is also confusing, so the terminology is not likely to be tested.

Both minor and major have tiny **target** lesions (probably dusky in the middle). Sometimes you have to use your imagination to envision the target. It may just look a little darker on the inside of the lesion than the outside. Lesions usually start on the hand and/or feet and THEN progress to the trunk. 0-1 mucous membranes involved (if more, it will likely be called SJS or TEN). IF you are tested on the terminology, pick minor if the patient is not toxic. Possible etiologies include HSV, Mycoplasma and Syphilis.

**IMAGE**: http://bit.ly/pwOmhQ

**IMAGE**: http://bit.ly/r2nHWI

**IMAGE**: http://bit.ly/V0lmqZ

**IMAGE**: http://bit.ly/VbOIZx

**MNEMONIC**: Imagine Stevens and Johnson as two very arrogant hunters. They went TARGET shooting one day in an area that said, "beware of BULLS." They learned their lesson the hard way when a BULL came out of nowhere and did some target practice of his own.

## NEONATAL LUPUS

The baby does NOT have lupus. Neonatal lupus occurs in children of mothers with SLE due to fetal exposure to maternal antibodies. It is rare. Findings may include increased LFTs, petechiae, rash, scaling, thrombocytopenia, **3rd degree AVB/bradycardia** or hydrops (death) due to AVB. Diagnose by sending **Anti-Ro** or anti-La antibodies (aka anti-SS-A or SS-B).

**IMAGES**: http://emedicine.medscape.com/article/1006582-media

**PEARL:** RASHES WITH CENTRAL CLEARING: Hives/urticaria, Rheumatic Fever ("jonEs" = E. Marginatum = MARGINs progress to give central clearing), Tinea (raised border/ringworm).

**PEARL:** RASHES WITH CENTRAL DARKENING/TARGET LESIONS: SJS/TEN ("target shooting, bull"), Brown recluse spider bite (see Emergency Medicine), Lyme Disease/Borrelia/Erythema Migrans

## URTICARIA/HIVES

Urticaria (hives) is a pruritic rash due to an allergic exposure. Pink center with a more erythematous border. Giving histamine blockers (both H1 & H2) may be helpful. Foods are the most likely cause of chronic urticaria.

**IMAGE**: http://www.webmd.com/skin-problems-and-treatments/picture-of-urticaria-wheals

**IMAGE**: http://www.webmd.com/skin-problems-and-treatments/picture-of-hives-urticaria

## SCLERODERMA

Scleroderma patient's have thickened skin with an ivory or waxy, appearance. Affects girls more frequently. The limited form is more common than the systemic form in children (located at 1 site only). Lesions may initially be painful and tender. Skin is often hard and may have a linear appearance. Treat with topical lubricants for limited cases. May have to use steroids or other immunosuppressives in more severe cases.

**IMAGES**: (bottom) http://dermatlas.med.jhmi.edu/derm/indexDisplay.cfm?ImageID=433473685

## DERMOID CYSTS (aka EPIDERMOID CYSTS)

Dermoid cysts (aka epidermoid cysts) are saclike growths present at birth. They are like teratomas in that they can contain hair and teeth. They are often associated with tufts or sinuses. They grow slowly and can get infected, so most of them should be REMOVED. Especially those in sensitive areas, including the face or nasal area.

**IMAGE**: http://bit.ly/V4FCbp

**IMAGE**: http://bit.ly/QJMShp

**IMAGE**: http://www.dermaamin.com/site/atlas-of-dermatology/4-d/351-dermoid-cyst-.html

## COMEDONAL ACNE

Think of comedonal acne as an OBSTRUCTIVE process that creates white heads and black heads. Treat with a RETINOID keratinolytic agent. May also prescribe benzoyl peroxide.

**PEARL**: Any answer with topical retinoic acid plus benzoyl peroxide twice daily is WRONG. Benzoyl peroxide inactivates retinoids, so one should be used at night, and the other in the morning.

## INFLAMMATORY ACNE

Inflammatory acne is differentiated from comedonal acne by its red base.

* **Minor cases**: If the acne is localized with small lesions, use a TOPICAL agent, such as Benzoyl peroxide, Clindamycin or Erythromycin.

* **Severe cases**: If large, nodular or in multiple areas, use ORAL antibiotics. First line is Tetracycline or Erythromycin. Second line agents include Doxycycline and Minocycline. These antibiotics provide a bacteriocidal and an anti-inflammatory effec. May also try oral contraceptive pills (OCPs) in females for their anti-androgen effects. If all else fails, use ISOTRETINOIN.

## ISOTRETINOIN

Isotretinoin is a miracle drug that fights sebum production and bacteria, while also decreasing inflammation and comedonal acne. But, it is **TERATOGENIC**, so obtain TWO negative pregnancy tests before starting the medications. Also, patients must use TWO forms of birth control starting 1 month before starting the medication and until 1 month after. In addition, they should have monthly pregnancy tests.

**PEARL**: Acne can begin as early as 8 years of age. If the boards present a 7-year-old child with what looks like acne, CONSIDER ANOTHER DIAGNOSIS! Consider exogenous steroid use, precocious puberty and TUBEROUS SCLEROSIS.

## (DOUBLE TAKE) APHTHOUS ULCERS

Aphthous ulcers are painful lesions found within the oral mucosa (buccal mucosa, lips and tongue) with a grayish-white base and a rim of erythema. Can occur in isolation or in association with Behcet's or Schwachman Diamond syndrome.

**IMAGE**: http://upload.wikimedia.org/wikipedia/commons/4/45/Aphthous_ulcer.jpg

**IMAGE**: http://upload.wikimedia.org/wikipedia/commons/2/2d/Aphtha2.jpg

# *TEETH ISSUES*

## TOOTH TIMELINE

Tooth appearance follows a timeline. All anterior teeth are present (nine of them) by about 12 months. Primary teeth are present by about age 2. Some children do not have teeth by 1 year of age, so reassurance is ok. For ABP questions, they will be more focused on **abnormal looking teeth**.

## PEG TEETH

Peg teeth refers to teeth that are smaller than usual. Sometimes they are tapered and look like fangs. Usually affects the lateral incisors. Associated with INCONTINENTIA PIGMENTI and HYPOHIDROTIC ECTODERMAL DYSPLASIA.

**IMAGE**: http://www.peglateralincisors.com/pli-e.jpg

**IMAGES**: http://www.peglateralincisors.com/

## HUTCHINSON TEETH

Hutchinson teeth are found in CONGENITAL SYPHILIS. These children have teeth that are smaller and more widely spaced. They also have notches on the biting surfaces.

**IMAGE**: http://en.wikipedia.org/wiki/File:Hutchinson_teeth_congenital_syphilis_PHIL_2385.rsh.jpg

**IMAGE**: http://www.lhup.edu/smarvel/Seminar/FALL_2004/McGovern/

## TETRACYCLINE TEETH STAINING

If tetracycline is used at a young age, teeth can end up having yellow, brown or blue band-like stains. Avoid tetracycline until patients are at least 8 years of age.

**IMAGE**: http://img.medscape.com/pi/emed/ckb/dermatology/1048885-1076389-2302tn.jpg

## FLUOROSIS

Fluorosis is the mottled discoloration of teeth noted due to excess fluorine use during tooth development (up to age 4).

**IMAGE**: http://img.medscape.com/pi/emed/ckb/dermatology/1048885-1076389-2303tn.jpg

# *VASCULAR & PIGMENTED LESIONS*

**PEARL/MNEMONIC**: HEMANGIOMAS are different than VASCULAR MALFORMATIONS (e.g., Port Wine Stains/capillary malformations). VASCULAR MALFORMATIONS tend to have much more associated morbidity. You might say that **VM**s are **V**ery **M**orbid in comparison.

**IMAGES**: (slideshow on birthmarks) http://www.webmd.com/skin-problems-and-treatments/slideshow-birthmarks

## HEMANGIOMAS

Hemangiomas are an abnormal build up of blood vessels. They eventually self-involute but are dangerous during PROLIFERATION PHASE. They are otherwise benign. They usually look red, but can appear blue if deep (CAVERNOUS HEMANGIOMAS). Proliferation is greatest during the first 6 months and lesions are largest around 1 year of age. Lesions start to involute around 2 years of age and disappear by 5-10 years of age. If in a benign area, they can be left alone. If in a more sensitive area (near the eyes, ears, nose, throat or spine), they may require treatment with **systemic** steroids. Second line therapies include pulsed dye laser therapy and surgery.

**COMPLICATIONS**: If located in the beard area, look for airway issues. If near the eye, it's okay as long as there is no problem with VISION. Those near the ears, nose and lips can be troublesome if they ulcerate. If in the lumbosacral area, there is concern for spinal dysraphism (incomplete fusion of a raphe, especially the neural folds/tube). High output congestive heart failure (CHF) can occur due to large, or multiple hemangiomas.

**IMAGE**: http://en.wikipedia.org/wiki/File:Capillary_haemangioma.jpg

**IMAGE**: http://bit.ly/Tkz347

## PHACES SYNDROME

A diagnosis of PHACES syndrome requires a large hemangioma in the face/neck area PLUS one of the following defects:

* **P**ost fossa malformation (DANDY WALKER)
* **H**emangioma. Often in the distribution of the Facial Nerve. Look for a large **segmental** hemangioma in on the **FACE**. Segmental refers to what looks like a nerve distribution (segmented by normal skin in between). Can be associated with STROKES.
* **A**rterial cerebrovascular anomaly
* **C**ardiac anomalies: Especially COARCTATION OF THE AORTA
* **E**ye anomalies: MICROPHTHALMIA, STRABISMUS
* **S**ternal defect
* **IMAGES**: http://www.baylorcme.org/hemangioma/presentations/metry/presentation_text.cfm

## (DOUBLE TAKE) KASABACH MERRITT SYNDROME

In Kasabach Merritt syndrome, there are large, congenital vascular tumors. They are not true hemangiomas but can cause a severe CONSUMPTIVE COAGULOPATHY (in the form of **thrombocytopenia** and the consumption of coagulation factors) and death.

**IMAGE**: http://img.medscape.com/pi/emed/ckb/hematology/197800-202455-5213tn.jpg

**IMAGE**: http://img.medscape.com/pi/emed/ckb/hematology/197800-202455-5214tn.jpg

**PEARL**: Look at the above images closely. Make sure you look closely at images so that you do not get this vascular tumor confused with hemihypertrophy.

**MNEMONIC:**

- >---< is used by many of us when recording CBC results.

↓---<ASSABACH = low platelets, risk of bleeding and death

## NEVUS SIMPLEX

A nevus simplex is a **S**almon colored lesion often called a **S**tork bite or **S**almon patch. These fade with time and are benign. Do not get this term confused with Nevus FLAMMEUS (aka PORT WINE STAIN).

**PEARL**: These BLANCH with pressure.

## PORT WINE STAINS (PWS) (aka NEVUS FLAMMEUS)

Port Wine Stains (PWS), aka nevus flammeus, are **CAPILLARY** malformations. They start as pink/flat lesions that become dark red-purple. They then progress to being thick/raised in adulthood. These **P**WSs are **P**resent at birth and are **P**ERMANENT. They are benign if noted in isolation. If noted on the face, they can be associated with glaucoma (increased intraocular pressure that can present as a red eye).

**IMAGE**: http://www.fungur.com/uploads/2010/05/Mikhail-Gorbachev-1.jpg

**IMAGES**: http://bit.ly/U1IWnE

**PEARL**: They grow in proportion to child.

**MNEMONIC**: Glaucoma is a concern if a PWS is noted in the facial area. Is that why Mikhail Gorbachev wears Glasses? Because he has that big FLAME on his head?

## STURGE WEBER SYNDROME (SWS)

The Sturge Weber Syndrome (SWS) includes the following findings: Port Wine Stain (PWS or NEVUS FLAMMEUS) + **EYE/TRIGEMINAL NERVE DISTRIBUTION** + INTRACRANIAL VASCULAR MALFORMATION (look for with **MRI**) +/- glaucoma +/- Seizures +/- cognitive deficits

**MNEMONICS**: "p**WS** = s**WS**"… Ever heard of a basketball player named Chris WEBBER? Think WEBBER = Sports = **ESPN (I know it's a stretch).**

* **E**YE - glaucoma

* **S**WS

* **P**WS

* **N**EUROLOGIC issues: Developmental delay, Seizures

## *CAPILLARY MALFORMATION ASSOCIATIONS*

## (DOUBLE TAKE) KLIPPEL-TRENAUNAY SYNDROME

Klippel-Trenaunay syndrome is associated with AV fistulae causing skeletal or limb OVERGROWTH (hemihypertrophy). Patients with Klippel-Trenaunay have Port Wine Stains and overgrowth of tissue, bones and soft tissue. Look for unilateral limb overgrowth and CHF.

* **IMAGES**: http://geneticpeople.com/wp-content/uploads/2011/03/bobbyreport001xb.jpg

* **PEARL**: Hemihypertrophy images on the pediatric exam should very quickly clue you in to a few disorders. Highest on your differential should be Beckwith-Wiedemann Syndrome, then Klippel-Trenaunay, then Russell-Silver Syndrome and then possibly Proteus Syndrome.

* **MNEMONIC**: From now on, say CRIPPLE-**T**. Think of these patients as having a CRIPPLING disorder in which they have one HUGE leg that prevents them from getting around.

* **NAME ALERT:** KLIPPEL-FEIL SYNDROME. Completely different disorder. Look for a Torticollis-like photograph (due to fused cervical vertebrae).

- **IMAGES**: http://emedicine.medscape.com/article/1264848-media

## (NAME ALERT) KLIPPEL-FEIL SYNDROME

Klippel-Feil Syndrome results in a torticollis-like appearance. Results from fused cervical vertebrae. Patients will likely have a short, webbed neck, limited range of motion at the neck and possibly other anomalies. Etiology is unknown. The "Name Alert" is because this is a completely different disorder than Klippel-Trenaunay Syndrome (limb overgrowth due to AV fistulae).

**IMAGES**: http://emedicine.medscape.com/article/1264848-media

## CONGENITAL MELANOCYTIC NEVUS

Congenital melanocytic nevi are commonly referred to as moles. Present at birth or within the first few months of life. Generally benign, but carry an increased risk of MELANOMA if there are multiple moles (more than three) or if they are > 20 cm. Associated with spinal dysraphisms and Dandy Walker Syndrome (fossa abnormality)

## MCCUNE ALBRIGHT SYNDROME (aka POLYOSTOTIC FIBROUS DYSPLASIA)

McCune Albright syndrome (aka Polyostotic Fibrous Dysplasia) findings include **IRREGULAR** café-au-lait **MACULES** (either > 3 cm or multiple), **PRECOCIOUS PUBERTY**, BONE PROBLEMS (fractures, cranial deformities) and possibly other endocrine issues (hyperthyroidism). Can cause fractures of long bones and bowing of arms.

**IMAGE**: http://en.wikipedia.org/wiki/File:Mccune-albrightsyndrome1.jpg

**MNEMONIC**: Call it **MACULE** Albright Syndrome from now on.

## TUBEROUS SCLEROSIS

Tuberous sclerosis is AUTOSOMAL DOMINANT. Look for at least **2** of the following features:

* ASH LEAF SPOTS: These are hypOpigmented lesions, which can be seen with a Woods Lamp. Need at least **3** on the body to help make the diagnosis.

- **IMAGE**: http://bit.ly/nwloFE
- **IMAGE**: http://www.nlm.nih.gov/medlineplus/ency/imagepages/2538.htm

* SHAGREEN PATCH (hypERpigmented plaque that can be rough/thick and papular)

- **IMAGE**: http://bit.ly/n02Q8f
- **IMAGE**: http://bit.ly/ojT5CF

* SEBACEOUS GLAND HYPERPLASIA

- **PEARL**: Often misdiagnosed as acne. LOOK FOR SPARING OF THE FOREHEAD
- **IMAGE**: http://en.wikipedia.org/wiki/File:TuberSclerosisCase-143.jpg

* PERIVENTRICULAR OR CORTICAL TUBERS: Usually associated with INFANTILE SPASMS or seizures

* CARDIAC RHABDOMYOMAS: Look for a kid with **arrhythmias**!

* RENAL ANGIOMYOLIPOMA

**MANAGEMENT** OF TUBEROUS SCLEROSIS: Most of the management has to do with seizures/infantile spasms and cardiac arrhythmias.

* **MNEMONICS**:

- Imagine a TUBULAR bazooka shooting out WHITE LEAVES. The leaves have DANCING (seizing) tics on them!
- ASH is typically GRAY/WHITE/HYPOPIGMENTED, whereas a "PATCH of GREEN" is typically DARKER/HYPERPIGMENTED.

* **MNEMONIC**: **ASHES** come from burned WOOD. Woods Lamp needed to see them.

## NEUROFIBROMATOSIS I (NF1)

Neurofibromatosis I (NF1) is an AUTOSOMAL DOMINANT disorder involving the SKIN, BONES and NERVOUS SYSTEM. Diagnose with at least **2** of the following:

* First degree relative has the disease

* Neurofibromas

* Lisch Nodules in the iris (they look like mini neurofibromas

- **IMAGE**: http://www.milesresearch.com/eerf/images/lisch1-300.jpg

* Optic nerve gliomas. This is the neurologic component

* 6 **REGULAR** café-au-lait macules. As they get older, SIZE **DOES** MATTER. If prepubertal, ≥ 5 mm, if postpubertal, ≥ 15 mm. 10 years of age is a good cutoff. These macules can be present at birth. Children can have an increase in the **size and number** as they age. Therefore, it is very important that they have regular followup, especially if there is a family history of the disorder. As a side note, children can also get pheochromocytomas or renal artery stenosis, so BP should be monitored regularly.

Scoliosis or bony abnormalities

* Axillary or inguinal freckling

* **MNEMONIC**: (FOR NF-1) **S**KIN + "**O**RTHO" + **N**EURO issues = **S.O.N. This is NF ONE, *SON (or daughter!)***!!!

## NEUROFIBROMATOSIS 2 (NF2)

(Low yield topic). Neurofibromatosis 2 (NF2) findings include nonmalignant tumors of the nervous system, especially acoustic nerve tumors (aka neuromas or schwannomas). These can cause tinnitus or even hearing loss. Can also have eye tumors, cataracts, retinal problems, spinal cord tumors and meningiomas. Look for a family history.

**PEARL**: Tuberous Sclerosis and Neurofibromatosis are both AUTOSOMAL DOMINANT. BUT, they both have a HIGH RATE OF NEW MUTATIONS. Do not exclude these from your differential if they mention that the patient's parents do not have the disorder.

## INCONTINENTIA PIGMENTI

Incontinentia pigmenti is a severe X-linked DOMINANT disease which results in DEATH for all MALES. If presented with this as an answer choice, make sure the ABP vignette refers to a FEMALE patient. There are four stages of this disorder: Inflammatory vesicular phase, followed by a verrucous phase, followed by the hyperpigmentation phase noted along the **lines of Blaschko**, and finally a phase in which the hyperpigmentation disappears. Can leave atrophy or hypopigmentation behind.

SYSTEMIC ASSOCIATIONS: **DELAYED DENTITION**, Mental retardation, Paralysis, **PEGGED teeth** and Seizures.

**IMAGE**: http://www.med-ars.it/various/blaschko_def.jpg

**IMAGE**: http://bit.ly/n5LAMT

**IMAGES**: http://bit.ly/nK0Nx7

**IMAGES**: http://bit.ly/oH27dE (bottom of page)

**IMAGE**: http://bit.ly/o6VQEu

**MNEMONIC**: As WOMEN age, they tend to have more "INCONTINENTs." Incontinentia = Female patient. Imagine a WOMAN on the ground having a SEIZURE. She becomes INCONTINENT of urine which streams down her PEGG legs and creates black and white LINEAR SKIN LESIONS. PEGG refers to PEGGED TEETH

## HYPOHIDROTIC ECTODERMAL DYSPLASIA

Hypohidrotic ectodermal dysplasia is a condition related to INCONTINENTIA PIGMENTI, but this can occur in boys. HYPOHIDROSIS (decreased sweating, which can lead to hyperthermia), HYPOTRICHOSIS (sparse hair, so no eyebrows/lashes), DELAYED TOOTH ERUPTION, DEFORMED/PEGGED TEETH

**IMAGES**: http://bit.ly/oAyEsN

**IMAGE**: http://en.wikipedia.org/wiki/File:Michael_Berryman_2007.png

# *INFECTIOUS SKIN CONDITIONS*

## (DOUBLE TAKE) ECTHYMA GANGRENOSUM

Ecthyma gangrenosum is usually a sign of a **PSEUDOMONAS** infection and possibly sepsis in an immunocompromised patient, especially **LEUKEMIA**! Look for a **neutropenic** patient with black, necrotic, ulcerative lesions with surrounding erythema and edema. Often located in the groin/diaper area.

**IMAGE**: http://www.nlm.nih.gov/medlineplus/ency/imagepages/2647.htm

## STREPTOCOCCAL INFECTIONS OF THE GROIN

Streptococcal infections of the groin or perineum are associated with pain with stooling, pruritus, redness and possible a fissure. Unlike zinc deficiency, there is **no desquamation**. If vaginal or vulvovaginitis, look for a history of vaginal discharge. Diagnose by culturing the area. Treat with amoxicillin, penicillin (PCN) or a first generation cephalosporin. Risk factors include abuse and previous instrumentation. Look for a history of recent antibiotics in case the discharge is due to Candida.

## (DOUBLE TAKE) CUTANEOUS CANDIDIASIS, A DIAPER DERMATITIS

Cutaneous candidiasis, a diaper dermatitis, can occur secondary to a contact dermatitis or recent antibiotic use. Beefy red rash with papular satellite lesions. This rash goes **into the inguinal folds**. Use a KOH prep to confirm diagnosis and treat with a topical antifungal, such as nystatin or clotrimazole.

**IMAGE** (includes satellite lesions): http://www.pediatrics.wisc.edu/education/derm/tutc/80.html

## BULLOUS IMPETIGO/STAPH SCALDED SKIN SYNDROME (SSSS)

Bullous impetigo, or Staph Scalded Skin Syndrome (SSSS), is a spectrum of the same disease.

* **IMPETIGO**: Look for honey colored, crusting lesions and bullae. Non-bullous impetigo and look similar but without vesicle/bullae (more oozing/crusting)

- **IMAGE**: http://bit.ly/rmVh8u

* **SSSS**: A very painful and red rash in which large, thin blisters are the result of an exotoxin. There is "**sheet-like**" skin loss/separation. Looks very superficial compared to impetigo. Obtain a BIOPSY to prove that it is SSSS and Stevens Johnson Syndrome (SJS) or Toxic Epidermal Necrolysis (TEN), both of which have deeper/dermal involvement.

- **IMAGES**: http://emedicine.medscape.com/article/788199-media
- **PEARL**: Lesions are **NOT** in eye or mouth, but may be **around** the eyes and mouth (as opposed to SJS/TEN, which may be IN the eyes and mouth).

## STAPHYLOCOCCUS EPIDERMIDIS

Staphylococcus epidermis is the most likely answer if you are presented with a premature baby that has a skin infection.

## CELLULITIS

Cellulitis is defined as a well-demarcated area of erythema, edema and induration secondary to an infection. May be associated with bullae. For treatment, start with **Cefazolin** as your first line agent.

## TINEA CORPORIS

In tinea corporis, a thin, circular lesion with **THIN SCALES,** a RAISED border and central clearing is noted. The ring of the "ringworm" looks like a worm.

**IMAGES**: http://dermatlas.med.jhmi.edu/derm/indexDisplay.cfm?ImageID=447741726

## TINEA VERSICOLOR

Tinea versicolor results in hypopigmentd OR hyperpigmented macules. It's caused by MALASSEZIA FURFUR. Treat with fluconazole, ketoconazole or selenium sulfide, but NOT griseofulvin (use that for T. capitis).

**IMAGE**: http://www.nlm.nih.gov/medlineplus/ency/imagepages/1890.htm

**IMAGE**: http://www.herbalgranny.com/wp-content/uploads/2009/09/Tinea-Versicolor.jpg

## PITYRIASIS ROSEA

Pityriasis rosea presents as oval, parallel lesions with THICK scales. Look for a herald patch (first lesion). Associated with winter and spring. Lesions are often in a "Christmas tree pattern." Treat with light exposure.

**IMAGES**: http://emedicine.medscape.com/article/762725-media

**PEARL**: Unlike secondary syphilis, there are no lesions on the palms/soles

## MOLLUSCUM CONTAGIOSUM

Molluscum contagiosum results in flesh-colored, pearly papules that are dome-shaped and **umbilicated**. Caused by the POX virus. NO treatment is needed, but sometimes cryotherapy or topical Cantharidin, Podophyllotoxin, Imiquimod or Potassium hydroxide.

**IMAGE**: http://www.nlm.nih.gov/medlineplus/ency/imagepages/1380.htm

**MNEMONIC**:

mollusc**UM**bilicated **P**apules
**O**
**X**

## HUMAN PAPILLOMA VIRUS (HPV)

Human papilloma virus (HPV) causes VERRUCA VULGARIS (warts). They can be on the hands, knees, feet and in the anogenital region. If genital, the condition is referred to as CONDYLOMA ACUMINATA (> 90% are from HPV 6 or HPV 11, which are NOT likely to induce cervical cancer). Lesions are NOT tender but easily bleed with minimal trauma. The risk of cervical cancer **is increased** depending on the subtype **(16, 18 are most commonly associated with cervical cancer)**. Anogenital warts can be due to maternal-fetal transmission and may not present unitl 3 years after birth! BUT, if you note anogenital warts AFTER 3 years of age, think SEXUAL ABUSE. Treat with self-applied topical podofilox or imiquimod. Treatment with cryotherapy or podophyllin is done by a physician.

**IMAGE**: http://dermatlas.med.jhmi.edu/data/images/verruca_vulgaris_1_090828.jpg

**IMAGE**: http://img.medscape.com/pi/emed/ckb/urology/435575-455021-4185tn.jpg (Acuminata)

**MNEMONIC:** Don't get confused with molluscum. hp**V** = **VV**arts/Warts = **V**erruca **V**ulgaris = **V**enereal **VV**arts/Warts. "**VV**arts on your hands or knees? It's probably from those darn **V**'s!"

**MNEMONIC:** The HPV 16 & HPV 18 strains are the two you should remember (associated with the highest risk of cervical cancer): Imagine an adolescent couple. Their birthdays are on the same day, 7/1 (Zodiac of CANCER). The boy is turning 18 and he's excited to finally VOTE. His girlfriend is turning 16 and she's excited because she'll finally get her DRIVER'S LICENSE now that she's celebrating her SWEET SIXTEENTH. As they go to blow out the BIRTHDAY CAKE candles, you notice that she has **VV**arts on her lips! It turns out he also has **VV**arts, but his are **V**enereal (anogenital).

**NAME ALERT**: An "**A**" in **A**cuminata looks like a flipped "**V**," which may help you remember that a diagnosis of Condyloma **A**cuminata represents an hp**V** infection. The "**L**" in Condyloma **L**ata, should remind you that you are dealing with syphi**L**is.

## CONDYLOMA LATA

Condyloma lata is found in secondary syphi**L**is = White-gray, coalescing papules. These are much more FLAT appearing than Condyloma Acuminata

**IMAGE**: http://en.wikipedia.org/wiki/File:Vaginal_syphilis_(disturbing_image).jpg

**NAME ALERT**: An "**A**" in **A**cuminata looks like a flipped "**V**," which may help you remember that a diagnosis of Condyloma **A**cuminata represents an hp**V** infection. The "**L**" in Condyloma **L**ata, should remind you that you are dealing with syphi**L**is.

## HERPES SIMPLEX VIRUS II (HSV II)

Herpes simplex virus II is an STD. The initial flare is often very painful. Pain may precede the presentation of lesions. Look for multiple, painful ulcers or vesicles on the labia or penis. Can have lymphadenopathy. The vesicles are CLUSTERED on an ERYTHEMATOUS BASE. Lesions can also be ULCERATIVE. Diagnose by obtaining a viral culture or HSV PCR. The Tzanck smear is not specific for HSV. Treat with ORAL Acyclovir x 7 days (not topical). Treat babies with **IV** Acyclovir.

**IMAGE**: http://www.coldsoreblisters.com/images/herpes1.jpg

**PEARL**: HSV can be associated with a very painful infection called a HERPETIC WHITLOW (typically of a thumb or finger) .

**IMAGE**: http://img.medscape.com/pi/emed/ckb/pediatrics_general/960757-964866-1733tn.jpg

## HERPES SIMPLEX VIRUS ENCEPHALITIS (HSV ENCEPHALITIS)

A question about herpes simplex virus encephalitis (HSV encephalitis) would likely mention fever, seizures and possibly the a CT finding in the **temporal lobe**. Treatment is STAT **IV acyclovir, followed by a lumbar puncture** to obtain fluid for PCR testing. EEG might show PLEDs (periodic lateralizing epileptiform discharges).

## HERPES SIMPLEX VIRUS GINGIVOSTOMATITIS

Herpes simplex virus gingivostomatitis presents with oral and perioral/vermillion border lesions/vesicles. Gingiva is friable and malodorous. There is associated lymphadenopathy.

**IMAGE**: http://img.medscape.com/pi/emed/ckb/pediatrics_general/960757-964866-1734tn.jpg

## (DOUBLE TAKE) ECZEMA HERPETICUM

Eczema herpeticum is a potentially life threatening disseminated herpes (HSV) infection occurring at sites of skin damage, including sites of eczema. Look for HSV Vesicles + Crusted Lesions. Even if a description is not given of a vesicular rash, have a high index of suspicion for a rash "**not improving with steroids and/or antibiotics**." Diagnose with viral culture for HSV, but do not delay treatment. A Tzanck smear can support the diagnosis. Treat by **STOPPING** topical steroids and/or immunosuppressants and starting **Acyclovir**.

**IMAGE**: http://bit.ly/nkSGQU

## (DOUBLE TAKE) BLUEBERRY MUFFIN SYNDROME

Blueberry muffin syndrome represents extramedullary hematopoiesis. Can be seen in congenital **viral** infections like **Rubella**, Coxsackie, Cytomegalovirus (CMV), Herpes Simplex Virus (HSV) and Parvovirus. Can also be associated with congenital Toxoplasmosis (a protozoa).

**IMAGES**: http://alturl.com/pfdqj

## SCABIES

Scabies presents as linear, papular, erythematous, **pruritic**, vesicular and crusting lesions most often seen in areas with CREASES (wrist, groin, webbing of fingers). May see burrows. Treat with permethrin overnight from head to toe for the **entire family**. Retreat if the patient is still having No need to retreat unless the patient is still having symptoms after 14 days and LIVE MITES are found because the persisting pruritus can be from residual inflammation. Try topical steroids or antihistamines for that interim.

**PEARL**: Unlike papular urticaria, lesions are not in crops

**IMAGES**: http://hardinmd.lib.uiowa.edu/dermnet/scabies.html

### PEDICULOSIS CAPITIS (aka HEAD LICE)

Pediculosis capitis (aka head lice) results in nits/ova of the lice at the hair shafts, especially in the occipital area. Treat with permethrin. More symptoms at night when lice tend to be more active. Itching is from the bites. Unlike scabies, **repeat permethrin again in 7-10 days** because eggs can hatch up to 10 days later.

**PEARL**: If an African American child is pictured, it is NOT lice

**IMAGES**: http://www.dermaamin.com/site/atlas-of-dermatology/15-p/1246-pediculosis-capitis-.html

### PEDICULOSIS PUBIS (aka PUBIC LICE or CRABS)

Pediculosis pubis (aka pubic lice or crabs) is an infection in the groin that results in red, crusted suprapubic macules and possibly bluish-gray dots. There is a STRONG ASSOCIATION with sexual abuse in children.

**IMAGES**: http://www.dermaamin.com/site/atlas-of-dermatology/15-p/1248-pediculosis-pubis-.html

## *THE "ERYTHEMA" RASHES*

### ERYTHEMA NODOSUM

For erythema nodosum, look for **PAINFUL**, shiny, red to bluish shin lesions in a patient with a history of a chronic disease or on certain medications. Associations include Crohns Disease, Ulcerative Colitis, Drugs (oral contraceptives and sulfa drugs), Infections (yersinia, EBV, Tuberculosis, fungal infections) and Sarcoidosis.

**MNEMONIC**: For this shiny skin finding, use CUDIS (kind of like CUTIS, which means skin) to help you remember the following associations: **C**rohns, **UC**, **D**rugs, **I**nfections and **S**arcoidosos.

**IMAGE**: http://www.health-writings.com/erythema-nodosum-children/

**IMAGE**: http://dermatlas.med.jhmi.edu/derm/IndexDisplay.cfm?ImageID=-731469363

### (DOUBLE TAKE) ERYTHEMA CHRONICUM MIGRANS

Erythema chronicum migrans is caused by BORRELIA BURGDORFERI, the spirochete that causes LYME DISEASE. Look for a large, flat lesion **(> 5 cm) that is annular and has a red border. It is located at the tick bite site in about 75% of patients. Classis description is a "bullseye" lesion.** The rash shows up 1-2 wks after the bite. Titers may still be negative during this period. Borrelia is transmitted via the Ixodes deer tick. IF the patient has an acute arthritis, disseminated erythema migrans, a palsy (BELL'S PALSY) or neuropathy, then treat with ORAL medication (**Doxycycline if >8 years old, or Penicillin or Amoxicillin if < 8 years old**). If the patient has **CARDITIS**, neuritis (encephalitis/meningitis) or **RECURRENT** arthritis, then treat with INTRAVENOUS medication **(PCN or Ceftriaxone)**. Arthritis is usually located at the large joints (especially the **knees**). Diagnosing using labs is often difficult. Obtain **Lyme antibody titers**. If positive, confirm with a Western blot. This is often a **clinical diagnosis**.

* **IMAGE**: (BULLSEYE LESION) http://web.princeton.edu/sites/ehs/bullseye.jpg

* **IMAGE**: (BELL'S PALSY) http://en.wikipedia.org/wiki/File:Bellspalsy.JPG

* **SIDE NOTES**
  - BELL'S PALSY: Unilateral facial nerve paralysis (CN VII). Often idiopathic.
  - The Jarisch-Herxheimer reaction results in fever, chills, hypotension, headache, myalgia, and exacerbation of skin lesions during antibiotic treatment of a bacterial disease (typically spirochetes). Due to large quantities of toxins released into the body. Classically associated with syphilis, but can also occur with Lyme disease. May only last a few hours

* **MNEMONICS:**
  - From now on, think/say borreLIYME. "Don't ever throw a borLIYME to MY GRANy!" Or, "Don't ever borre-LIE to MY GRANy." Borrrelia = borreLIYME. MY GRANy = Migrans.
  - Imagine that BULL'S EYES are made of 2 bright neon-green LIMES! This should remind of you of the classic description.
  - Imagine squeezing LYME into a CAN = **C**arditis, **A**rthritis, and **N**euritis.

## ERYTHEMA MARGINATUM

Erythema marginatum is a **transient**, erythematous, macular and light colored described as "serpentiginous," in which the **MARGIN**s progress as the center clears. It is part of the Jones criteria for Rheumatic Fever.

**IMAGE**: http://www.dermaamin.com/site/atlas-of-dermatology/5-e/470-erythema-marginatum-of-rheumatic-fever-.html

**MNEMONICS**: **E**rythema **MARGIN**atum. The **E** in **E**rythema is part of the **E** in jon**E**s, and the name **MARGIN**atum should remind you to look for an interesting description of the rash's **MARGIN**s.

## (DOUBLE TAKE) ERYTHEMA INFECTIOUSUM

Erythema infectiousum IS an INFECTIOUS rash!!! Caused by Parvovirus B19. It is also called Fifth Disease Look for erythematous facial flushing of the cheeks (sometimes described as "slapped cheeks" appearance). The extremities will have diffuse macular (or morbilliform) erythema (especially on the extensor surfaces) referred to as "lacy" or "reticular." Diagnose with Ig**M titers** (there is no culture or rapid antigen available).

**PEARLS:** The rash occurs AFTER the slapped cheeks rash (often a week later). Patients may also have knee or ankle pain. Parvovirus B19 infection can result in **APLASTIC CRISIS**. Intrauterine exposure can result in **hydrops fetalis**.

**MNEMONIC**: infectio**5**u**M** = FIFTH disease = "Fiver fingers." Imagine a cheek being SLAPPED with FIVE fingers covered by a white LACY/LACEY glove with a red M on the back of it (extensor surface). **M** = ig**M** titers

**MNEMONIC**: ParVoVirus B19: From now on, say/think "par**V**o**V**irus **V**19." **V = Roman numeral 5!**

## ERYTHEMA TOXICUM NEONATORUM

Erythema toxicum neonatorum is seen in up to 50% of newborns and consists of **erythematous macules** with raised central lesions (papules **or** vesicles). Usually seen at birth or by DOL 2. It is a benign rash with an unknown etiology. It usually disappears by DOL 7. Diagnose by noting eosinophils on microscopy.

**IMAGE**: http://www.nlm.nih.gov/medlineplus/ency/imagepages/2198.htm

**MNEMONIC**: Although the name "TOXICum" suggests otherwise, this is a NON-toxix rash resulting in nontoxic looking babies.

**MNEMONIC**: This is an **E**arly, **E**rythematous, "**E**osinophilled" rash called **E**rythema tox**EEE**cum.

### ERYTHEMA MULTIFORME

See the Stevens-Johnson syndrome section for more information on erythema multiforme. Look for target lesions.

## *THE NEWBORN RASHES*

### MILIARIA RUBRA

Look for very superficial vesicles that are easily ruptured in a case of miliaria rubra. Occurs due to obstruction of sweat glands. Also called "prickly heat rash."

**IMAGES**: http://www.skinsight.com/infant/miliariaRubra.htm

**MNEMONIC**: Miliaria sounds like malaria, which is usually found in hot countries where you sweat!

### MILIA

Milia are small, pearly inclusion cysts which look like little white heads. There's NO associated erythema. If on the nose, can be very easy to confuse with SEBACEOUS HYPERPLASIA.

**IMAGE**: http://newborns.stanford.edu/PhotoGallery/Milia1.html

**IMAGE**: http://www.nlm.nih.gov/medlineplus/ency/imagepages/2677.htm

### SEBACEOUS HYPERPLASIA

In sebaceous hyperplasia, pinpoint white-yellow papules appear on the nose and central face. NO associated erythema. Results due to maternal androgen exposure. Benign.

**IMAGE**: http://www.dermatlas.com/data/images/milia_1_040329.jpg

**IMAGE**: http://newborns.stanford.edu/images/sebac-hyp2.jpg

### TRANSIENT NEONATAL PUSTULAR MELANOSIS

Transient neonatal pustular melanosis is more common in **African-American kids**. This is a benign rash with NO associated erythema. Starts in utero and **PRESENT AT BIRTH**. Resolves within a few days. Can leave hyperpigmented macules for a while. Diagnose by examining contents and looking for **PMNs** on Tzanck smear.

**IMAGE**: http://bit.ly/Y1GaX1

**IMAGE**: http://www.aafp.org/afp/2008/0101/afp20080101p47-f3.jpg

**MNEMONICS:** transient neonatal PUStular melanosis should remind you of the PMNs on Tzanck smear in the PUS-like contents of these PUStules. MELANosis should make you think about dark skinned individuals (AA kids) and the dark macules that can be left behind.

### NEONATAL ACNE (aka NEONATAL CEPHALIC PUSTULOSIS)

Neonatal acne (aka Neonatal Cephalic Pustulosis) occurs within the **first month** of life. Resolves by 4 months of age. Look for inflammatory pustules on the cheeks and forehead without comedones. This is a benign rash that requires no treatment.

**IMAGE**: http://www.skinsight.com/infant/neonatalAcneBenignCephalicPustulosis.htm

**MNEMONIC**: NEONATal = FIRST MONTH OF LIFE!

### INFANTILE ACNE

Infantile acne looks like typical pubertal acne, but it is found in babies. Onset is usually around 2-3 months of age and it is due to androgenic stimulation. There can be COMEDONES (whiteheads and blackheads). The rash can resolve in a few weeks or it can take up to a year.

**MNEMONIC**: INFANTile = Infants. Don't choose this if the baby is 4 weeks old.

**IMAGE**: http://www.dermnetnz.org/common/image.php?path=/acne/img/s/acne-infant.jpg

### LIVEDO RETICULARIS (aka CUTIS MARMORATA)

Livedo reticularis (aka cutis marmorata) presents as a mottled, reticulate patterned rash. Benign. May be described as a lacy rash. Resolves by 1 month.

**IMAGE**: http://www.dermatlas.com/data/images/cutis_marmorata_1_060523.jpg

**PEARL**: If the baby is healthy and without any concerning symptoms, choose this. If not, consider sepsis in your differential.

## *ALOPECIA & HAIR FINDINGS*

### ALOPECIA AREATA

In alopecia areata, there are round/well-circumscribed area(s) of alopecia. Alopecia can be on the scalp or in other areas. Hairs at the periphery of the areas are short, **pluckable** and may resemble an **exclamation point!**

**IMAGE**: http://bit.ly/RgZPzd

**IMAGE**: http://bit.ly/SoAmlH

### ALOPECIA TOTALIS

Alopecia totalis is the loss of all hair on the HEAD.

**IMAGE**: http://www.dermaamin.com/site/atlas-of-dermatology/1-a/88-alopecia-totalis-.html

### ALOPECIA UNIVERSALIS

Alopecia universalis is the loss of all hair on the entire BODY. Usually a **SYSTEMIC** etiology such as hypothyroidism, a nutritional deficiency or even lupus (SLE).

**IMAGE**: http://www.pediatrics.wisc.edu/education/derm/tutc/86m.jpg

## (DOUBLE TAKE) ZINC DEFICIENCY

Breastfeeding helps with zinc absorption. If a child is begins having medical problems once weaned from breast milk, consider zinc deficiency in your differential. Zinc deficiency causes a **SCALY and EXTREMELY ERYTHEMATOUS** dermatitis in the perioral and perianal area (**around the natural orifices**) that can DESQUAMATE. The rash is sometimes described as erosive and eczematous. Can also be associated with ALOPECIA and poor taste.

* **MNEMONIC**: Poor taste, huh? Have you ever had Zinc lozenges? They are disgusting! It's probably a good thing that you have hypogusia when you are eating Zinc lozenges!

* **IMAGES**: http://www.dermaamin.com/site/atlas-of-dermatology/25-z/505-zinc-deficiency-.html

* **IMAGE**: http://bit.ly/Y1IxJf

* **PEARLS:**

- CROHNS DISEASE: If a Crohns patient is suffering from diarrhea, they may have zinc deficiency since Zn is lost in the stool.
- STRICT VEGETARIANS & VEGANS: Susceptible to multiple nutritional deficiencies, including deficiencies in IRON, ZINC, CALCIUM and VITAMIN B12. Vegans avoid all animal derived products (including milk and eggs). B12 deficiency can result in megaloblastic anemia, vitiligo, peripheral neuropathy and even regression of milestones
  - **MNEMONIC**: Did you know giraffes are vegetarian? Imagine a giraffe standing in Time Square reaching its long neck into the sunroof of a FUZZY CAB that has green, grass-like seats and fuzzy floor mats. FUZZY CAB = FeZi CaB12!

## (DOUBLE TAKE) ACRODERMATITIS ENTEROPATHICA

Acrodermatitis Enteropathica is an inherited condition (autosomal recessive) in which there is a Zinc transport defect. Can result in **alopecia**, diarrhea, failure to thrive (FTT) and the **rash** of zinc deficiency.

**IMAGES**: http://emedicine.medscape.com/article/912075-media

## (DOUBLE TAKE) BIOTIN/BIOTINIDASE DEFICIENCY

Biotin, or Biotinidase, deficiency may present with a RASH + ALOPECIA + **NEUROLOGIC SIGNS** (ataxia, coma, etc.). Patients may also have lactic acidosis. Treat with Biotin.

**MNEMONIC**: Imagine the TIN MAN from the Wiz of Oz walking with an ATAXIC gait as he SCRATCHES his arm. Notice that he has NO HAIR!

## TELOGEN EFFLUVIUM

Telogen effluvium is a form of acute hair shedding that occurs diffusely. Often related to a psychologic or medical stressor. Treat with REASSURANCE because the will grow back.

## TINEA CAPITIS (aka RINGWORM)

Tinea capitis (ringworm) results in broken hair that looks like "**black dot alopecia.**" There is often inflammation and this condition can be associated with a kerion (the raised spongy lesion associated with this infection). Treat with GRISEOFULVIN. You do not need any baseline labs.

**IMAGES**: http://emedicine.medscape.com/article/787217-media

**IMAGES**: http://www.healthhype.com/itchy-scalp-causes-and-treatment.html

## TRICHOTILLOMANIA

Trichotillomania is a body-focused repetitive behavior in which patients pull out their own hair (may be on a location other than the scalp). Look for loss of hair in an irregular pattern (not a nice circle). Also, the irregularly shaped patches will contain incomplete hair loss in which you will see hair of **differing lengths.**

**IMAGE**: http://bit.ly/nySzrr

**IMAGE**: http://en.wikipedia.org/wiki/File:Trichotillomania_1.jpg

## (DOUBLE TAKE) ESSENTIAL FATTY ACID DEFICIENCIES

Essential fatty acids include **LINOLEIC ACID** and alpha-linolenic acid. Deficiency results in alopecia, a scaly dermatitis and **thrombocytopenia**. Treat with IV lipids.

**MNEMONIC:** Imagine a fish whose red SCALES are shaped like HAIRY PLATELETS. As the fish struggles to find food, it becomes SKINNIER and skinnier (malnourished) and the hairy platelets begin to fall off. What's left is a SKINNY (fat free), BALD and THROMBOCYTOPENIC fish!

## APLASIA CUTIS CONGENITA

In aplasia cutis congenita, there is a congenital absence of the skin in an area. Usually in a single location (most often the scalp), but can be in multiple areas. After the lesion heals and scars, a BALD SPOT is left behind. Aplasia cutis can be associated with underlying spinal dysraphisms and underlying skull defects.

**IMAGES**: http://emedicine.medscape.com/article/1110134-media

**IMAGE**: http://en.wikipedia.org/wiki/File:Aplasia_cutis_congenita.jpg

**PEARLS:** Look for the HAIR COLLAR SIGN. This is a hairless area with a collar of dense hair at the edges. If given a picture of a scalp with the hair collar sign, get an MRI

**IMAGE**: http://bit.ly/nu3VmS

# NEONATOLOGY

## *WEIGHT, LENGTH & HEAD CIRCUMFERENCE*

### NEWBORNS WEIGHT

Newborns should double their birth weight by 4-6 months and **triple** their birth weight by 1 year of age. For a full-term baby, expect weight gain of 20-30 g/day. For premies, expect weight gain of 15-20 g/day.

* **PEARL**: 50th Percentile for WEIGHT at birth is 3.25 kg

* LARGE FOR GESTATIONAL AGE (LGA): Top 10th percentile for weight, or > 3900 grams at birth.

* SMALL FOR GESTATIONAL AGE (SGA): Bottom **10th** percentile for weight, or < **2500 grams at birth.**

* LOW BIRTH WEIGHT (LBW): < 2500 grams

* VERY LOW BIRTH WEIGHT (VLBW): < 1500 grams

* EXTREMELY BIRTH WEIGHT (ELBW): < 1000 grams

- **PEARL**: A baby that is ELBW **can** be "AGA" if it is a premie

### NEWBORNS LENGTH

Here are some great shortcuts to help you calculate the expected length for newborns:

* 1-year-old: 1.5x the birth length

* 4-year-old: 2x the birth length

* 13-year-old: 3x the birth length

* Mid-parental height: Expect the adult height to be +/- 5 cm (2 inches) of the mid-parental height. The mid-parental height is calculated as follows:

- Boys = (Mom's Height + Dad Height + 5 inches)/2
- Girls = (Mom's Height + Dad Height - 5 inches)/2
- NOTE: If using centimeters, replace "5 inches" with "13 cm" in the formulas above.

* **PEARL**: 50th Percentile for LENGTH at birth is 50 cm.

* **PEARL/EXAMPLE:** The most average baby you can find would be 50 cm in length at birth and by 1 year of age s/he would be 75 cm. By 4 years old, s/he would be about 100 cm (2 x 50), and by 13 s/he would be 150 cm (3 x 50). 150 cm / 2.5 = 60 inches = 5 feet.

### INTRAUTERINE GROWTH RESTRICTION = INTRAUTERINE GROWTH RETARDATION = IUGR

The term intrauterine growth restriction, aka intrauterine growth retardation or IUGR, is not the same thing as "small for gestational age." The term is used to describe intrauterine fetal growth that has deviated from the norm. IUGR babies' description can sound like that of an IDM = hypoglycemia, polycythemia, poor feeding, and hypothermia.

### HEAD CIRCUMFERENCE – Macrocephaly, Hydrocephaly & Microcephaly

To know if you're dealing with possible macrocephaly, hydrocephaly or microcephaly, you first have to know what a baby's head circumference should be. A normal newborn's head circumference is approximately 35 cm. After that, expect the Head Circumference to increase by 1 cm/month for 6 months and then ½ cm/month from 6-12 mo. The means at 12 months, a child should have a HC of about 44.

**PEARL**: 50th Percentile for HC at birth is 35 cm

* MACROCEPHALY: > 97 percentile for age. By definition, macrocephaly refers to a normal child. If the child's development is in fact normal, just measure the parents' heads to see if it is FAMILIAL MACROCEPHALY! HC is usually **normal at birth**.

- **PEARL**: If there are developmental issues or other issues with the child, HYDROcephaly is likely the main problem.

* HYDROCEPHALY: Look for a **BIG HEAD AT BIRTH** + NEUROLOGIC ISSUES

- **PEARL**: Large" anterior fontanelles are found in both macrocephaly and in hydrocephaly. The finding is very nonspecific. If the ABP mentions a **BULGING FONTANELLE**, think HYDROCEPHALY.
- **PEARL**: Neither macrocephaly nor hydrocephaly result in papilledema. Both of these conditions are usually due to chronic issues, and papilledema is usually the result of an ACUTE problem.

* MICROCEPHALY: HC is usually **normal at birth** (~35 cm) and then falls off the curve.

## *NUTRITION, BREAST MILK & FORMULA*

### NEONATAL POTASSIUM REQUIREMENTS

Neonatal potassium requirements are 2 mEq/kg/day.

### NEONATAL SODIUM REQUIREMENTS

Neonatal sodium requirements are 3 mEq/kg/day.

### PROTEIN INTAKE

High protein intake can result in a **high osmotic load**. Osmotic load is affected by chloride, phosphorus, potassium and sodium (NOT calcium). Consider a high osmotic load as a source of fluid losses, renal impairment, and weight loss.

* FULL TERM BABY: 2 – 2.5 grams/kg/day for the first 6 months of life

* PREMATURE BABY: **3.5** gr/kg/day. Premature babies have higher protein requirements in order to avoid a negative nitrogen balance (which can result in weight loss).

### NEONATAL CALORIC REQUIREMENT

The neonatal caloric requirement is 100 kcal per day for each of the first 10 kg + 50 kcal/day for each of the second 10 kg + 20 kcal/day for each kg beyond that (e.g., a 16 kg patient needs 1300 kcal/day). So, 1500 for the first 20 kg then 20 cal/kg after that.

**PEARL**: This is a GENERAL GUIDE. If a child is falling off of his curve, or is at the 3rd to 5th percentile for weight, increase the caloric intake.

**MNEMONIC**: **Use 100/50/25 instead**. It's easier to remember! 100/2 = 50. 50/2 = 25. Hence, 100/50/25! SIMILAR TO THE HOURLY IV FLUID RATE MNEMONIC RATE AND THE NUMBERS ARE THE EXACT SAME (100/50/25) FOR THE DAILY FLUID REQUIREMENT MNEMONIC!

## EXCLUSIVELY BREAST FED BABIES

In exclusively breast fed babies, look for Vitamin D and Vitamin K deficiencies.

**PEARL**: ALL BABIES/KIDS are now supposed to get a total of 400 IU (international units) of Vitamin D per day in their diet. For formula fed babies, this is usually attainable through the formula. Breastfed babies need supplementation.

**PEARL**: Patients with CF (CYSTIC FIBROSIS) or RICKETS need > 1600 IU per day!

## BREAST MILK

Breast milk contains arachidonic acid, DHA, whey, casein, colostrum, hind milk, etc. It's a lot to remember, so memorize the following and move on!

* ARACHIDONIC ACID (AA) & DOCOSAHEXAENOIC ACID (DHA): Help with neurologic development. Greatest in COLOSTRUM. Not as much in mature milk.

* WHEY: The primary protein in breast MILK.

* CASEIN: The primary protein in FORMULA.

* COLOSTRUM: The milk produced at the end of pregnancy and early after delivery. Only small amounts are expressed in the first few days until the more mature milk finally comes in.

- Yellow color is from carotene.
- Stimulates passage of meconium.
- High in PROTEIN (immunoglobulins, especially IgA).

* HIND MILK: Last bit of milk expressed during breast-feeding. It is highest in CALORIES and FAT.

* FROZEN BREAST MILK: Good for 3-6 months. Once thawed, use within 48 hours.

* CONTRAINDICATIONS TO BREAST-FEEDING: Mother with herpes simplex virus (HSV), HIV, tuberculosis (TB), on chemotherapy, on HYPERthyroid medications, on metronidazole, on sulfa drugs or on Tetracycline. Breast-feeding is also usually contraindicated if the baby has an **INBORN ERROR OF METABOLISM**. An inverted nipple may be a contraindication depending on the degree of inversion. Breast shells may be needed.

**PEARL**: Candidiasis, mastitis and fibrocystic disease are NOT contraindications.

**PEARL**: Breastfeeding is NOT a contraindication for Hepatitis B. For mothers who are CMV **carriers** (not recent converters), they may also breastfeed.

**MNEMONICS:**

- COLOSTRUM: Although it is supplied to babies very EARLY in life, it has tremendous LONG-TERM protective benefits/ingredients (AA, DHA, IG's/IgA aka protein).
- MATURE MILK is the regular, everyday milk that provides the regular, everyday ingredients to a baby (fat, lactose, "energy," etc.).
- HIND MILK: HIND milk has a high FAT and CALORIC content, like the unusually oversized be**HIND** of an appropriately overweight/fat new mom. (Sorry for the un-PC mnemonic. Hopefully it helps).
- WHEY: Breast milk is WHEY better, and FORMULA comes IN a CASE (CASE-IN). (Thanks to PBR member samori01 for bringing this one to my attention!)

## FORMULA

When using formula, there is no need to sterilize bottles/water unless using well water. Use COLD tap water (less lead). Caloric content is approximately 20 kcal/oz. 1 oz = 30 cc or ml.

## IRON SUPPLEMENTATION

Premature infants need iron supplementation at 2 months. Full term infants need supplementation after 4-6 months when their liver stores run out (**even breast-fed babies**). Supplementation DOES NOT have to be in the form drops, and it does NOT cause constipation. Iron-fortified cereal and iron-fortified formula are just as good. If using formula, make sure it contains at least 12 mg/L.

**MNEMONICS:** Use this 12 lb IRON to help you remember iron should be in formula (12 mg/L). If that doesn't work, imaging making a 12-EGG omelet on a hot IRON.

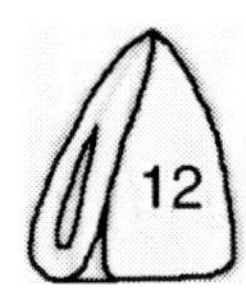


## WHOLE MILK

* Whole milk should NOT be given to babies until they are 1 year of age. It (can cause anemia, electrolyte abnormalities, such as hypocalcemia, and renal damage

* DEFINITELY give whole milk from 1 - 2 years of age. The fat content helps with nervous system development.

* After 2 years of age, CHANGE to 2% milk in order to LIMIT the fat content.

# *PREMATURE INFANTS*

## ESTIMATING GESTATIONAL AGE BY PHYSICAL EXAM

For the pediatric board exam, you should be able to properly identify a **premature child** by estimating gestational age through a physical exam!

**SHORTCUTS:** At a gestational age of < 35 wks, hair is all over the body +/- face. There are 0-3 creases on the sole and none at the heel. Ear cartilage still slowly unfolds. There is NO BREAST BUD.

* In males, there are NO PALPABLE TESTES and very few rugae.

* In females, there is a PROMINENT CLITORIS, but the labia majora is small.

**SHORTCUT:** If a vignette mentions the presence of any of the following, it is likely that the child is FULL TERM:

* MANY CREASES ON SOLES
* BREAST BUDS
* PALPABLE TESTES
* MANY RUGAE
* SPRINGY EAR CARTILAGE

## CALCULATING GESTATIONAL AGE

Although a child is considered full term at 37 weeks gestation, for patients who are considered premature, use 40 WEEKS to calculate corrected gestational age. For example, if a child was 8 weeks ago at 36 weeks gestation, s/he should be meeting the developmental milestones of a 4-week-old child (corrected gestational age is 4 weeks). Corrections are no longer needed after 2 years.

## PREMATURE INFANT NUTRITION

Nutritional requirements for premature infants include high amounts of calcium, phosphorus, Vitamin D (due to a lack of bile salts needed to absorb the fat soluble vitamins), and Vitamin E. Therefore, premies are at much higher risk of Rickets. Vitamin E supplementation can help prevent hemolytic anemia.

* CALORIC REQUIREMENT: 120 kcal/kg/day. Use 24 kcal/oz premature formula, rather than the usual 20 kcal/oz formula.
* CARBOHYDRATES: Glucose polymers are preferred over Lactose.
* FATS: Intake should consist of at least 50% Medium Chain Triglycerides (MCTs). This is because premies lack certain lipases needed for fat digestion.
* PROTEIN: 3.5 grams/kg/day
* Catch up growth happens in the first 2 years so should be at a normal by 2 years of age.

## TOTAL PARENTERAL NUTRITION (TPN)

Total Parenteral Nutrition (TPN) monitoring should focus on frequent electrolyte checks and adjustments because high potential morbidity from abnormalities. Lipid and protein content is dictated by weight, so frequent monitoring is not needed.

## RETINOPATHY OF PREMATURITY (ROP)

According to the AAP, **all kids < 1500 g and < 32 wks should be screened for Retinopathy of Prematurity (ROP). The greatest risk is for kids born at < 29 weeks gestation or < 1200 g**. Check for ROP at 32 weeks gestation, at discharge or at **4 wks after the patient is born**. Whichever criterion is later.

**PEARL**: AGE and WEIGHT should guide your suspicion for ROP, NOT exposure to oxygen.

# *NEONATAL JAUNDICE, HYPERBILIRUBINEMIA & HEMOLYTIC DISEASE OF THE NEWBORN*

## NEONATAL JAUNDICE

Causes of neonatal jaundices include breastfeeding jaundice, human milk jaundice and physiologic jaundice. Descriptions are below:

* BREASTFEEDING JAUNDICE: Jaundice that occurs within the fist few days of life and is due to insufficient intake and dehydration. DO NOT withhold breast milk. Treat by working on technique.
* HUMAN MILK JAUNDICE: Usually occurs after about 1 week of feeding with breast milk, but can occur as early as DOL 3. Complicated etiology (innate steroids in mom's milk decrease induction of conjugating enzymes in the liver, resulting in indirect hyperbilirubinemia). For diagnostic and treatment purposes, HOLD

BREAST MILK for one day and give formula instead. May then resume breast milk. The indirect bilirubin will come down and likely stay down.

- **MNEMONIC**: If the term has "FED" or "FEEDING" in it, the problem lies in technique and is therefore related to hydration. If the term has "MILK" in it, the problem arises from an intrinsic issue with the milk itself, which should then be held for a day.

* PHYSIOLOGIC JAUNDICE: This is a **diagnosis of exclusion**. Look for a healthy infant with jaundice on **DOL 2 through 5** with no pathologic explanation.

## HYPERBILIRUBINEMIA

Jaundice on DOL 1 is always BAD and can quickly result in dangerous hyperbilirubinemia (consider G6PD deficiency and TORCH infections in your differential). Obtain a total and direct (conjugated) bilirubin level. Phototherapy is CONTRAINDICATED in DIRECT hyperbilirubinemia because it can cause "bronze baby syndrome".

**PEARL**: Use of alcohol, heroin, phenobarbital, phenytoin and tobacco are all associated with a DECREASED risk of hyperbilirubinemia.

## RISK FACTORS FOR DEVELOPING HYPERBILIRUBINEMIA

Risk factors for developing hyperbilirubinemia include G6PD deficiency, asphyxia, temperature instability, sepsis, acidosis, albumin < 3, ABO incompatibility, Rh disease, sibling with history of phototherapy, bruising, cephalohematoma and exclusive breastfeeding that is not going well. SO ALMOST ANYTHING IS A RISK FACTOR. Focus instead on identifying neonates that are at high and medium risk of developing hyperbilirubinemia, as well as memorizing the bilirubin levels at which phototherapy should be initiated.

* **HIGHEST RISK BABIES:** 35 – 37 + 6/7 weeks WITH risk factors

* MEDIUM RISK BABIES: 35 – 37 + 6/7 weeks WITHOUT risk factors. OR, > 38 weeks WITH risk factors.

* LOW RISK BABIES: >38 weeks and well

* START PHOTOTHERAPY ON HIGH RISK NEONATES IF:

* **TOTAL** bilirubin **> 8 at 24 hours**

* **TOTAL** bilirubin **>11 at 48 hours**

* **TOTAL** bilirubin **>13 at 72 hours**

**SHORTCUTS FOR WHEN TO START PHOTOTHERAPY IN MEDIUM & LOW RISK NEONATES:**

* MEDIUM risk: The bilirubin threshold increases by 2 (so approximately 10, 13, and 15).

* LOW risk: The bilirubin threshold increases by 4 (so approximately 12, 15 and 18).

**PEARL**: NONE OF THE ABOVE-MENTIONED RISK FACTORS AND BILIRUBIN THRESHOLDS APPLY TO PREMATURE NEONATES < 35 WEEKS.

**PEARL**: If there is ANY concern that the child will develop hyperbilirubinemia, have the family follow up at 48 hours.

## (DOUBLE TAKE) RHESUS DISEASE (aka RH DISEASE)

When checking for Rhesus Disease (aka RH Disease), look for an Rh- mom in her SECOND pregnancy: Maternal IgM antibodies are made during the FIRST pregnancy and are too large to cross over into the fetal circulation. During the SECOND pregnancy, IgG antibodies are present, which are small enough to cross (can cause ERYTHROBLASTOSIS FETALIS if they cross early in pregnancy). Rh- mom's are supposed to get RHOGAM at 20 wks, and then again after delivery if baby is found to be Rh+.

* KLEIHAUER BETKE TEST: Check MATERNAL blood to see if there are FETAL red cells present.

## (DOUBLE TAKE) ABO INCOMPATIBILITY

ABO incompatibility usually occurs in mothers with an "O" blood type. Naturally occurring "anti-A" or "anti-B" IgG antibodies may be present. This can result in hemolytic disease of the newborn in a FIRST TIME pregnancy. So for hemolysis in a G1P1 baby, consider ABO incompatibility as the etiology.

## (DOUBLE TAKE) GLUCOSE-6-PHOSPHATE DEHYDROGENASE DEFICIENCY (G6PD DEFICIENCY)

Glucose-6-Phosphate Dehydrogenase Deficiency (G6PD Deficiency) is an X-linked recessive disorder (so look for a male patient!) resulting in jaundice, dark urine and anemia due to hemolysis from oxidative injury. In newborns, this will present with jaundice in a MALE baby AFTER 24 HOURS. In other children, can result in hemolysis after ingestion of fava beans, malaria medications, trimethoprim-sulfamethoxazole, ciprofloxacin or nitrofurantoin. Look for HEINZ BODIES (purple granules noted in the red cells on microscopy). Associated with African-American and Mediterranean descent.

**PEARL**: Do not test for the deficiency during the acute hemolytic phase (wait a few weeks) because a false negative result can occur due to the build up of G6PD in reticulocytes.

**MNEMONIC**: Imagine a bottle of HEINZ BODIES ketchup that has an **X** located on the exact spot that you have to hit it to make the broken RBCs come out of the bottle.

# *MISCELLANEOUS*

## FULL TERM

For the boards, assume 37 - 42 week gestation is full-term.

## NEONATE

The term "neonate" applies from Birth to 1 month of age.

## INFANT

The term "infant" applies from 1 month – 12 months of age.

## APNEA

Apnea is defined as $\geq$ 20 seconds between breaths. If less, it's called PERIODIC BREATHING. Differential diagnosis is wide, including cardiac issues, infections, metabolic disorders, neurologic problems and prematurity. Treat with THEOPHYLLINE or CAFFEINE.

**TERMINOLOGY:**

* PRIMARY APNEA: Reversible with stimulation

* SECONDARY APNEA: Not reversible with stimulation, so intubate.

**PEARLS:** Start your workup with NONINVASIVE testing, such as an echocardiogram. Apnea of Prematurity is NOT associated with an increased risk of Sudden Infant Death Syndrome (SIDS). Low birth weight IS associated with an increased risk of SIDS.

## SUDDEN INFANT DEATH SYNDROME (SIDS)

The risk of Sudden Infant Death Syndrome (SIDS) is increased with smoking, low socioeconomics, low birth weight, younger parents, co-sleeping and tummy sleeping.

## ANURIA

For anuria lasting > 24 hours after birth, first gently attempt to obtain a catheter urine specimen. If unsuccessful, obtain a RENAL ULTRASOUND.

## ANEMIA

For newborns, anemia is considered any hemoglobin < 13

## APT TEST

When bloody NG aspirate is noted, an APT test is done to identify whether or not the blood is from the baby or the MOTHER. The test is based on the difference between maternal and fetal hemoglobin.

## NEONATAL HYPOGLYCEMIA

The definition for hypoglycemia is different in the neonatal population. Neonatal hypoglycemia is defined as a glucose level **< 40 (NOT 60)!** May be caused by use of a beta-adrenergic tocolytic (can inhibit glucagon and allow insulin to continue working, and thus cause hypoglycemia). Symptoms of hypoglycemia include apnea, tachypnea, cyanosis, shaking, seizures and poor feeding. Treat with a bolus of **3 ml/kg of D10 solution**. For an average sized newborn, that is about 10 ml of D10. Glucagon may not work in low birth weight babies because of a lack of muscle mass (and therefore low glycogen stores).

## SHOCK-LIKE SYMPTOMS

If shock-like symptoms are noted within 10 days of life, consider bacterial sepsis and inborn errors of metabolism high in your differential. Also consider congenital adrenal hyperplasia (CAH), ductal dependent complex congenital heart disease (CHD), intentional/non-intentional trauma and viral sepsis.

## SEPTIC WORKUP

For babies, a septic workup includes lots of FLUIDS, NO IMAGING. Blood, CSF and urine cultures. Obtain imaging only if there is a suspected focus of infection (e.g., get a CXR for respiratory symptoms).

## CRYING

2-3 hours of crying per day is normal.

## COLIC

Colic is defined by LOTS OF THREES! The infant will have a normal exam and normal lab values. Look for a baby who has been crying >3 hours/day at least 3 times/week for at least 3 weeks. The age should be between 3 weeks – 3 months. This should resolve by 3-4 months of age. Recommend that the parents recruit help from family/friends so that they can get breaks.

**PEARL**: If a child presents like this after 4 months of age, do an aggressive work-up.

## SLEEP

Babies sleep 18 hours/day when first born.

**PEARL/SHORTCUT:** After that, approximate the average number of hours a baby should be sleeping per day by subtracting 1 hour for each month they age. For example, a 6 month old baby should sleep about 12 hours per day.

## SUN SAFETY

The vast majority of a person's lifetime sun exposure occurs before the age of 18, so sun safety is important almost immediately. Both UVA and UVB exposure lead to aging of the skin, skin cancers and sunburns.

* Infants < 6 months of age: Have poor perspiration and are not mobile enough to get themselves out hot spots like you. So, avoid all direct sunlight for this age group.

* Infants > 6 months of age: These kids should use SPF 15 sun block (blocks UVA & UVB) and be kept out of the midday sun.

## AUTOMOBILE AND CAR SEAT SAFETY

Automobile and car seat safety guidelines vary depending on a child's age and weight.

* REAR-FACING CAR SEAT: From birth to 24 months of age, infants should be seated in the back seat at a 45-degree angle. They should be seated in the middle and face the rear of the car. This should be continued until the child is over 24 months or until they outgrow the manufacturer's weight and height limits. After that, the car seat may face forward.

* FORWARD FACING CAR SEAT: When they graduate from the rear-facing seat, they can be in a front-facing car seat with a harness until they again outgrow the manufacturer's weight and height limits (usually around 4 years of age).

* BOOSTER SEAT: Around 4 years of age or 40 lbs, kids outgrow the car seat and will need a belt-positioned booster seat. They stay in that until a regular seatbelt fits.

* SEATBELT: Children may use a regular seatbelt, while still sitting in the back seat, once it fits properly. Usually when they reach 4 feet 9 inches around the ages of 8 – 12.

* FRONT SEAT: Children may sit in the front seat once they become teenagers (regardless of weight).

## VERY LOW BIRTH WEIGHT (VLBW)

On admission to the ICU, 10% Dextrose in Water (D10W) is fine for a Very Low Birth Weight (VLBW) baby. Using D5 will provide insufficient calories.

**PEARL**: Electrolytes do not need to be added for the first 72 hours because of high free water losses.

## PREGNANCY INDUCED HYPERTENSION (PIH)

Pregnancy Induced Hypertension (PIH) is associated with LOW weight, LOW length and LARGE head.

## NALOXONE

Do NOT use Naloxone in a baby with respiratory depression if the mother was on chronic opioids or ADDICTED to opioids. Can precipitate withdrawal and SEIZURES. If the mom was given opioids for PAIN CONTROL during labor, using naloxone for infant respiratory depression is acceptable.

## FAILURE TO THRIVE (FTT)

Always look for dietary and psychosocial issues within the family first when dealing with Failure To Thrive (FTT) cases. Then look for organic causes, including cardiac, malabsorptive, metabolic and renal issues.

**PEARLS: RENAL TUBULAR ACIDOSIS** is a VERY COMMON reason for FTT. Any child growing less than 5 cm/year needs a workup and a diagnosis, even if the overall height is "normal" for age.

## ARTHROGRYPOSIS MULTIPLEX

Arthrogryposis Multiplex is characterized by contractures due to idiopathic fetal akinesia (decreased fetal movement).

**IMAGE**: http://www.pediatric-orthopedics.com/Topics/Muscle/Diseases/Arthrogryposis_Multiplex/Arthrogryposis.jpg

## CEPHALOHEMATOMA

A cephalohematoma is a firm, localized and tense subperiosteal hematoma of the calvaria.

**PEARL:** It is subperiosteal, so it does NOT cross suture lines.

**MNEMONIC**: It has the word HEMATOMA in it. Like any hematoma, it is firm and tense. NOT boggy and soft like a caput.

## CAPUT SUCCEDANEUM

Caput Succedaneum is a swelling of the scalp, from urine wall or vaginal wall pressure, which results in a soft and boggy lesion on the head.

**PEARL:** CAN cross suture lines.

**MNEMONIC**: "Caput" is Latin for hat. Hats are placed on the scalp and cross multiple suture lines!

## UMBILICAL CORD

Keep the umbilical cord/baby below mom's waist for 30 seconds at birth, but DO NOT MILK IT. The cord should fall off by TWO weeks. If it has not fallen off by 4 weeks, consider LEUKOCYTE ADHESION DEFECT or other problems in which there is LEUKOPENIA (low WBCs).

## CORD CATHETERS

Cord catheters should be placed at L3-L5 or T6-T10. Complications include hepatic infection (omphalitis), decreased femoral pulse, emboli to the liver, hemorrhage, perforation of a bigger artery... but does NOT CAUSE NEC.

**MNEMONIC**: "3 & 5 x2 = 6 & 10." Remembering one set of numbers will help you remember the other. For example, if you remember L3 – L5, simply multiply the digits by 2 to get T6-T10

## SINGLE UMBILICAL ARTERY

Obtain a RENAL ultrasound and look for VATER/VACTER-L syndrome if a single umbilical artery is noted.

* **(DOUBLE TAKE) VACTER-L (aka VACTERL or VATER) SYNDROME**: VACTER-L (aka VACTERL or VATER) syndrome is an acronym. VACTER**L** is now used instead of VATER because it stand for **V**ertebral anomalies, **A**nal atresia/imperforate anus, **C**ardiac defects (especially VSD), **T**racheoesophageal fistula, **R**adial hypoplasia and **R**enal anomalies, and **L**imb abnormalities. These children have a normal IQ. When associated with hydrocephalus, this can be an X-Linked disorder.

**PEARL**: May present with single umbilical artery.

**MNEMONIC**: Imagine Darth VACTER cutting off his own son's ARM (radial hypoplasia and limb abnormalities) and then using the ARM as a light saber to create ANAL ATRESIA and a TE Fistula."

**IMAGE**: (go to 5:30 in the video) http://www.youtube.com/watch?v=C-Del3ohVbY

**IMAGE**: http://www.youtube.com/watch?v=uXQxstTrhTI

## NECROTIZING ENTEROCOLITIS

Necrotizing Enterocolitis is often located at the ileocecal junction. Look for thrombocytopenia, bloody stool, an erythematous abdomen, poor feeding, PNEUMATOSIS INTESTINALIS +/- air in biliary tree. Associated with infection and hypoxic injury (apnea, asphyxia, RDS, etc). Do not feed these kids for **3 weeks**.

* **MNEMONIC**: NECKrotizing enterocolitis usually occurs at the NECK of the colon (ileocecal junction).

* (DOUBLE TAKE) TYPHLITIS: If you are presented with a clinical description in an older child with leukopenia or neutropenia that sounds like NEC, choose typhlitis as the answer.

## HYPOSPADIAS

Hypospadias is associated with Silver Russell Syndrome, Laurence-Moon-Biedle Syndrome, Opitz Syndrome and Beckwith-Wiedemann Syndrome. It can be associated with renal anomalies, but as a general rule, the more distal the lesion, the less chance of renal problems.

**MNEMONIC**: If you pee all over the floor because of hypospadias, you will likely be called a **SLOB**. **S**ilver Russell Syndrome, **L**aurence-Moon-Biedle Syndrome, **O**pitz Syndrome and **B**eckwith-Wiedemann Syndrome.

## UNDESCENDED TESTICLE

Risk of malignancy is increased for BOTH testicles even if only one testicle is undescended. The increased risk persists even after orchiopexy (surgical procedure done to move undescended testicle into the scrotum). Risk for infertility increases with increased time spent in the canal. If still undescended at at 6 months of age, refer for orchiopexy to decrease chances of infertility.

# DEVELOPMENTAL MILESTONES

PEARL: For the American Board of Pediatrics (ABP) Initial Certification Board Exam, KNOW THE DEVELOPMENTAL MILESTONES WELL! This is a CRITICAL SECTION. Be able to identify a child's age based on milestones reached, and also be able to identify the specific category in which a child is delayed. Questions could be in matching or multiple-choice format. Rote memorization of this section is difficult. Many residents avoid this section and rely on their clinical training to get them through these questions. BAD IDEA! You must know small details in order to get these questions correct. This guide will help you memorize key milestones through the use of mnemonics. Please note that the milestones in the mnemonics DO NOT ENCOMPASS every milestone listed in the tables. Lastly, there are many sources available for your review on this topic. The timelines do vary since milestones are achieved on a spectrum in reality. Memorize 1 source and do your best.

## *DEVELOPMENTAL MILESTONES THROUGH ADOLESCENCE*

### DEVELOPMENTAL MILESTONES SCREENING TOOLS

Developmental milestones evaluation includes the evaluation of language, gross motor, fine motor, cognitive and social development. For the pediatric boards, the ABP requires you to know about more screening tools for developmental milestones than just the Denver.

*** PEDS: Developmental Milestones (PEDS:DM):** Newer tool. Parent and physician versions are available. Both screening and an assessment levels are available. May be used in children 0-8 years of age to assess the major developmental categories (gross motor, fine motor, expressive language, receptive language, self-help and social-emotional). For older children, an also be used to screen for reading and math problems.

*** BAYLEY NEURODEVELOPMENTAL SCREEN (BINS):** Used in children 3-24 months of age. Often used in premature infants. Assesses neurologic processes, neurodevelopmental skills and developmental accomplishments reflexes via DIRECT OBSERVATION of tone, movement, symmetry, object permanence, imitation and language.

*** CHILD DEVELOPMENT INVENTORIES or CHILD DEVELOPMENT REVIEW:** Used in children 3-73 months of age. Screens major developmental areas via a PARENT QUESTIONNAIRE.

*** DENVER DEVELOPMENTAL SCREENING TEST II (DDST-II):** For use at 1 month to 6 years of age. Used as a primary screen for possible developmental problems and to monitor children at risk for developmental problems. Some also use this to confirm suspected problems with this objective tool, but the sensitivity is lower than other more detailed (and time consuming tools).

* **AGES AND STAGES QUESTIONNAIRE (ASQ):** For use at 4 – 60 months of age. Used to screen for developmental delays in the first 5 years of life through a parent or caregiver QUESTIONNAIRE. Covers 5 developmental areas (communication, gross motor, fine motor, problem solving, and personal-social interactions).

* **TOOLS RECOMMENDED BY THE AAP:** http://www.earlychildhoodmichigan.org/articles/7-03/DevScrTools7-03.htm#recommendations

## DRAWING SHAPES

Age 3 = circle, 4 = cross, 5 = square, 6 = triangle and 7 = diamond

**MNEMONICS: Multiple the above ages by TWO. The product creates a shape similar to one a child can draw at that age!**

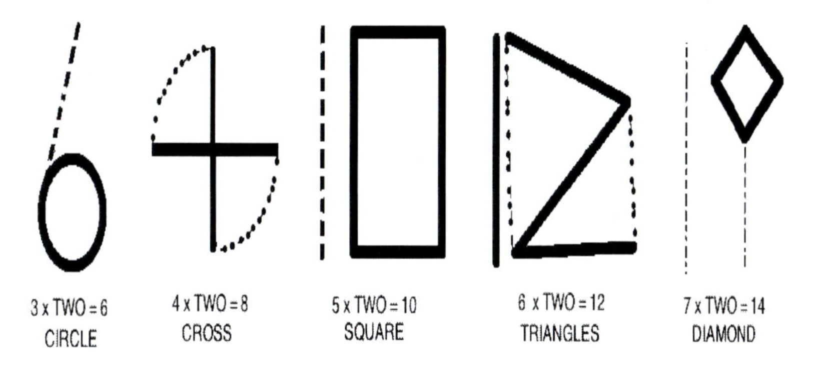


**MNEMONIC**: Remember the intro peg for **3**? It's a STOOL. Imagine a **3**-year-old drawing a circle that looks like the top of a STOOL!

**MNEMONIC**: A **4**-year-old can draw a CROSS. There are 4-points on a CROSS/PLUS sign. Also, a plus sign (**+**) looks like a "**4**."

## DEVELOPMENTAL MILESTONES CHART, BIRTH TO 2 MONTHS AGE

Developmental milestones chart for birth to 2 months of age:

| AGE | LANGUAGE | GROSS MOTOR | FINE MOTOR | COGNITIVE | SOCIAL | RED FLAGS |
|---|---|---|---|---|---|---|
| Birth | | Moro reflex. Fixates. Briefly follows objects. | | Alert with voice/bell. | Sleeps ~18 hours/day. | |
| 1 mo | | Lift head briefly while prone and look at object. | | | | |
| 2 mo | COO. | Lift head & shoulders. Partial head lag (45°) with pull-to-sit at 2-3 mo. Visually track from one side to the other. | Palmar grasp reflex disappears by 2-3 mo. | | Social smile. Cry 2-3 hours/day up until 2 months old. | Can't fixate. Poor visual attention. |

**MNEMONICS:** COO rhymes with TWO. Lifts TWO body parts when prone (head & shoulders). Tracks in TWO directions (TWO sides or 180° at "1.8 mo").

**MNEMONICS FOR CRYING BY AGE:** By three months of age, babies cry approximately 1 hour/day. Use 1-2-3 and 3-2-1 to remember that by at months 1, 2, and 3, babies cry 3, 2, and 1 hours per day.

## DEVELOPMENTAL MILESTONES CHART, 4 MONTHS AGE

Developmental milestones chart for 4 months of age:

| AGE | LANGUAGE | GROSS MOTOR | FINE MOTOR | COGNITIVE | SOCIAL | RED FLAGS |
|---|---|---|---|---|---|---|
| 4 mo "4attle" | Laugh. Squeal. | While prone, lifts **hands**, bears weight on **forearms**. Roll front to back. Bear wt on feet while being held by hands. Moro disappears by 3-6 months. Head lag disappears by 4 months with pull-to-sit. Sits in a tripod position. | **Grasp** a rattle ~3-4 months (not shake or reach) and place in mouth. | Turn to sound ~3 mo. Anticipates being lifted and reacts to peek-a-boo (but doesn't play it). Smiles with joy. | Minimally imitates social interaction. ~70% infants are sleeping through the night. Cry ~1 hour/day by 3 months. | Not tracking from side to side. Unsteady head control. |

**MNEMONICS:** Spell Rattle with a **4** (**4**attle) for 4 months of age.

**MNEMONIC STORY:** Go over the following story over and over again! Imagine a 4-month-old baby wearing a shirt that has 4attles all over it. He loves his RATTLE, and his MOUTH. You see him lying on his belly holding a very thin RATTLE that is shaped like a **4** in his right hand. He LIFTS HIS HEAD and while PUTTING WEIGHT ON HIS ARMS. Then, he TURNS in the only direction he can, from FRONT TO BACK (Just like the rattle! If you lay a 4-shaped rattle on its pointy side, or "face down," it will easily end up on its flat back. Once on its back, it is much more difficult to roll from back-to-front!). You now cover the infant's eyes and shake the rattle on his left side. He TURNS TO THE SOUND. You then place the rattle in his left hand and note that he is ABLE TO GRASP it well (**but not reach for it**). He SMILES WITH JOY, then LAUGHS and SQUEALS in delight as he SHAKES THE RATTLE and then MOUTHS it. When **YOU** DO PEEK-A-BOO, he SQUEALS in delight, but he cannot yet return the favor. You STAND him up, but he can only IMITATE a real stand with his stiff legs as you hold him (similar to a 4-shape RATTLE that cannot stand on its own or it will topple over!). You lay him back down, but then note that he's crying and getting fussy because he is ready for a nap. As you reach to pick him up, he stops crying because he can ANTICIPATE what you are about to do, but he still cannot participate and reach for you to help you lift him.

## DEVELOPMENTAL MILESTONES CHART, 6 MONTHS AGE

Developmental milestones chart for 6 months of age:

| AGE | LANGUAGE | GROSS MOTOR | FINE MOTOR | COGNITIVE | SOCIAL | RED FLAGS |
|---|---|---|---|---|---|---|
| 6 mo "sitting in a high chair" | Babbles consonants, different cries, imitates speech. | Rolls in both directions. Sits using hand support. Stand with assist (not cruise). Transfers from one hand to other. **Low crawl.** Sit in a high chair. Mouth and bang objects (on floor/table). Bounce when held standing. Shake rattle. Radial-palmar grasp. | **Reach.** | Turn directly to sound or voice. Open mouth for food. Imitate familiar actions. Smile at mirror (does NOT recognize self). | Sleeps ~12 hours/night plus 2 one-hour naps. | Not turning to sound. |

**MNEMONICS:** **S**ix = **S**IT = **S**WITCH hands (aka TRANSFER). Once a baby can TRANSFER objects, he also has the motor skills to coordinate a REACH. If a **S**ix-month-old is strong and coordinated enough to sit without support, he can also ROLL IN **2** DIFFERENT DIRECTIONS, CRY IN **2** DIFFERENT/UNIQUE WAYS and take 2 SEPARATE NAPS during the day. Also, his legs are no longer stiff like the 4attle of a 4-month-old, so he can now BOUNCE with excitement when being held on his feet.

**MNEMONIC STORY:** Go over the following story over and over again! Imagine a **S**ix-month-old **S**ITTING in a HIGH CHAIR. He SHAKES objects and then BANGS THEM ON THE HIGH CHAIR'S TABLE. The table is made out of a MIRROR. He looks at the MIRROR and SMILES when he sees the **other** baby (not yet able to recognize self). He **BABBLES** and thinks the other baby in the mirror is talking, so then he begins to **IMITATE** the other baby's **SPEECH**.

## DEVELOPMENTAL MILESTONES CHART, 9 MONTHS AGE

Developmental milestones chart for 9 months of age:

| AGE | LANGUAGE | GROSS MOTOR | FINE MOTOR | COGNITIVE | SOCIAL | RED FLAGS |
|---|---|---|---|---|---|---|
| 9 mo "cat" | Nonspecific "mama" or "dada" | Bangs two blocks. Claps. Sit without support. | Develops parachute reflex around 6-9 mo. | **Plays** peek-a-boo. Turns to name. | Stranger anxiety (9-12 mo). Recognizes objects & people. Plays pat-a-cake | No babbling |

**IMAGE**: (PARACHUTE REFLEX) http://www.winfssi.com/images/parachute.jpg

**MNEMONICS:** **9**eek-a-boo, **9**at-a-cake and **9**arachute reflex.

**MNEMONIC STORY:** 9 = CAT with 9 lives! 9-month-old infants are like CURIOUS LITTLE CATS. Image a 9-month-old baby being like a curious, yet scared little CAT. It can RECOGNIZE OBJECTS/PEOPLE, but typically has ANXIETY WHEN APPROACHED BY STRANGERS. It's a SCAREDY CAT! She says "Meow-Meow" NONSPECIFICALLY to her owners, like a baby says "MA MA" NONSPECIFICALLY to both mom and dad. Now imagine this black cat being scared of the growing 9-month-old baby boy that also lives in its home. She is scared of the boy because he likes to BANG BLOCKS TOGETHER, or CLAP with the cat's head in the middle!

## DEVELOPMENTAL MILESTONES CHART, 12 MONTHS AGE

Developmental milestones chart for 12 months of age:

| AGE | LANGUAGE | GROSS MOTOR | FINE MOTOR | COGNITIVE | SOCIAL | RED FLAGS |
|---|---|---|---|---|---|---|
| 12 mo "eggs" | Points. Shakes head. "Mama" and "Dada" are said specifically. Speaks 1 other word. | Few steps. Pulls to stand. Cruises with 1 hand. Possibly takes a few apprehensive steps. Waves bye-bye with one hand. Crawls on hands/knees. | Pincer grasp. Drink from cup held by others. Points. Feeds self finger foods. Removes blocks or objects from a container. | Assist with dressing (offers feet). Responds to music. Object permanence (crawls to last place toy was seen). Follows 1-step commands with 1-finger gestures. | Knows family members' names. Cries if you cry. Shows affection. Raises arms to be picked up (anticipate + participate). ~90% infants sleep through night. | |

**MNEMONIC STORY:** Peg for 12 is EGGS (as in a dozen eggs!). Imagine a weird little 12-month-old kid who LOVES EGGS. He always wears his favorite 12-egg omelet T-shirt. Imagine that his DADA has a CUP filled with a dozen egg yolks. Baby PULLS TO STAND and CRUISES WITH 1 HAND on a table all the way to DADA. He then takes a **ONE** APPREHENSIVE **STEP** to and then DRINKS FROM THE CUP full of eggs being **held by** his dad! To make things worse, the dad is a little odd too. He loves wearing wigs that make him look like he has an egg-white-colored PERM (object PERManence)!

* NOTE: "**ONE** APPREHENSIVE **STEP"** refers to 1-step commands as well as the ability of a 12-month-old to take a **few** apprehensive steps.

## DEVELOPMENTAL MILESTONES CHART, 15 MONTHS AGE

Developmental milestones chart for 15 months of age:

| AGE | LANGUAGE | GROSS MOTOR | FINE MOTOR | COGNITIVE | SOCIAL | RED FLAGS |
|---|---|---|---|---|---|---|
| **15 mo "pay-check"** | 4-6 words | Give and take an object. Stoop and recover. Crawl up stairs. | Drink from cup by self. Take cube from cup. Stack 2 cubes. Scribble a vertical line (15-18 mo). | Follow 1-step commands **without** gesture (15-18 mo). Knows at least ONE body part. Asks by pointing or vocalizing. | 1 nap per day | |

**MNEMONIC STORY:** Peg for 15 is PAYCHECK or TAX DAY on April 15th. Imagine an IRS tax agent WALKING into your home with his pad of papers and an oversized red crayon. He SCRIBBLES jargon that you can't understand and GIVES it to you, and you TAKE it. While your attention is diverted because you are trying to read his scribbled mess, he takes a drink from *your* coffee CUP. He even takes out your single CUBE of sugar because he thinks your coffee is too sweet. He takes that cube and STACKS IT ON ANOTHER CUBE. He then notices your wallet on the ground so he STOOPS/RECOVERS as he picks up you hard earned money. You then notice what he has done and you POINT TO ONE INAPPROPRIATE BODY PART and **ASK** HIM TO kiss your… FOOT. You use several other FOUR-letter words. You then crumple his paper into a RED BALL and GIVE it back to him. You then grab your head in frustration and scream the word "OUT!" (He understands you even though you did NOT point.) After all of that madness, you are EXHAUSTED SO YOU CRAWL UPSTAIRS and get into bed for your ONE NAP of the day.

* **NOTES:** Pointing to a body part refers to knowledge of at least one major body part. **ASK** refers to asking by pointing. "FOUR-letter words" refers to a 4-6 word vocabulary. Screaming "OUT," refers to the tax man's ability to understand a one-step command without the need for a gesture.

## DEVELOPMENTAL MILESTONES CHART, 18 MONTHS AGE

Developmental milestones chart for 18 months of age:

| AGE | LANGUAGE | GROSS MOTOR | FINE MOTOR | COGNITIVE | SOCIAL | RED FLAGS |
|---|---|---|---|---|---|---|
| 18 mo "voter" | 10-15 words. Asks for mom or dad. Says "hi," "bye," and "please." | Takes off hat, mittens and socks. Carries a doll. Runs stiffly. Walks up stairs **with help** (18-24 mo). Toss ball underhand. Climb into an adult chair. | Self-feeding with a spoon, or sp∞n (messy). Turn pages (3 at a time). Stack 4 cubes. Scribble. | Recognizes **self** in mirror/photos and objects in a book. Object permanence (now seeks out hidden objects). Trial and error. Recognizes named body parts. Follows 1-step commands without gestures (15-18 mo). Understands 75-100 words. | Share toys with **parents**. Play alone. Imitates simple chores (vacuuming or **sweeping**). | Not walking. No words. |
| 21 mo | 30-50 words | | | | | |

**MNEMONIC**: 18 months = age at which a SPOON is used (which is usually shaped like an "8"). SP∞N (note the 18 on its side)

**MNEMONIC STORY:** Peg for 18 is VOTING (age in years when you can VOTE). Now imagine an 18-year-old African-American with dwarfism who lives with his parents. It's an election year and he sees a commercial of "Uncle Sam" asking him to vote with ONE FINGER (one-step command). He looks at a book of presidential candidates and TURNS 3 PAGES AT A TIME because that's all he can do with his stubby fingers. He lives in the cold and windy city of Chicago and decides to vote for Obama. He goes to the polling site carrying a BOBBLE HEAD DOLL of Obama. Once there, he NEEDS HELP GOING UP THE BUILDING'S STAIRS and goes ONE STEP AT A TIME. He's being helped by a greeter who has been instructed to limit her vocabulary to 10-15 WORDS so as not to influence any voters. The dwarf is then helped into the actual booth up FOUR BLOCKS/CUBES being used as a stairs. He TAKES OFF HIS MITTENS and notices a mirror in front of him. He RECOGNIZES himself and makes himself look as nice as he can for his first voting experience. Then his stomach growls so he grabs some pudding from his coat because he doesn't want to disturb other voters. He eats the pudding using a FIGURE 8-SHAPED SP∞N. It's too much for his tiny stomach, so he throws it away using an UNDERHAND FLIP/THROW. The pudding flies out of the booth and makes a mess on the floor. He looks at the ballot and is able to UNDERSTAND THE 75 – 100

WORDS/topics and eventually figures out how to vote for the first time using TRIAL & ERROR. As he's about to leave, he is asked to help SWEEP up his mess of pudding. Carrying his bobble head doll, he goes home. As a doll carrying 18 year old dwarf, he's somewhat of an outcast. He usually only PLAYS BY HIMSELF, but at least he has his parents. He doesn't like anyone else, but he at least SHARES his doll WITH HIS PARENTS when it's their turn to go vote.

## DEVELOPMENTAL MILESTONES CHART, 2-YEAR OLD

Developmental milestones chart for a 2-year old:

| AGE | LANGUAGE | GROSS MOTOR | FINE MOTOR | COGNITIVE | SOCIAL | RED FLAGS |
|---|---|---|---|---|---|---|
| 2 yo | 50-100 words. 2-3 word phrases. Says "me" and "mine." Starts using pronouns (2-3 yo). Speech is 50% intelligible to strangers. | Take off all clothes. Toss big ball overhead. Roll ball. Up/down single steps alone. Walk backwards. Broad jump (2-3 yo). Kick a big ball. Run well. | Drink from a straw. Put, arms in sleeves. Stack 6 cubes. Put cube into performance block. Open boxes, cabinets & drawers. Use a spoon (less mess). Washes & dries hands with help. | Follows 2-step commands. Points to named objects or photos. Sorts objects by color. Points to 5 body parts. Refers to self by name. Verbalizes feelings, wants & toileting needs. | Parallel play (plays alongside others, but not with them). Comforts distressed peers. Takes turns to play. | No phrases. |

**MNEMONICS:** Half, or 1/**2** of their words are intelligible to strangers. TWO-year-old children follow TWO-step commands. The peg for TWO is a light bulb. Imagine a 2-year-old child drinking out of a STRAW made out of a light bulb. Or try to associate the TWO ends, or TWO holes, of a straw with TWO years of age.

**MNEMONIC STORY:** Imagine a 2-year-old child in his "Terrible two's" RUNNING WELL all around the house after TAKING OFF ALL OF HIS CLOTHES. He then goes to the kitchen, uses his TWO feet to perform a BROAD JUMP over the pet dog, and then KICKS the dog's food dish to make a mess. He then proceeds to make a bigger mess by OPENING ALL OF THE BOXES, CABINETS and DRAWERS in the kitchen. His mom finally catches him and she starts to cry. He COMFORTS HER. She then HOLDS HIS HAND and as they go up 6 CUBE-SHAPED STAIRS (stack 6 cubes), which he is able to navigate with ALTERNATING FEET. His mom then helps him WASH/DRY his hands, feet, face, nose, ears (5 BODY PARTS that he KNOWS he shouldn't have gotten dirty). As she helps him finally get some clothes on, he PUTS HIS ARMS THROUGH THE SLEEVES of a coat.

## DEVELOPMENTAL MILESTONES CHART, 3-YEAR OLD

Developmental milestones chart for a 3-year old:

| AGE | LANGUAGE | GROSS MOTOR | FINE MOTOR | COGNITIVE | SOCIAL | RED FLAGS |
|---|---|---|---|---|---|---|
| 3 yo | 3-8 word sentences. Asks short questions. Uses plurals. Says "I, you, he, she & it." 75% of language is intelligible. May ask "why?" | Stands on 1 foot for 1-2s. Hops on 1 foot 2-3x. Broad jump (2-3 yo). Regular-sized ball thrown overhead. Uses toilet with help. Walks **up** steps with **alternating** feet. Jump down from 1 step. Tiptoes. Kicks small ball. Rides tricycle. | Draws a circle & a line. Stacks 8-10 cubes. Can button & unbutton large buttons. Puts on shirt, shoes & shorts. Feeds self. Holds a glass in 1 hand. Uses crayon well. Washes hands by self. Draw a head. | Daytime continence. Remembers yesterday. Knows simple adjectives (hungry, tired). Understands "now, soon & later." Can match circle & squares. Can ID 1 color. Knows first & last name. Counts to 4. Understands "routines." | Some pretend play with imaginary friends. Imitates housework and helps with tasks. Makes others laugh. Plays spontaneously with random children. | Lacks 3-word phrases.<br><br>Speech is unintelligible to strangers.<br><br>Trouble with comprehension. |

**MNEMONIC STORY:** Peg for 3 is a STOOL (with 3 legs). Now imagine a fictitious 3-year-old character named CRAYALO. Crayalo is a 3-legged orphaned stool. Her legs are made out of 3 RED crayons (ID one color) and she has a CIRCULAR seat (can draw a circle). Crayalo comes to life, but only when she's alone. She is playful and can DRAW USING HER CRAYON LEGS REALLY WELL. She can also RUN UP 8-10 CUBE-SHAPED stairs with ALTERNATING HER FEET and without any help. COMING DOWN STAIRS IS AN ISSUE because she's afraid of heights. Crayalo lives in a home with a 3-year-old girl who is very respectful of Crayalo whenever she climbs up the stool to WASH/DRY HER HANDS BY HERSELF, or to GRAB A GLASS IN ONE HAND from the cabinet before sitting to drink and EAT BY HERSELF. The girl can also do some other BIG GIRL things by herself, like using BIG CIRCULAR BUTTONS (shaped like the top of the stool), putting her feet into the BIG CIRCULAR HOLES OF SHORTS, putting her feet in VELCRO® SHOES (can't yet tie laces) that are shaped like SKATES (8 is the number of blocks she can stack), and putting her head through the BIG CIRCULAR HOLE of a shirt. Even though she can't talk to the girl, Crayalo loves it when the girl PRETEND PLAYS with her IMAGINARY FRIEND, the stool.

**MNEMONICS:** Uses more pronouns. Use the picture below to remember pronouns are now being commonly used at the age of 3 (note the 3 prongs).

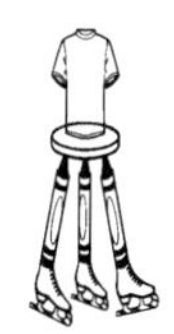

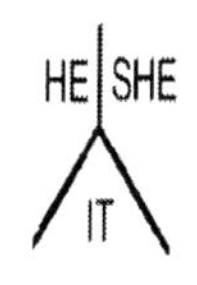

## DEVELOPMENTAL MILESTONES CHART, 4 YEAR OLD

Developmental milestones chart for a 4 year old:

| AGE | LANGUAGE | GROSS MOTOR | FINE MOTOR | COGNITIVE | SOCIAL | RED FLAGS |
|---|---|---|---|---|---|---|
| 4 yo | 3-5 word sentences clearly. Uses past tense. Asks questions like, "how, what, when, where & why?" 100% intelligible. Rhymes. Uses articles. | Balances on 1 foot for 4 seconds. Goes up AND down stairs easily and with alternating feet. Uses handrails on the way down. | Draws a plus sign (+) or cross. Draws a simple stick-person with 4 body parts. Dresses self completely without help. Brushes teeth without help. | Reads a Tumbling E chart (4-5 yo). Tells a story/recent event. Tries to bargain & change game rules. Knows age/city. Recognizes 3-4 colors. Understands more adjective (big, little, & tall). Follow 3-step commands. Wants to know what's next. Knows opposites. | Plays interactively with peers. Prefers peer play. Separates from parents for a bit without crying. Plays dress up. **Participates in pretend play with imaginary and real friends.** Shares when asked to do so. Enjoys rules based games. Plays hide & seek. Plays tag. No longer takes naps. | No peer interaction.<br><br>Trouble navigating stairs. |

**MNEMONIC STORY:** Imagine a COMPASS that is sitting on a stack of SHARED newspapers at the airport (shares with **strangers** and now plays with peers). The compass is lying on an ARTICLE about CROSS-DRESSING men (play dress up) who CHEAT (change rules) on their wives. The compass has FOUR needles made out of "E" looking TOOTHBRUSHES. The needles make a CROSS. Imagine that each needle is a different color (FOUR COLORS). The compass can obviously answer the question of WHERE! Let this remind you that this is the age when children ask DIRECT questions, like who, what, where, why, and **WHEN**. **WHEN** reminds us that the child can now speak in PAST tense. Now imagine a child decides to STAND ON 1 FOOT FOR 4 SECONDS on the compass (just long enough to go around the compass once before falling off). Lastly, try to imagine that the BRUSHES are taken off of the compass, enlarged, and made into HANDRAILS to allow a 4-year-old child to walk DOWN STAIRS WITH ALTERNATING FEET USING HANDRAILS.

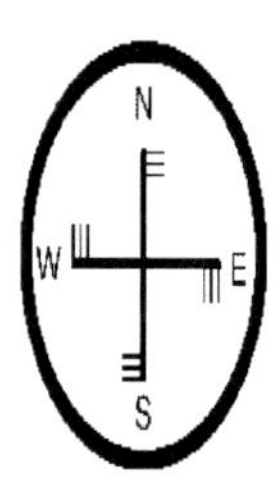


(Please note that the figure above is a special COMPASS with 4 brush-shaped needles. It is NOT the Hindu symbol called the Swastika.)

**MNEMONIC**: By remembering what a child can draw at 3 and 4 years of age (circle and line at 3, and then a cross at 4), you can assume that the child can now draw a stick figure with FOUR body parts. In the image below, the 4 parts are the head, arms, legs and torso.

## DEVELOPMENTAL MILESTONES CHART, 5 YEAR OLD

Developmental milestones chart for a 5 year old:

| AGE | LANGUAGE | GROSS MOTOR | FINE MOTOR | COGNITIVE | SOCIAL | RED FLAGS |
|---|---|---|---|---|---|---|
| 5 yo | | **Walks backwards.** **Skips.** | Has kindergarten & school skills: Holds a pencil properly. Prints letters (not own name). Draws a person with 6 body parts. Draws a square. Ties a knot and laces (5-6 yo). | Names all of the primary colors. Defines words. Able to tell fantasy from reality. | Plays board game. | |

**<u>MNEMONIC</u>**: Look at the 5 made out of a square font below. It looks like a "hangman" drawing. It contains a stick figure that is more complicated than before. Note the HANDS & FEET (2 new body parts for a total of 6 BODY PARTS that the child can now draw. You cannot see the FIVE toes because there are SHOES on its feet. Imagine the LACES ARE TIED. Keep in mind this is a school-going age, and that is around the time that children learn how to tie their laces. It's also the time they learn how to WALK BACKWARDS TO AVOID SCHOOL! At recess, they like to SKIP around as they show off to their friends, and after school they like to sit and play BOARD GAMES as they wait for their parents. The peg for 5 is a GLOVE or a HAND. Use that peg to imagine a child reaching out to catch a RAINBOW with his GLOVED FIVE FINGERS, and then NAMING ALL THE PRIMARY COLORS of the rainbow before releasing it.

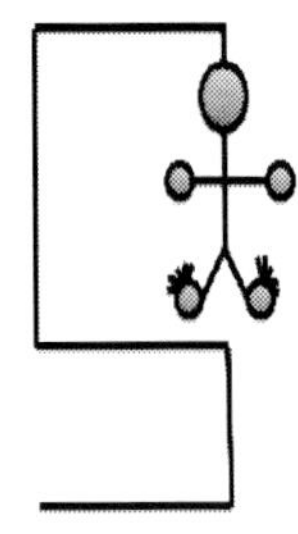

## DEVELOPMENTAL MILESTONES CHART, 6 YEAR OLD

Developmental milestones chart for a 6 year old:

| AGE | LANGUAGE | GROSS MOTOR | FINE MOTOR | COGNITIVE | SOCIAL | RED FLAGS |
|---|---|---|---|---|---|---|
| 6 yo | | Rides a bicycle (tricycle at 3, bicycle at 6!) | Draws a triangle. Writes own name with a mature pencil grasp. Uses a knife to spread (can cut at ~7 yo). Ties own laces. Prepares a bowl of cereal. | Knows right from left. Counts to 10. | | |

**MNEMONICS/STORY:** 6 = ∆'. A 6 year old can draw a **TRI**angle and ride a **BI**cycle (TRIcyle at 3, Bicycle at 6!). Imagine that the BICYCLE has a 6-year-old child's FULL NAME (write full name) on the back (like a license plate). The child can turn the bike to the RIGHT to go home, or he can turn it to the LEFT to go to the candy store. He goes left (able to tell RIGHT from LEFT).

# COGNITION

## COGNITIVE AND CONCRETE THINKING

Cognitive development refers to the development of the ability to **THINK AND REASON**. Children (6 to 12 years old) develop the ability to think in **CONCRETE** ways (concrete operations), such as how to combine (addition), separate (subtract or divide), order (alphabetize and sort), and transform objects and actions (change things such as 5 pennies = 1 nickel). Cognitive reasoning begins to develop in early adolescence and is complete by approximately 21 years of age.

| AGE | COGNITIVE | SOCIAL |
|---|---|---|
| 6-12 yo | Self conscious over small things (hair). Values winning. Early **concrete** thinking. Routines become automatic. Begin making plans to meet specified goals. | More independence. Minor rebellion as children start to talk back to parents. Fear the idea of death and rejection. Able to appreciate others' point of view. Attachment to non-parental adults (as in teachers). |
| 10-13 yo | Concrete thinking still predominates. Needs explicit discussions. Impulsive without good cognition = BAD COMBINATION! | Increasing emphasis on peer relations. Major rebellion against parents. |
| 14-16 yo | Cognition begins to develop. Still very impulsive. | Begin to explore relationships with the opposite sex. Peers are the most influential group in effecting change and for emotional support. Sexual ORIENTATION is developed by 16 years of age (parent's orientation and difficult upbringing does NOT influence sexual orientation). |
| 17-21 yo | Body image and GENDER ROLE defined. Idealistic (in college)! Cognitive development is now complete. Life goals tend to be the center of patient discussions. Still some impulsiveness, but now with cognition of consequences. | Individual relationships become extremely important. |

* **NOTE:** GENDER ROLES refers to the social and behavioral norms for a man or woman in a given culture (what the role of a man or woman is, etc.). SEXUAL ORIENTATION refers to whether or not a person is heterosexual or homosexual.

# EMERGENCY MEDICINE & TOXICOLOGY

## MENTAL STATUS CHANGES

If the child is mobile, mental status changes may indicate medication or toxic ingestion.

**PEARL**: Did the question mention a FAMILY MEMBER'S medical history? Ingestions are often NOT observed and may be related to a family member's disease (e.g., hypoglycemia from a diabetic medication, or hypotension from antihypertensive medications).

## PUPILS

Normal pupil size is 3 mm.

**PEARL**: If any other pupil size is mentioned, it is probably a hint.

## MIOSIS

Miosis means small pupils. Can be caused by clonidine, cholinergics, opioids, organophosphates, phenothiazine, pilocarpine and sedatives/barbiturates. Technically, pupils should be 2 mm or less.

**MNEMONICS:** CoPS know to look for small pupils o o

* C – **Clo**nidine & CHOLinergics
* **o** – **o**pi**o**ids & **O**RGAN**o**phosphates. Does the **ORGAN** remind you of a **COLON**-ergic?
* P – **Pheno**thiazine, pil**o**carpine (Know this cholinergic!)
* S – **S**edatives & barbiturates

## MYDRIASIS

Mydriasis (large pupils) can be caused by antihistamines, antidepressants, anticholinergics and aympathomimetics. Technically, pupils should be 5 mm or more.

**MNEMONIC**: Think of how BIG your eyes got the first time you realized the benefit of AAA Savings!!! "AAA $avings." If you have a AAA® card, you can save on cars, hotels and more!

* A – **ANTI**histamine
* A – **ANTI**depressants
* A – **ANTI**cholinergics (**Atropine**)
* S – **S**ympathomimetics (Amphetamines, Cocaine, Phencyclidine)

## DIAPHORESIS

Diaphoresis (aka sweating or perspiration) can be caused by sympathomimetics, organophosphates, aspirin & phencyclidine.

**MNEMONIC**: Sweaty/diaphoretic skin needs **SOAP**!

* **S**ympathomimetics
* **O**rganophosphates (and CHOLINERGICS in general)
* **A**spirin
* **P**hencyclidine (PCP)

| Toxidrome | BP | HR | RR | Temp | Pupils | Bowel Sounds | Diaphoresis |
|---|---|---|---|---|---|---|---|
| Anticholinergics | ~ | Up | ~ | Up | Large | Down | Down |
| Cholinergics | ~ | ~ | ~ | ~ | Small | Up | Up |
| Opioids | Down | Down | Down | Down | Small | Down | Down |
| Sympatho-mimetics | Up | Up | Up | Up | Large | Up | Up |
| Sedative-hypnotics | Down | Down | Down | Down | ~Small | Down | Down |

## NYSTAGMUS

Nystagmus can be caused by alcohol, lithium, anticonvulsants and **PCP**. PCP can cause vertical OR horizontal nystagmus.

## SYRUP OF IPECAC

Syrup of Ipecac is no longer recommended for toxic ingestions.

## CHARCOAL

The typical dose of charcoal is 1 gram/kg. Used for ingestion of anticholinergics, aspirin, acetaminophen and theophylline. If the ingestion was less than 1 hour ago, give charcoal. If not, charcoal may still be appropriate for aspirin, anticholinergics and theophylline.

**PEARLS:** MULTIPLE and/or LATE doses are allowed for medications in which there is delayed transit time, or decreased GI motility (anticholinergics, ASPIRIN and theophylline). For ASA and theophylline, absorption also occurs in the systemic circulation.

**MNEMONIC**: **AAAA = A, A, A**, and an **A**sthma medication = **A**nticholinergics, **A**spirin, **A**cetaminophen, and an **A**sthma med (Theophylline). Also, note that all of these medications come in **TABLET** form.

**PEARL/MNEMONIC/SHORTCUT:** Charcoal is NOT useful for ingestion of MLL = METALS, LIQUIDS and LITHIUM.

- "METALS" = Iron & Lead. Can you imagine charcoal powder mixing well with an iron or a lead pipe? Not really.
- "LIQUIDS" = Ethanol, Organophosphates, Acids, Bases and Cyanide. For the test, assume they "are absorbed, or act, too quickly."

## GASTRIC LAVAGE

Gastric lavage is indicated for pill/solid ingestions that occurred within the 60 minutes of presentation. CONTRAINDICATED in hydrocarbon, acid and base ingestions.

## AMPHETAMINES

Amphetamines may cause HYPERTENSION, EUPHORIA, agitation, arrhythmias, psychosis and seizures. Treatment is supportive.

## COCAINE

Cocaine may cause AGGRESSION, VIOLENCE, EUPHORIA, overconfidence and HTN (treat with nitroprusside). Death is not dose dependent. May try haloperidol for treatment.

## PHENYLCYCLIDINE (PCP)

**Phenylcyclidine, aka PCP, may cause vertical** and/or horizontal nystagmus. It also causes ataxia, MYDRIASIS (or normal), lid edema, muscle rigidity, RHABDOMYOLYSIS, paranoia, agitation, HTN, violence and hallucinations (auditory or visual). Treat with a cooling blanket +/- haloperidol. Do NOT use restraints.

## BARBITURATES (like phenoBARBITal)

Barbiturates, such as Phenobarbitol, may cause RESPIRATORY DEPRESSION, hypotension and psychosis.

## OPIOIDS

Use of opioids may result in respiratory depression, miosis, euphoria, INDIFFERENCE to pain, hypotension, hypothermia, urinary/stool retention and pulmonary edema.

**PEARL/SHORTCUT:** Osmolar gaps (aka osmolal gaps) are present in ALL of the alcohols about to be discussed (ethanol, methanol, ethylene glycol and isopropyl alcohol). Anion gaps are also present in ALL of them EXCEPT FOR ISOPROPYL ALCOHOL!

## ALCOHOL (ETHANOL)

Alcohol (ethanol) may cause ataxia, normal but sluggish pupils, nystagmus, perspiration, flushing, hypokalemia, HYPOGLYCEMIA, POSITIVE ANION GAP, nausea and vomiting. Higher intake results in stupor and possibly a coma. Binge drinking in teens is an INDICATOR OF DEPRESSION, but it is NOT a risk factor for future alcoholism. Treat the patient's hypoglycemia and hypokalemia. Chronic abusers may have withdrawal symptoms of (dilated pupils, tremulousness, tachycardia, HTN, hallucinations, confusion and even seizures).

**PEARLS:** Alcohol is the most common substance of abuse among adolescents now. It is no longer tobacco. Marijuana is the most common ILLICIT substance of abuse in adolescents. Hypoglycemia can result due to inhibition of gluconeogenesis by ethanol.

**MNEMONIC**: Imagine a **drunken** bum passed out at the **GAP®**.

## ETHYLENE GLYCOL INGESTION

Ethylene Glycol is the really bad one. It is found in ANTIFREEZE**.** Urine fluoresces with a Woods lamp and has CALCIUM OXALATE CRYSTALS on microscopy. There is a POSITIVE ANION GAP. Can result in HYPOCALCEMIA. Symptoms include tachycardia, HTN, tachycardia and eventually cardiorespiratory failure from acidosis and hypocalcemia. Will also eventually result in renal failure. Treat with **ethanol, fomepazole or emergent hemodialysis**.

* **PEARL**: Presentation can look just like ETOH so look for clues such as "playing in garage," or a picture of calcium oxalate crystals. Also, the Osmolar Gap (OG) is VERY high (> 20). Normal is < 10 mOsm/kg. OG = Measured – Calculated Osmolality.

* **MNEMONICS:** For a problem this bad, it's okay to give a kid ETHANOL for treatment. When you see the words ethylene glycol, think eth**X**lene gl**X**col to remind you of the calcium oxalate crystals in the urine. Regarding the images, this could get a little confusing, so also imagine an X on a long can of beer to remember that the crystals associated with ethylene glycol are tubular in shape. Regarding the hypocalcemia, assume it's from the loss of the calcium in the calcium oxalate crystals!

*** (DOUBLE TAKE) CALCIUM OXALATE CRYSTALS**

There are two types of calcium oxalate crystals. One is square and looks like it has an X on it. That one very common. Calcium oxalate stones are the most common ones in humans. and the other is elongated and looks like a rod. The long one is the one usually associated with **ethylene glycol poisoning**.

- **IMAGE**: http://alturl.com/mg3xv (This one is related to ethylene glycol and kind of looks like a long can of beer, or needles).
- **IMAGE**: http://meded.ucsd.edu/isp/1994/im-quiz/images/caloxal.jpg
  - **MNEMONIC**: calcium oxalate crystals look like a big X on a crystal.

## METHANOL INGESTION

Methanol toxicity can result from ingestion of **W**indshield washer fluid. Patients present with visual complaints (snowstorm appearance), abdominal pain and a POSITIVE ANION GAP. Treatment is the same as ethylene glycol.

**MNEMONICS:** "M ~ W." Also, imagine how terrible your driving VISIBILITY would be in a SNOWSTORM if someone drank all of the WINDSHIELD washer fluid from your car!

## ISOPROPYL ALCOHOL

Isopropyl alcohol ingestion results in a positive ETHANOL reading on toxicology screens. Alcohol dehydrogenase breaks it down into ACETONE (SO KETONES WILL BE PRESENT). Found in mouthwash and rubbing alcohol. Used as an ETOH substitute for its intoxicating effects. It is mainly a CNS depressant. Breath may have a fruity odor from the acetone.

**PEARL**: **There is NO anion gap**. If ETOH is positive and there is NO anion gap, choose ISOPROPYL ALCOHOL poisoning. If an anion gap is PRESENT, choose ETHANOL, METHANOL or ETHYLENE GLYCOL. This is not intuitive given the presence of ketones, but knowing this could be key.

## MARIJUANA (MJ)

Look for poor interpersonal skills and inattentiveness (poor memory) when Marijuana (MJ) use is suspected. May have mydriasis or gynecomastia. Withdrawal is not common, but may result in insomnia, nystagmus, or tremors. Drug testing can be positive for up to ONE week for a ONE use (actually it's 5 days), TWO weeks for frequent use (actually it's about 10 days), and 1 month for daily use.

**PEARL**: ETOH is the most common "substance of abuse," but MJ is the most common "illicit" drug used. Tobacco has fallen out of favor in recent years.

## NICOTINE/TOBACCO/SMOKING

Nicotine from smoking or chewing tobacco increases alertness, relaxation and memory. Decreased appetite. Children exposed to second-hand smoke are more likely to have reactive airway disease, upper and lower respiratory tract infections (URIs and PNAs), otitis media and pneumonias.

## ACETAMINOPHEN INGESTION

In cases of acetaminophen ingestion, obtain an acetaminophen level at **4 hours** post-ingestion and plot on the nomogram. Treat with N-acetylcysteine (NAC). NAC can be gien by IV or PO. It acts as a glutathione precursor. If the number of tablets is known and the dose exceeds **140 mg/kg**, treat with NAC immediately. Otherwise, you may WAIT until the level returns to plot it on the nomogram before giving the NAC, but it must be given within 8 hours of ingestion. So, if the level will not come back until after 8 hours post ingestion, give NAC now.

**IMAGE**: (NOMOGRAM) http://www.ars-informatica.ca/content/acetaminophen2.gif

**PEARL**: Do not base treatment on liver function testing because they may not be abnormal until 3 days post-ingestion. Charcoal (+/- gastric lavage) may be given immediately if the ingestion occurred within 60 minutes of presentation.

**SHORTCUTS:** Using the 4-hour Rumack-Mathew Nomogram, acetaminophen levels > 150-200 means some liver damage, > 300 means severe damage.

**MNEMONIC**: Peg for 14 = Chocolate Heart for Valentine's Day. Peg for 4 = Car with 4 doors. imagine a troubled 14-year-old boy who is trying to commit suicide on VALENTINE'S DAY. He makes a big WHITE CHOCOLATE HEART in which the main ingredient is ACETAMINOPHEN. He then goes to his CAR and eats the entire WHITE CHOCOLATE HEART over FOUR hours. (WHITE CHOCOLATE HEART = 14 ~ 140 mg/kg. CAR = Nomogram at 4 hours post-ingestion)

## CHOLINERGICS

Cholinergics include ORGANophosphates (pilocarpine and pesticide ingredients), neostigmine and bethana**CHOL**. Do not get cholinergics and anticholinergics confused! Symptoms of exposure include diarrhea, urination, miosis, bronchospasm, emesis, lacrimation, salivation, lethargy, wheezing and bradycardia. Treat with ATROPINE or PRALIDOXIME. Use benzodiazepines for CNS symptoms. Without treatment, symptoms can persist for weeks.

**MNEMONICS:** "COLON"-ergic. A COLON is an ORGAN (ORGANophosphates). Now look at the **DUMBELLS made out of COLON"-ergic ORGANs** including a heart and a "CHOLIN" below to remind you of the major DUMBELLS symptoms associated with COLON-ergics: **D**iarrhea, **U**rination, **M**iosis, **B**ronchospasm, **E**mesis, **L**acrimation, **L**ethargy, **S**alivation.

## ANTICHOLINERGICS

Use of anticholinergics may result in fever, blurred vision, DRY MOUTH/EYES, hyperthermia, psychosis, DECREASED BOWEL SOUNDS/MOTILITY, URINARY RETENTION and tachycardia. These patients also have MYDRIASIS. Medications with anticholinergic effects include diphenhydramine, **ATROPINE**, pralidoxime, TRICYCLIC ACIDS (TCAs) and JIMSON WEED. Treat with CHARCOAL.

**PEARL**: Due to decreased bowel motility, it is OKAY to give charcoal even after 60 minutes for an anticholinergic ingestion! For other medications, though, 60 minutes is the limit.

**MNEMONICS:** HOT as a hare, BLIND as a bat, DRY as a bone, RED as a beet, MAD as a hatter while the BOWEL and BLADDER lose their tone, and the HEART runs alone. In AA**A**$, the 3rd A is for **A**nticholinergics or **A**tropine. For Jimson WEED, imagine Harold and Kumar smoking "WEED" at a White Castle® that has plenty of food, but NO DRINKS (DRY mouth, slow passage of foods in the GI tract because the food is "dry").

## TRICYCLIC ACID (TCA) TOXICITY

Tricyclic Acid (TCA) toxicity can reultt in all of the anticholinergic symptoms. Distinguishing features include tachycardia, labile blood pressure, coma, seizures, ANION GAP ACIDOSIS, a **wide QRS COMPLEX**, QT prolongation and possible ventricular fibrillation. Treat with **sodium BICarbarbonate**. If the patient presents within 1 hour of ingestion, may also try gastric lavage and charcoal.

**MNEMONICS:** Treatment is sodium BICarbonate ≈ BIC ≈ BIKE ≈ bicycle, like a tricycle and triCYCLic acids. Or, imagine a depressed patient who is on a TCA. In order to lift his spirits, you play some loud music with some heavy BASE (bicarb).

**MNEMONICS:** "tri-C-A" = **C**ardiac abnormalities (HR, BP) & arrhythmias, **C**omplex long, **C**onvulsions and **A**cidosis (or **A**rrhythmias).

## SALICYLATES

ASPIRIN (ASA) and OIL OF WINTERGREEN are considered salicylates. Patients' initial tachypnea is a response to a PRIMARY ANION GAP METABOLIC ACIDOSIS ("mudpile**S**"). After that, patients develop a PRIMARY RESPIRATORY ALKALOSIS. The major causes of death are cerebral edema, pulmonary edema, electrolyte disturbances and eventual cardiovascular collapse. Salicylates also decreased gastric motility. A level >100 is bad. Measure level at 2-6 hours. Treat with gastric lavage, charcoal, hydration, bicarbonate to alkalinize the urine and/or hemodialysis. Gastric decontamination is paramount!

**PEARLS:** Consider **MULTIPLE** doses of charcoal (even AFTER 60 minutes) because of the decreased GI motility. Can also try LATE gastric lavage due to decreased GI motility.

**MNEMONIC**: Imagine mixing a purely white powder (ASA) with a purely black powder (charcoal) over and over again until you have a perfect gray color.

## IBUPROFEN OVERDOSE

Ibuprofen may act similar to ASA but consequences are not as severe. May only gie nausea, vomiting, abdominal pain with some nystagmus or headache. If the patient is asymptomatic at 4 hours post-ingestion, DISCHARGE.

## IRON OVERDOSE

Iron overdose results in CAPILLARY LEAK, hypotension, ANION GAP ACIDOSIS (mudp**i**les) AND possible cardiovascular collapse. There may also be associated hematemesis, scarring of the gastric pylorus and hepatic toxicity. On presentation, obtain ABDOMINAL X-Rays to look for undigested pills. There are 4-5 stages of toxicity (GI, Latent, Metabolic/Cardiovascular, Hepatic and then Delayed). Obtaining serum iron levels at about 3-5 hours post-ingestion helps determine the risk of systemic toxicity. Treatment requires CHELATION with DEFEROXAMINE. Do NOT pick charcoal because it does NOT bind well to iron. Chelation will result in pink urine, often referred to as "Vin Rose Urine." Continue giving deferoxamine until the urine is no longer pink.

**MNEMONICS:** DE-FERO. The word says it all. To DE-iron, or DE-FERROus the body. Again, do not choose charcoal. Imagine pouring powdered charcoal onto a shiny iron. It does NOT stick, but it does make a mess on your ironing board.

**MNEMONIC**: The mnemonic below summarizes the stages of iron overdose.

* I = Indigestion (nausea, vomiting, abdominal pain and possible hematemesis)
* R = Recovery
* ♥ = Cardiovascular collapse and liver damage
* N = Narrowing of pylorus or other parts of GI tract and causing obstruction from strictures/scars

## (DOUBLE TAKE) LEAD TOXICITY

Lead toxicity often results because kids love its sweet taste. Sources of lead toxicity are paint in old homes and dust from home renovations. Because lead blocks iron from being incorporation into heme, lead toxicity can look very much like iron-deficiency anemia (MICROcytic). To differentiate this from iron deficiency, look for BASOPHILIC STIPPLING of RBC's, constipation, neurologic issues and LEAD LINES (obtain imaging of long bones). Check a lead level at 12 and 24 months of age. Even mild elevations (level of 10-20) can cause neurologic problems with learning and behavior. For mild elevations, make changes in the home and educate the family. For levels > 45, CHELATION therapy is needed for lead poisoning with **edetate disodium calcium (EDTA)**, dimercaprol or d-penicillamine. Always confirm an elevated capillary sample with a VENOUS sample because these are more reflective of RECENT exposure.

**PEARLS:** Iron levels should be normal and TIBC should be normal or low.

**IMAGE**: (Lead Lines) - http://www.alsehha.gov.sa/dpcc/images/lead_lines.gif

**IMAGE**: (Lead Lines) - http://en.wikipedia.org/wiki/File:Lead_PoisoningRadio.jpg

**IMAGE**: (Basophilic Stippling) - http://bit.ly/oeWPKR

**MNEMONIC**: The treatment for lead is "LEADetate."

## CLONIDINE & PHENOTHIAZINES OVERDOSE

Overdose from clonidine and phenothiazines typically results in hypotension, but may cause transient, or rebound hypertension. Both cause miosis. Clonidine can also produce some anticholinergic effects, cogwheel rigidity and dystonic reactions

* **(DOUBLE TAKE) DYSTONIC REACTIONS**: Dystonic reactions involve sudden muscle contractions and usually some **odd postures**. Can be related to clonidine, phenothiazine, metoclopramide, promethazine and athetoid cerebral palsy. Treat **medication-induced** dystonic reactions with **diphenhydramine**.

## CALCIUM CHANNEL BLOCKER) OVERDOSE

Calcium channel blocker overdose results in hypotension. Treat with CALCIUM to "unblock" the blocker effect!

## DIGOXIN TOXICITY

Digoxin toxicity leads to anorexia, weakness, n/v, poor appetite, **V-tach**. Increased chance of dig toxicity if LOW K+, Mg, poor renal function or hypoxia. Treat with BB, Digibind if life threatening correction of electrolytes.

## THEOPHYLLINE

At theophylline levels > 20, nausea, vomiting, hypOtension, HYPERcalcemia, HYPOkalemia, metabolic Acidosis and seizures will occur. Treat with CHARCOAL until the level normalizes and give beta-blockers for arrhythmias (if the blood pressure is not too low). Use benzodiazepines for seizures. May also choose to do emergent hemodialysis or whole bowel irrigation if levels continue to increased.

**PEARL**: Multiple and late doses of charcoal are allowed because of increased GI transit time and charcoals systemic absorption of theophylline.

## CARBON MONOXIDE (CO)

Carbon monoxide (CO) can cause ERYTHEMATOUS/FLUSHED skin, flu-like symptoms, ATAXIA, confusion, lethargy, somnolence, GI symptoms and headache. Can be associated with burns/fires. The CO reversibly binds to hemoglobin creating carboxyhemoglobin (obtain a level). Treat with 100% O2 if mild. If the carboxyhemoglobin level is > 25%, the patient has GI symptoms or the patient has neurologic symptoms, then treat with 100% O2 in a hyperbaric chamber. If answer choices include "100% O2,"

**PEARLS**: Pulse oximetry may read as NORMAL (don't let them trick you!). May present as a drunken-appearing family, or a family with a prolonged viral syndrome. Presentation is similar to cyanide poisoning. If the patient is not getting better on 100% O2, go ahead and treat for cyanide poisoning with Sodium Thiosulfate or Nitrate.

**MNEMONIC**: If you've ever had a CO monitor, it was probably located where it is HOT AS HECK! For example, in the basement laundry room where your skin is already HOT, RED & FLUSHED!

## METHEMOGLOBINEMIA

Methemoglobinemia is an enzyme deficiency (NADH methemoglobin reductase) that results in increased methemoglobin levels (thus reducing O2 carrying capacity). Symptoms include BLUE or CYANOTIC SKIN without evidence of respiratory distress, hypotension/shock and tachycardia. Blood becomes CHOCOLATE COLORED (less oxygen). Certain exposures increase the methemoglobinemia (associated with sulfa drugs and **well water** being used with formula). Treat with methylene blue, oxygen and removal of the offending agent.

**PEARL**: Pulse oximetry is NORMAL.

**MNEMONIC**: "**METH** HEMOGLOBINEMIA" is treated with **METH**ylene **BLUE** for the **METH**-emoglobin kid that looks **BLUE** but has a **NORMAL PULSE OXIMETRY**. Or, imagine a PALE, BLUE teen that abuses METH.

## HYDROCARBON INGESTION

Hydrocarbon toxicity can result from ingestion of furniture polish/cleaner, gasoline, kerosene and lighter fluid. Some of these hydrocarbons fall under the term "MINERAL SPIRITS." Symptoms of hydrocarbon ingestion may result in cough, tachypnea, nausea and vomiting. Can eventually lead to hypoxia and ARDS. Symptoms occur quickly, so if there are no respiratory symptoms and a CXR is normal at 6 hours post-ingestion, patients may be discharged from the ER. Otherwise, admit for supportive therapy (which could include intubation.)

## HYDROCARBONS INHALATION

Hydrocarbon **inhalation**, often from inhalant abuse, can result in symptoms including ataxia, weakness (can get rhabdomyolysis), nystagmus, possible hallucinations and slurred speech. May cause respiratory acidosis OR alkalosis. Can appear drunk, and often feel hung-over afterwards. Chronic use can result in GI symptoms and encephalopathy. Can lead to death from **arrhythmias**, aspiration or asphyxia.

## ACID OR BASE INGESTION

NEVER do a gastric lavage for an acid or base ingestion. These are CAUSTIC/CORROSIVE substances and lavage can lead to emesis and subsequent aspiration. Symptoms may include drooling, chest pain, dysphagia, odynophagia, oral ulcers or tachycardia. If an acid or base ingestion is suspected, you MUST ORDER AN UPPER ENDOSCOPY (aka EGD) to look for esophageal burns.

**PEARL**: ACIDS are bitter and may leave esophageal **&** stomach burns (not neutralized).

**PEARL**: BASES are tasteless (e.g., baking soda) and cause "liquefaction necrosis" and injury of the esophagus (neutralized in the stomach). Drain cleaners often contain sodium hydroxide (quick, often associated with oral lesions and may cause airway obstruction due to edema).

## FOREIGN BODY INGESTION

Coins are most common in cases of foreign body ingestion.

* **PENNY: REMOVE IT (old pennies contained corrosive zinc)**

* Other coin in **PROXIMAL esophagus**: REMOVE IT.

* Other coin in **DISTAL ESOPHAGUS**: LEAVE IT and reimage 24 hours later. If it is still has **NOT REACHED THE STOMACH AFTER 24 HOURS**, REMOVE IT. If it is in the stomach, leave it because it will usually pass within about 5 days.

* **BUTTON BATTERY** = REMOVE IT!

## (DOUBLE TAKE) RABIES VIRUS

Rabies is caused by a VIRAL infection in which the virus is transmitted through bites, scratches and contact with mucous membranes of infected animals, such as BATS, dogs, foxes, **RACCOONS** (most common in US), skunks, and WOODCHUCKS. If the history suggests a possible exposure (wild/aggressive animal), treat with standard wound care **PLUS** Human Rabies Immunoglobulin (HRIG) **PLUS** the 5 vaccine doses. If the animal is a pet, observation of the patient and animal is allowed without giving HRIG.

**PEARL**: Rabbits, rats and squirrels (rodents) are **NOT** associated with rabies.

**PEARL: For the boards, if the word "BAT" is mentioned (alive, escaped, whatever!), treat as an exposure!**

## BROWN RECLUSE SPIDER

The brown recluse is a small spider. The bite is usually painless. Symptom onset is usually within hours, but most bites are minor. Some can result in a hemorrhagic blister, a necrotic ulcer or a TARGET LESION. The spider does carry a potentially deadly toxin, but systemic symptoms are rare. Discharge from ER if there are no systemic symptoms. For minor bites, treatment is supportive.

**MNEMONICS: RECLU**se. **RECLU** spelled backwards spells **ULCER**! See below for a visual mnemonic to remind you of the potential target lesion.

## BLACK WIDOW

If a black widow spider bite is suspected, look for a single, small **puncture wound**. Quick onset of symptoms (by 6-8 hours), which may include HTN and generalized muscular pain (abdomen, back, chest). Treat with antihypertensives and pain medications.

## COMMON BITES

Common bite marks can be differentiated by shape and depth of a wound. A half-moon or semicircular lesion is likely a HUMAN bite, a tear-lesion is likely from a DOG, and PUNCTURE wounds are likely from a CAT. Cat bites carry the worst prognosis due to the deepest penetration. Amoxicillin-Clavulanate provides adequate coverage for staph and pasteurella.

## BURN TREATMENT

Antibiotics are NOT mandatory in the treatment of burns. Tetanus booster is a consideration.

* SUPERFICIAL: Red, painful skin. Treat with soap, cool water and analgesics.

* PARTIAL THICKNESS: Look for blisters and erythematous skin that blanches with pressure. These burns are also PAINFUL. Once the blisters rupture (do not actively rupture them), treat with debridement, topical antibiotics and a nonadherent dressing.

* FULL THICKNESS: Look for white, dry, leathery skin that is PAINLESS. These burns require immediate referral to a burn center and will likely require surgical excision.

* DRESSINGS: If asked about dressings, use cool, wet dressings for small burns to avoid hypothermia from fluid losses. This can also help to prevent hypovolemic shock.

* BURN CENTER CARE: Refer for full thickness burns and any burns on more than 10% of the body (by surface area), and any burns in sensitive areas (face, mouth, perineum, hands and feet). For burns on more than 15% of the body, add an EXTRA 4 ml/kg to maintenance fluids.
* ELECTRICAL BURNS: Can cause deep tissue and muscle injury that is not obvious to the eye. Electrical burns are frequently in the perioral area (electrical cord bite). Provide anticipatory guidance prior to the child becoming mobile (around or prior to the 6 month visit).
* **PEARLS:** Almost 20% of pediatric burns are from abuse. Classic patterns include well-demarcated burns and stocking or glove patterned burns. Sparing of the folds and splash pattern burns are NOT necessarily from abuse (may be unintentional).
* **MNEMONIC**: Use the "rule of 9's" to help estimate the body surface area affected. An arm is approximately 9% of a child's BSA. A leg is approximately 18% of a child's BSA. A palm is approximately 1% of a child's BSA. These BSA percentage of the head changes significantly with age. The head is approximately 19% of a newborn's BSA and only 9% of a teenager's. For more specific calculation, ER's use a Lund & Browder chart (low yield fact): http://archive.student.bmj.com/issues/06/09/education/images/view_19.jpg

## NEAR DROWNING

**Near drownings usually require INPATIENT monitoring because they can result in ARDS after an asymptomatic period. This can lead to DEATH from hypoxic ischemic brain injury and cerebral edema.**

**PEARLS:**

* GOOD PROGNOSIS: Good pulses on EMS arrival, or s/he required < 10 minutes of CPR. If the patient was in the water for less than 60 seconds, had NO loss of consciousness and did not require CPR, then patient does NOT need to be hospitalized.
* POOR PROGNOSIS: Cold on EMS arrival (< 90° F), CPR needed for > 10 min, > 25 minutes under water, apnea, coma or pH < 7.1. Rewarm these patients to < 90° F and consider intubation to provide PEEP for possible ARDS.

## POOL SAFETY

Recommend a four-sided fence around pool with a locked gate for greatest pool safety. Most drowning victims are preschoolers and 16-18 year olds (end of high school). Toddler swimming lessons have NOT been shown to prevent drownings. In-ground pools result in more drownings than aboveground pools. Boys are more likely to drown than girls. African Americans are more likely to drown than Caucasians. Infants drown in tubs, preschoolers drown in pools, and teens most often drown in fresh water. The most common cause of death in epileptic patients is DROWNING.

## HYPOTHERMIA

Hypothermia is defined as a core body temperature less than 95° F. If temperature is < 90° F, provide core rewarming with bladder irrigation, warm and humidified air and warm IV fluids. Once areas of the body are warmed, blood can pool in those areas resulting in HYPOTENSION. Therefore rewarming should be done carefully/slowly. Passive rewarming includes the use of a warm blanket and warm baths.

## HEAD INJURY

Head injury can result in a basilar skull fracture or papilledema. Try to remember the GCS scale.

* **Basilar skull fracture = "clear rhinorrhea or otorrhea." Bruise over mastoid or rcoon eyes**. Could have a facial nerve palsy. No abx. Good prognosis. **Confirm with a CT** (not XR)

* Nasal septum or ear's pinna with hematoma/swelling = ENT CONSULT to avoid saddle nose or cauliflower

* Papilledema is a LATE finding of increased ICP and may take weeks or months to develop

* **If GCS < or = 8, intubate!** 8 = coma. **GCS is ONLY for head trauma**... it is not for METABOLIC dz

* GCS = Eyes (4), Verbal (5), Motor (6). Lowest = 3 = not opening eyes, no sounds, not moving. 15 = opens eyes spontaneously, oriented and talks appropriately, obeys motor commands. 8 might = opens eyes to painful stim (2), makes inappropriate sounds (3), abnormal flexion to painful stim (3). Extension would be GCS 2 for motor = bad. Longer time in a coma = worse prognosis.

## POST CONCUSSION TREATMENT

Post concussion treatment varies depending on whether there was loss of consciousness and/or amnesia:

* GRADE 1: NO LOC + NO amnesia **+ Confusion**. Reevaluate every 5 minutes for neurologic changes. May return to sports if mentation is clear and there are no symptoms for 20 minutes.

* GRADE 2: NO LOC **+ AMNESIA + Confusion**. Patient should have medical follow-up at 24 hours. May return to sports once the patient is asymptomatic for 1 week.

* GRADE 3: **LOC + AMNESIA + Confusion**. Take these patients to the ER for further evaluation and probable imaging. May return to sports once the patient is asymptomatic for 2 weeks.

- PROGNOSIS: Worse if concussion was associated with amnesia, prolonged confusion, prolonged recovery or repeated trauma.

* IMAGING: Required if loss of consciousness (LOC) > 1 minute, or if there are still neurologic symptoms at presentation in the ER.

* **MNEMONIC**: Instead of "LOC," think "LAC" for the 3 differentiating factors (**L**OC, **A**mnesia and **C**onfusion). Grades 1 through 3 add one finding per grade, and they happen to go in the reverse order of C-A-L (Gr 1 = Confusion, Gr 2 = Confusion and Amnesia, Gr 3 = Confusion, Amnesia and LOC). Regarding timings for return to play, they are 0 weeks, 1 week and then 2 weeks of symptom-free time.

## ENDOTRACHEAL TUBES & VENTILATION

* PEDIATRIC ENDOTRACHEAL TUBE SIZE (aka ET TUBE or ETT SIZE): In general, keep (Age/4) + 4 in mind. It's doubtful that you will be asked about cuffed ET tubes, but if so, CUFFED size = (Age/4) + 3. Babies' ETT size is based on weight. 2.5 mm if < 1.5 kg, 3.0 mm if < 2.5 kg and 3.5 mm if > 2.5 kg.

* VENTILATION: Tidal Volume (aka TV) = 7 cc/kg

* MEDICATIONS: Naloxone (for acute opioid exposures), Atropine (for bradycardia), Valium, Epinephrine and Lidocaine may be given via the ETT.

- **MNEMONIC**: The word "NAVEL" should help you remember the medications that can be given through the ETT. If that doesn't work, imagine an ETT that goes so deep that it almost pokes out of an intubated baby's NAVEL!

* CRASHING PATIENT: If you are presented with a rapidly deteriorating patient that is intubated patient, consider EXTUBATION followed by bag & valve mask ventilation as the next step. If the patient's condition is deteriorating more gradually, evaluate for displacement of the ETT, obstruction of the ETT, pneumothorax and equipment failure.

- **MNEMONIC**: DOPE = Displacement, Obstruction, Pneumothorax and Equipment failure. For rapidly deteriorating patients, extubation and bag & valve mask ventilation addresses D, O, and E.

## IMPAIRED PERFUSION/HYPOVOLEMIA

Since poor capillary refill is an early finding in shock, pulse oximetry is unreliable in cases of impaired perfusion, or Hypovolemia. Hypotension is the LATE finding. Try to get IV access for 90 sec or 3 tries, but then MOVE ON to an intraosseous line (IO).

## CARDIOPULMONARY RESUSCITATION (CPR)

Cardiopulmonary Resuscitation (CPR) is a low-yield topic because guidelines are always changing.

* SINGLE RESCUER CPR FOR BABIES: Provide compressions and breaths at a ratio of 30:2 to minimize transition times. Also, COMPRESSIONS are more important than breaths.

* DOUBLE RESCUER CPR FOR BABIES: Provide compressions and breaths at a ratio of 15:1 (15 compressions for every breath).

* ADOLESCENTS: 30:2 regardless of the number of rescuers.

* **PEARL**: Guidelines have changed, but the key is to remember that it's becoming more and more important to focus on high quality chest compressions to get the blood flowing rather than focusing on breaths.

# VITAMIN & NUTRITIONAL DISORDERS

**PEARL**: For the vitamins and nutritional disorders sections, familiarize yourself with the proper names of vitamins, but know that the ABP has yet to institute their strict use. In general, they still tend to use "Vitamin E" instead of "Tocopherol." Keep in mind that you are not dealing with one test-question writer, so there is room for variability. In this chapter, the terms likely to be used will be presented, followed by the less likely associated term in parenthesis.

## *FAT-SOLUBLE VITAMINS*

### FAT-SOLUBLE VITAMINS

Fat-soluble vitamins include Vitamin A, Vitamin K, Vitamin E and Vitamin D.

**PEARL**: Keep deficiencies of these particular vitamins in mind when dealing any disease that is associated with malabsorption.

**MNEMONIC**: Used the word **FAKED** to remind you of the **F**at-soluble vitamins of **A**, **K**, **E** and **D**.

### VITAMIN A (aka RETINOL)

* **MNEMONIC**: Vitamin **A** = Retinol. Have you ever heard of Retin-**A**®?
* DEFICIENCY: Deficiency can lead to bitot spots on the eye and blindness.
* EXCESS: Can lead to PSEUDOTUMOR CEREBRI

- **(DOUBLE TAKE) PSEUDOTUMOR CEREBRI (aka IDIOPATHIC INTRACRANIAL HYPERTENSION or BENIGN INTRACRANIAL HYPERTENSION)**

The symptoms associated with **increased intracranial pressure** (ICP) predominate for Pseudotumor Cerebri (aka Idiopathic Intracranial Hypertension or Benign Intracranial Hypertension). These include headache, nausea, vision changes and bulging fontanelle. Late findings are papilledema and loss of vision. The disease is **not** benign, and although CT or MRI of the head will not show herniation, a lumbar puncture will show increased intracranial pressure. It can be associated with **excessive vitamin A**, isotretinoin, tetracycline and thyroxine. Treatment may require steroids and/or shunt.

- **PEARL**: PSEUDO-tumor. Don't get confused by the word "tumor." This condition is NOT associated with a mass.
- **MNEMONIC**: (image)

### VITAMIN K DEFICIENCY (aka PHYTONADIONE DEFICIENCY)

Vitamin K deficiency (Phytonadione deficiency) can result in HEMORRHAGIC DISEASE OF THE NEWBORN due to a coagulopathy. Vitamin K is needed for use with coagulation factors 2, 7, 9, & 10. This is especially a concern in babies because there is a lack of normal gut flora needed to make the Vitamin K. Deficiency is also more common in breastfed babies that did not receive Vitamin K at birth (watch out for a home birth). Look for a bleeding circumcision site or umbilical cord site.

## EARLY VITAMIN K DEFICIENCY

Early Vitamin K deficiency occurs within 3 days of birth.

## LATE VITAMIN K DEFICIENCY

Late Vitamin K deficiency can occur any time between 3 days and 3 months! Look for a breastfed baby with on antibiotics and diarrhea (decreased gut flora to make Vitamin K due to antibiotic use, and decreased absorption due to diarrhea).

* TREATMENT: For bleeding patients, give Vitamin K + FFP (Fresh Frozen Plasma contains plenty of functional Vitamin K dependent coagulation factors).

* **PEARL**: Oral Vitamin K at birth does NOT prevent HEMORRHAGIC DISEASE OF THE NEWBORN.

## VITAMIN E DEFICIENCY (aka TOCOPHEROL DEFICIENCY)

Vitamin E deficiency (Tocopherol deficiency),can result in hemolytic anemia, impaired reflexes, jaundice, neurologic problems, peripheral edema or thrombocytosis.

* **PEARL**: When seen, it is almost always seen in association with a patient getting TPN.

* **MNEMONICS:**

- TocophErol
- Vitamin EE deficiency leads to a knEE reflex deficiency

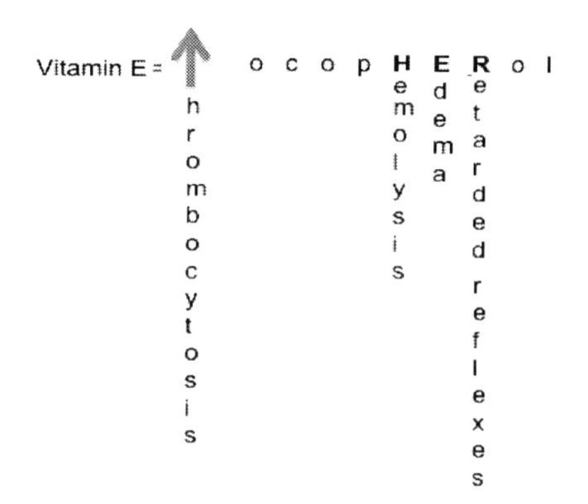


## VITAMIN D (ERGOCALCIFEROL, CHOLECALCIFEROL) EXCESS

An excess of Vitamin D (aka ergocalciferol or cholecalciferol) results in a hypercalcemia-induced diuresis. Look for hypERcalcemia, hypERphophatemia, polyuria and polydipsia. Patients may also have nausea, vomiting, nephrolithiasis and an elevated BUN and/or creatinine. Treat with IV fluids and furosemide (helps remove the excess calcium).

## VITAMIN D DEFICIENCY (Related terms: ERGOCALCIFEROL, CHOLECALCIFEROL)

A deficiency of Vitamin D (aka ergocalciferol or cholecalciferol) can result in RICKETS.

**PEARL**: When concerned about Vitamin D Deficiency, MEASURE 25-Vitamin D, but when treating, be sure to do so with oral Vitamin D 1,25 (active form).

**MNEMONIC**: Vitamin D gets absorbed in the gut and goes to the LIVER (one organ). It then gets activated in the KIDNEYS (two organs). One organ, one number (25). Two organs, two numbers (1,25).

## (DOUBLE TAKE) RICKETS

Findings of Rickets may include widening of wrist and ankle physes (growth plates), bowed legs, pain, decreased growth rate, anorexia, enlarged costochondral junctions (rachitic rosary), pigeon chest, delayed suture/fontanelle closure, frontal bossing (thick skull), or bad tooth enamel. There is no singular lab pattern for Rickets. It can be due to Vitamin D deficiency secondary to one of multiple different disorders and you **MUST** learn the lab patterns associated with the different disorders:

*** NORMAL (or LOW) CALCIUM + LOW PHOSPHORUS**

**This pattern represents FAMILIAL HYPOPHOSPHATEMIC RICKETS (aka "VITAMIN D RESISTANT RICKETS"). It is an X-linked DOMINANT renal disorder**. There is a defect of phosphate reabsorption in the proximal tubule AND a defect of the kidney to convert 25-Vitamin D to 1,25 Vitamin D → Treat with oral phosphate supplementation and avoid hypOcalcemia by giving the active/oral form of Vitamin D (1,25). Labs = Normal or low calcium, LOW serum phosphorus, **HIGH ALKALINE PHOSPHATASE, normal Vitamin D 25,** PTH IS NORMAL since calcium is usually normal.

- **PEARL**: For the exam, they will probably keep it simple and avoid giving you a low calcium level

*** NORMAL CALCIUM + LOW PHOSPHORUS**

This represents **INITIAL** VITAMIN D DEPLETION. Low Vitamin D results in low phosphorus reabsorption. There is a compensatory increased PTH that temporarily normalizes calcium.

- **PEARL**: For the test, they probably want you to focus on FAMILIAL Hypophosphatemic Rickets. The differentiating lab would be low Vitamin D level (25) in early Vitamin D depletion, versus normal in Familial Hypophosphatemic Rickets.

*** LOW CALCIUM + LOW PHOSPHORUS**

This represents SEVERE VITAMIN D DEFICIENCY resulting in poor absorption of calcium and phosphorus from the gut. PTH should be high.

*** LOW CALCIUM + NORMAL PHOSPHORUS**

This represents the initial Vitamin D repletion stage of healing Vitamin D Deficiency Rickets

*** LOW CALCIUM + HIGH PHOSPHORUS**

This represents hypoparathyroidism, phosphorus overload, or pseudohypoparathyroidism. All result in perceived (or real) low PTH in the body

*** NORMAL CALCIUM + HIGH PHOSPHORUS**

This represents renal disease, growth hormone excess, or a high phosphorus diet.

## (DOUBLE TAKE) RICKETS OF PREMATURITY

Premature babies have a high Vitamin D, Calcium and phosphorus requirement. Failure to administer these may result in Rickets of Prematurity. Treat with Vitamin D + Calcium + Phosphorus. Have to give all three (Vitamin D alone is insufficient).

## (DOUBLE TAKE) LIVER DYSFUNCTION

Liver dysfunction can result in decreased bile salts in the gut → Vitamin D absorption issues → Vitamin D deficiency.

# ***WATER-SOLUBLE NUTRIENTS***

**MNEMONIC:** The following inappropriate (aka gutter) humor should help you remember the names of the water-soluble B vitamins:

* The Really Nasty Prince Farted Constantly!
* Thiamine (B1), Riboflavin (B2), Niacin (B3), Pyridoxine (B6) and Cyanocobalamin (B12).

## THIAMINE (B1) DEFICIENCY

Thiamine (Vitamin B1) deficiency can result in WET BERIBERI (a type of high output cardiac failure due to peripheral vasodilation). It can also result in various neurologic symptoms, including ataxia, delirium/encephalopathy, neuropathy and nystagmus (called WERNICKE ENCEPHALOPATHY). Wernicke's can eventually lead to Korsakoff Syndrome.

## RIBOFLAVIN (B2) DEFICIENCY

Riboflavin (Vitamin B2) deficiency can result in anemia, angular stomatitis, cheilosis, glossitis ("tongue = riboFLAVor") and seborrheic dermatitis.

**PEARL**: Phototherapy can result in decreased riboflavin (B1) levels. Keep this in mind with premature hyperbilirubinemia children.

**MNEMONIC**: Imagine this drawing represents a premature baby with protective glasses on (for phototherapy). The left lens have "2" written on them to remind you of the B2 deficiency. There is a rash at the scalp (seborrheic dermatitis), and there are steep angles at the edges of the lips to remind you of angular stomatitis. He's bleeding from the angles of his mouth to help you remember of the possible anemia.

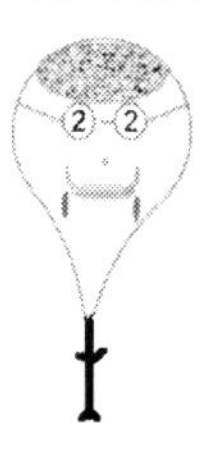


## NIACIN (B3) DEFICIENCY

Niacin (Vitamin B3) deficiency results in PELLAGRA, which is identified by a constellation of findings; including diarrhea, dermatitis and dementia. The condition is often found in communities in which corn is a cornerstone of their diet.

**IMAGES**: http://mizzouderm.com/genetic--metabolic.html

**MNEMONICS**: **NIA**cin deficiency results in pellagra. From now on, refer to pellagra as peNIAgra to help you remember it is associated with a NIAcin deficiency. Also, remember that a corNIAgra diet often leads to this. The "**P**iarrhea, **P**ermatitis and **P**ementia" of **D**ellagra are often called the 3 D's of Pellagra.

**(NAME ALERT) HARTNUP DISEASE (AKA PELLAGRA-LIKE DERMATOSIS)**: This is NOT a vitamin deficiency. It is a disorder of neutral amino acid transport and absorption, particularly tryptophan. Patients present with what looks like the 3 D's of **P**ellagra (diarrhea, dementia, dermatitis).

## PYRIDOXINE (B6) DEFICIENCY

Pyridoxine (Vitamin B6) deficiency can result in neuropathy, rashes, seizures and tongue swelling. Isoniazid (INH) can lead to pyridoxine deficiency. This is why your medical school probably taught you that patients on isoniazid therapy for tuberculosis should always be placed on pyridoxine supplements.

**PEARL**: Know that pyridoxine is the same thing as B6 (higher yield than others).

## (DOUBLE TAKE) FOLATE (B9) DEFICIENCY

Folate (Vitamin B9) deficiency results in a **macrocytic** anemia and macroglossia. In pregnant women, deficiency can lead to neural tube defects (such as spina bifida) in the unborn fetus.

**PEARLS:** Look for a diet poor in veggies, or a history of **goat milk** ingestion.

**PEARL**: Both B12 and Folate deficiency can lead to hypersegmented neutrophils and neural tube defects. Both can be masked by treatment with the other. Use the symptoms, diet history and blood testing to get the diagnosis right the first time!

**MNEMONIC**: FOLATE deficiency form FOLIAGE (veggies) deficiency!

## (DOUBLE TAKE) B12 DEFICIENCY (aka CYANOCOBALAMIN DEFICIENCY)

Cyanocobalamin (Vitamin B12) deficiency can present as a macrocytic anemia in a patient with neurologic issues (paresthesias, ataxia and decreased reflexes). B12 deficiency can also result in neural tube defects (such as spina bifida) in the unborn fetus if a pregnant woman is deficient. Common causes of B12 deficiency include diet (VEGANS) and an inability to absorb B12. Look for INTRINSIC FACTOR DEFICIENCY (Pernicious Anemia), or an inability to absorb B12 due to gastric surgery (diagnosed with a Schilling test). B12 is absorbed in the terminal ileum, which means it may be lacking patients with Short Bowel Syndrome, bacterial overgrowth or Inflammatory Bowel Disease (IBD). Treat with IM B12 injections.

**PEARLS:** B12 will likely be used on the exam, but KNOW that B12 = CYANOCOBALAMIN. In a B12 deficient patient, if a macrocytic anemia is noted and mistakenly treated with folate supplements, the B12 deficiency will be MASKED as the megaloblastic anemia improves. This can eventually lead to **irreversible** neurologic damage.

## VITAMIN C DEFICIENCY & EXCESS

* **VITAMIN C DEFICIENCY**: Can lead to SCURVY and OSTEOPENIA.

- **SCURVY**: Look for poor wound healing, skin spots, spongy/ bleeding gums or bone pain.
- **OSTEOPENIA**: In infants, osteopenia related to Vitamin C deficiency can lead to brittle bones and fractures.
- **PEARL**: If you encounter a patient with bone issues and bleeding gums, look for Vitamin C deficiency as the etiology.

* **VITAMIN C EXCESS:** May result in nephrolithiasis. In G6PD patients, may result in hemolysis.

## (DOUBLE TAKE) ZINC DEFICIENCY

Breastfeeding helps with zinc absorption. If a child is begins having medical problems once weaned from breast milk, consider zinc deficiency in your differential. Zinc deficiency causes a **SCALY and EXTREMELY ERYTHEMATOUS** dermatitis in the perioral and perianal area (**around the natural orifices**) that can DESQUAMATE. The rash is sometimes described as erosive and eczematous. Can also be associated with ALOPECIA and poor taste.

* **MNEMONIC**: Poor taste, huh? Have you ever had Zinc lozenges? They are disgusting! It's probably a good thing that you have hypogusia when you are eating Zinc lozenges!

* **IMAGES**: http://www.dermaamin.com/site/atlas-of-dermatology/25-z/505-zinc-deficiency-.html

* **IMAGE**: http://bit.ly/Y1IxJf

* **PEARLS**:

- CROHNS DISEASE: If a Crohns patient is suffering from diarrhea, they may have zinc deficiency since Zn is lost in the stool.
- STRICT VEGETARIANS & VEGANS: Susceptible to multiple nutritional deficiencies, including deficiencies in IRON, ZINC, CALCIUM and VITAMIN B12. Vegans avoid all animal derived products (including milk and eggs). B12 deficiency can result in megaloblastic anemia, vitiligo, peripheral neuropathy and even regression of milestones
    - **MNEMONIC**: Did you know giraffes are vegetarian? Imagine a giraffe standing in Time Square reaching its long neck into the sunroof of a FUZZY CAB that has green, grass-like seats and fuzzy floor mats. FUZZY CAB = FeZi CaB12!

## (DOUBLE TAKE) ACRODERMATITIS ENTEROPATHICA

Acrodermatitis Enteropathica is an inherited condition (autosomal recessive) in which there is a Zinc transport defect. Can result in **alopecia**, diarrhea, failure to thrive (FTT) and the **rash** of zinc deficiency.

**IMAGES**: http://emedicine.medscape.com/article/912075-media

## (DOUBLE TAKE) BIOTIN/BIOTINIDASE DEFICIENCY

Biotin, or Biotinidase, deficiency may present with a RASH + ALOPECIA + **NEUROLOGIC SIGNS** (ataxia, coma, etc.). Patients may also have lactic acidosis. Treat with Biotin.

**MNEMONIC**: Imagine the TIN MAN from the Wiz of Oz walking with an ATAXIC gait as he SCRATCHES his arm. Notice that he has NO HAIR!

## COPPER DEFICIENCY

Copper deficiency is rare. It can result in fatigue, neuropathies and kinky hair.

## (DOUBLE TAKE) STRICT VEGETARIANS & VEGANS

Strict vegetarians and vegans may be susceptible to multiple nutritional deficiencies, including deficiencies in IRON, ZINC, CALCIUM and VITAMIN B12. Vegans avoid all animal derived products (including milk and eggs). B12 deficiency can result in megaloblastic anemia, vitiligo, peripheral neuropathy and even regression of milestones.

**MNEMONIC**: Did you know giraffes are vegetarian? Imagine a giraffe standing in Time Square reaching its long neck into the sunroof of a FUZZY CAB that has green, grass-like seats and fuzzy floor mats. FUZZY CAB = FeZi CaB12!

## ***NUTRITIONAL DEFICIENCIES***

### KWASHIORKOR

Kwashiorkor is a PROTEIN DEFICIENCY. These children have pitting edema, large abdomens, hepatomegaly and can have an associated rash.

**IMAGE**: http://www.cs.stedwards.edu/chem/Chemistry/CHEM43/CHEM43/Leukotr/Kwashiorkor.GIF

**MNEMONICS:** These kids are sometimes referred to as "sugar babies" because of a carbohydrate heavy (rice) diet. Also, "kwashi" sounds like "cushy," which should help remind you that these babies have cushy, spongy skin with pitting edema.

### MARASMUS

Marasmus is deficiency of all nutrients. These children do NOT have edema. They are thin, anorexic-looking children with severe muscle wasting. Looks like a thin, old man.

**IMAGE**: http://upload.wikimedia.org/wikipedia/commons/7/71/Starved_child.jpg

**MNEMONIC**: Ever notice how skinny Marvin the "MARSHIAN" is?

### (DOUBLE TAKE) ESSENTIAL FATTY ACID DEFICIENCIES

Essential fatty acids include **LINOLEIC ACID** and alpha-linolenic acid. Deficiency results in alopecia, a scaly dermatitis and **thrombocytopenia**. Treat with IV lipids.

**MNEMONIC**: Imagine a fish whose red SCALES are shaped like HAIRY PLATELETS. As the fish struggles to find food, it becomes SKINNIER and skinnier (malnourished) and the hairy platelets begin to fall off. What's left is a SKINNY (fat free), BALD and THROMBOCYTOPENIC fish!

# GASTROENTEROLOGY

## *LIVER DISEASE*

### CONGENITAL HEPATIC FIBROSIS

Congenital hepatic fibrosis is associated with POLYCYSTIC KIDNEY DISEASE and can lead to varices and portal hypertension.

### HEPATOMEGALY

Hepatomegaly is defined as a palpable liver > 1 cm below the costal margin, or a liver that crosses the midline. A palpable liver edge is normal, and is especially notable in a newborn.

### GALLBLADDER HYDROPS

Gallbladder hydrops refers to RUQ pain from acute swelling or distension of the gallbladder in the absence of any gallstones. May be associated with FASTING, HENOCH-SCHONLEIN PURPURA, KAWASAKI SYNDROME, STREP PHARYNGITIS and TPN., Sepsis, Obesity, hemolysis

**MNEMONIC**: Imagine a Las Vegas "STREPPer" who has breasts that look like giant GREEN WATER BALLOONS. She's riding a KAWASAKI motorcycle through a drive through to get a BURGER IN A BAG. She gets on the road her speedometer only reads FAST or slow. She goes FAST, but when she sips her BURGER IN A BAG, she flies off the bike and gets BRUISES on her BUTT and LEGS.

* **KEY:** STREPPer = Strep pharyngitis, GREEN WATER BALLONS = Gallbladder, KAWASAKI = Kawasaki Syndrome, BURGER IN A BAG = TPN and the BRUISED BUTT & LEGS = HENOCH-SCHONLEIN PURPURA.

### HEPATOBLASTOMA

A hepatoblastoma is a malignant, liver neoplasm in infants and children. Usually presents as an abdominal mass by 3 years of age. ALPHA-FETOPROTEIN (AFP) levels are high. Prognosis is poor.

### PRIMARY SCLEROSING CHOLANGITIS (PSC)

Primary sclerosing cholangitis (PSC) is a chronic, cholestatic liver disease resulting from autoimmune inflammation leading to fibrosis of the intrahepatic and extrahepatic biliary tree. It is diagnosed with CHOLANGIOGRAPHY (aka ERCP or Endoscopic Retrograde Cholangiopancreatography), which looks for beading and stenosis of the biliary ducts. ULCERATIVE COLITIS and elevated pANCA levels are both frequently associated with PSC. "PSC often = UC."

**PEARL**: Since the biliary tree is affected, **look for GGT elevation**. Bilirubin levels are elevated in the advanced stages.

## HEPATOBILIARY IMINODIACETIC ACID SCAN (aka HIDA SCAN or CHOLESCINTIGRAPHY)

A hepatobiliary iminodiacetic acid scan (aka HIDA scan or cholescintigraphy) utilizes a nuclear medicine tracer which is injected into an IV. The gallbladder should then be visible within one hour post-injection. If the gallbladder is not seen, there is either CHOLECYSTITIS or CYSTIC DUCT OBSTRUCTION.

**IMAGE**: http://www.nature.com/nrc/journal/v4/n9/images/nrc1429-f1.jpg

## TRANSAMINITIS

Mild transaminitis is common with many viral infections. If transaminases are in the thousands, the diagnosis is likely VIRAL HEPATITIS.

**PEARLS:** If the ALT is higher than the AST, that is also suggestive of VIRAL HEPATITIS. AST > ALT usually means there is an alcoholic hepatitis. This would only be presented in a teen patient.

## ALKALINE PHOSPHATASE

Alkaline phosphatase levels are elevated in biliary and bone disease (e.g., biliary obstruction, bony tumors/metastases, PAGET'S DISEASE).

**PEARL**: A Gamma-Glutamyl Transpeptidase (GGT) level will guide you to the source of an elevated alkaline phosphatase. GGT is elevated in hepatic disease. It is normal in diseases of the bone.

## BILIARY OBSTRUCTION

After a biliary obstruction, AST becomes elevated first, followed by alkaline phosphatase.

# ***JAUNDICE***

## JAUNDICE

Jaundice may be a due to hyperbilirubinemia from common neonatal jaundice etiologies (discussed in the Neonatology section), hemolytic jaundice (discussed in the Hematology section) or may be due to liver or biliary diseases (discussed in this chapter).

**PEARL**: When evaluating a patient with jaundice, if a hepatobiliary etiology is suspected, look at the transaminase and alkaline phosphatase levels to help differentiate CHOLESTATIC DISEASE from HEPATOCELLULAR JAUNDICE. In hepatocellular jaundice, there will be an associated transaminitis. In cholestatic jaundice, there will be a marked elevation in alkaline phosphatase.

## CHOLESTASIS

Cholestasis may present in the neonatal period. Look for possible acholic (pale or gray) stools, hepatomegaly and an elevation in the DIRECT (or CONJUGATED) bilirubin. A **HIDA scan** will show hepatic uptake without biliary excretion due to an obstructive process.

## BILIARY ATRESIA

In cases of biliary atresia, look for an elevation in the direct bilirubin in a neonate. If found, obtain an abdominal ultrasound followed by a HIDA scan. If left untreated, can result in liver failure or KERNICTERUS (due to unconjugated bilirubin crossing the blood-brain barrier).

## CHOLEDOCHAL CYSTS

Choledochal cysts are congenital cystic dilations of the biliary tree. Along with **jaundice**, other symptoms may an **abdominal mass, RUQ abdominal pain**, nausea, vomiting and pancreatitis. These patients carry in increased risk of cancer. Treatment is surgical removal of the cyst.

**IMAGE**: (Various types of choledochal cysts) http://alturl.com/oxk49

## PROGRESSIVE FAMILIAL INTRAHEPATIC CHOLESTASIS (PFIC)

There are 3 types of Progressive Familial Intrahepatic Cholestasis (PFIC). In PFIC I, look for a **direct hyperbilirubinemia and severe pruritus**. Bilirubin is formed, but not properly.

**PEARLS:** Associated with the Amish community. Since the biliary tree/piping is normal in PFIC I and II, the GGT will be normal. In PFIC III, there is a mutation that causes damage to the biliary epithelium, resulting in very high GGT levels.

**MNEMONIC:** "pfic thrEE has the HIGH ggTEE!"

## ALAGILLE SYNDROME (aka ARTERIOHEPATIC DYSPLASIA)

Alagille Syndrome (aka arteriohepatic dysplasia) is a genetic disorder in which jaundice is noted in the newborn period. Look for a child with LIVER and HEART disease. Here are some associations: **paucity of bile ducts (aka intrahepatic biliary atresia or hypoplastic biliary ducts)**, **pulmonary stenosis, a triangular face** (underdeveloped mandible), hypercholesterolemia with xanthomas, eye abnormalities and acholic stools.

**IMAGE**: http://alturl.com/auydt

**MNEMONIC**: (image #1) The GREEN ALLIGATOR has a TRIANGULAR FACE that is green because it is filled with BILE. His head is shaped like a HEART to remind of the PULMONARY STENOSIS. He also has funny shaped EYES and XANTHOMATOUS lumps all over his face. Also, notice what he's eating. It's a LIVER!

**MNEMONIC**: (image #2) This cute little guy is named Alagille. He has a TRIANGULAR head and a funny shaped JAW that is in the shape of a HEART to remind you of PULMONARY STENOSIS.

**MNEMONIC**: (image #3) That's Al the Green Alligator.

## IDIOPATHIC NEONATAL HEPATITIS

Idiopathic neonatal hepatitis is a diagnosis of exclusion! Must do a workup first. If nothing is found except for enlarged hepatocytes on biopsy, it is likely this. Will likely resolve by 8 months of age.

## VIRAL HEPATITIS

**PEARL**: When evaluating a vital hepatitis, always note whether they give you A**g** or A**b**. Knowing how to interpret the presence of antigens and antibodies is key for this section.

**MNEMONIC**: There are many types of viral hepatitis. A, B, C, D, E. To remember which ones are transmitted fecal-orally, imagine that the A is the beginning of the GI tract, and the E is the end. So Hep A and Hep E are transmitted fecal-orally, while the rest (B, C, D) are transmitted through B-C-D (Blood, Cex/Sex & Drugs).

## HEPATITIS A

Incubation period Hepatitis A is approximately 4-6 weeks. May be asymptomatic in children (much worse in adults). Jaundice can be relapsing for up to 1 year.

- **PEARLS:** Look for a child with elevated transaminases, recent travel and what sounds like a viral syndrome. DO NOT OBTAIN AN IgG LEVEL FOR DIAGNOSIS. It is persistent for life. Diagnose with IgM. Also, keep in mind that many viral syndromes can cause a mild transaminitis. In viral hepatitis, look for elevation in the many hundreds to thousands.

## HEPATITIS B

Hepatitis B SURFACE ANTIGEN persistence beyond 6 months means there is a chronic infection. HEPATITIS B "E" **ANTIGEN** (HBe**Ag**) presence means there is high replication resulting in a high viral load and **high infectivity**. The WINDOW PERIOD is the period after the Hep B surface **Ag** presents and before the Hep B surface Abs are made. Diagnosis during this time can be difficult **UNLESS anti-HBc** (IgM, anti-core antibody) is sent. If this test is positive and everything else is negative, then this is the WINDOW PERIOD. If this test is positive and IgG is also positive, this represents a PAST INFECTION (not immunization). If only IgG is positive, that represents prior immunization.

- **IMAGE**: http://alturl.com/imrqp
- **PEARL**: Vertical transmission is a very common mode of transmission.
- **MNEMONIC**: HB**e**Ag reflects high "**E**nfectivity"

## HEPATITIS C

Hepatitis C is the most common blood borne infection in the US and the most common etiology of chronic viral hepatitis. Mostly asymptomatic in kids. Later with liver CA and Cirrhosis.

## GILBERT'S SYNDROME (aka GILBERTS SYNDROME)

In Gilberts Syndrome, there is a glucoronyl transferase deficiency (therefore conjugation of bilirubin is decreased). Most common inherited cause of **indirect** hyperbilirubinemia. The **intermittent indirect** hyperbilirubinemia is mild (usually with levels < 3), extremely common and BENIGN. It is autosomal recessive and so may be seen in multiple family members. Usually noted at times of illness and physiologic stress (dehydration, fasting and even in vigorous exercise).

**MNEMONIC**: gIlBerts. The "**I**" represents Indirect. The "**B**" represents both **B**ilirubinemia and the fact that this is a **B**enign condition.

## CRIGLER-NAJJAR SYNDROME

In Crigler-Najjar Syndrome, there is a glucoronyl transferase deficiency (therefore conjugation of bilirubin is decreased). Results in **indirect** hyperbilirubinemia. Rare.

* TYPE 1: **NO DIRECT BILIRUBIN** (zero) because of a complete lack of glucoronyl transferase. There is severe INDIRECT hyperbilirubinemia and jaundice within the in first days. Requires lifelong phototherapy.

* TYPE 2: Glucoronyl transferase function partially exists, so patients do well with the partial conjugation indirect to direct bilirubin. They do not need phototherapy.

**MNEMONIC**: Najjar sounds like NINJA. Imaging two ninja's born to man named Craig. The $1^{st}$-born carries 1 sword (Type 1) and is always seen **wearing protective goggles** because he'll need life-long phototherapy to protect himself.

**MNEMONIC IMAGE**:

## DUBIN JOHNSON SYNDROME

In Dubin Johnson Syndrome, there is a mild **DIRECT** hyperbilirubinemia. Benign.

**MNEMONIC**: **D**u**B**in Johnson = **D** for **D**irect, **B** for **B**enign.

## REYE'S SYNDROME (aka REYES SYNDROME)

Reyes syndrome is an acute non-inflammatory encephalopathy associated with liver function abnormalities. Rare. Look for a recent viral URI or varicella infection in the setting of aspirin use. Symptoms will include encephalopathy, possible coma, abnormal LFTs, possibly an elevated PT and hyperammonemia.

**PEARL**: This disease only occurs once in a patient's life.

## (DOUBLE TAKE) WILSONS DISEASE

Wilsons Disease (aka Wilson's Disease) is an autosomal recessive disorder resulting in excess copper accumulation, especially within the liver and brain. Accumulation in the liver can lead to **hepatomegaly**, spider nevi, esophageal varices and a Coombs-negative hemolytic anemia. Accumulation in the brain can lead to **neurologic changes** including tremors, poor school performance, ataxia, abnormal eye movements and spasms. On eye exam, a Kayser-Fleischer ring may be visible. Copper levels in the serum are **low**, but high in the **tissues**. Diagnose by LIVER BIOPSY. Treat with PENICILLAMINE, a copper chelator.

**PEARLS: Diagnose** by LIVER BIOPSY. Abnormal eye movements and a Fleisher ring may be present, but there is no visual disturbance. Kayser-Fleischer rings are seen in 90% of symptomatic patients, and almost

100% of patients with neurologic manifestations. **Screen** family members with CERULOPLASMIN levels. Ceruloplasmin is made in the liver and is the primary copper carrying protein. If the level is **LOW**, that suggests Wilsons Disease because excess copper is not being incorporated into ceruloplasmin, and is therefore still in the TISSUES. Therefore, supportive labs may include a low serum ceruloplasmin, high tissue copper levels, high urine copper levels and low serum copper levels.

**IMAGE**: http://www.gourmandizer.com/wilsons/kf2.html

**IMAGE**: http://upload.wikimedia.org/wikipedia/commons/0/00/Kayser-Fleischer_ring.jpg

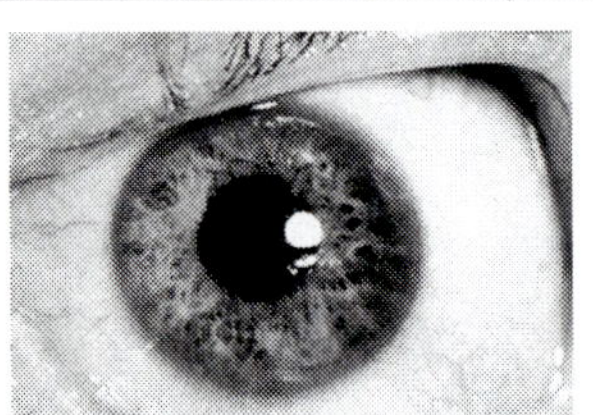

**MNEMONICS:** Treat with a "**COPPER PENNY-**cillamine." Also, ever heard of Wilsons? The leather company that makes baseball gloves? See the image below to note the strong resemblance in color to a COPPER PENNY. This should help you remember that Wilsons Disease has to do with copper, and that it is treated with "PENNY-cillamine."

## CHOLECYSTITIS

Cholecystitis is an inflamed and thickened gallbladder wall, usually due to gallstone obstruction of the cystic duct. Presents with fever and RUQ abdominal pain (Murphy's Sign). A RUQ mass is sometimes palpable. Unlike adults, many children also have JAUNDICE. Fatty meals exacerbate the pain, which may radiate to the right scapula or shoulder. Acalculous cholecystitis occurs in the absence of gallstones and may be associated with hemolysis, TPN sepsis, prolonged fasting or obesity. Diagnosis is usually made by ultrasound, though a HIDA scan may be needed in some cases to confirm.

**NAME ALERT:** CholeCYSTitis refers to the inflammation of the gallbladder (cyst = bladder, the same way a cystitis refers to an inflammation of the bladder). CholeLITHiasis simply means there are stones in the gallbladder. CholANGItis is a medical emergency which classically presents with Charcot's triad of fever, RUQ pain and leukocytosis from an infection in the biliary tract.

## CHOLELITHIASIS

Cholelithiasis is a term for gallstones in the gallbladder. May present with similar symptoms as cholecystitis. Patients are more likely to have jaundice and icterus. Unlike cholecystitis, hepatosplenomegaly may be present. Cystic fibrosis, TPN use or a history of Ceftriaxone use increase the risk of having cholelithiasis.

## ICTERUS

Icterus is a yellowing of the sclera.

**PEARL**: If a patient presents with yellow-orange skin but DOES NOT HAVE ICTERUS, s/he likely has had excessive beta carotene ingestion (apricots, carrots, sweet potatoes, pumpkins, etc.).

## ***ABDOMINAL DISCOMFORT & PAIN***

### CLASSIC FUNCTIONAL ABDOMINAL PAIN OF CHILDHOOD

Classic functional abdominal pain of childhood is a periumbilical, crampy abdominal pain that does not radiate in an otherwise healthy, prepubertal child. May be recurrent.

### CONSTIPATION

For cases of constipation, use stool softners, fiber supplements, and osmotic agents/laxatives to "clean them out." Then focus on good bowel habits, appropriate fiber intake and avoidance of withholding through positive reinforcement.

**PEARL**: Avoid suppositories and enemas in simple constipation as they may be traumatic.

### FECAL OVERFLOW ENCOPRESIS

Patients with fecal overflow encopresis may present with LLQ pain. May use more aggressive measures, such as enemas and suppositories for initial "clean out."

### HELICOBACTER PYLORI

* PEPTIC ULCER DISEASE (aka H. pylori induced PUD)

If a patient is diagnosed with an ulcer of any type that is found to be positive for *H. pylori*, treatment will require a proton pump inhibitor (PPI) and antibiotics. Possible regimens include:

- PPI + Amoxicillin + Clarithromycin
- PPI + Amoxicillin + Metronidazole

* NODULAR GASTRITIS: The most common etiology is *H. pylori*. An EGD with biopsy (samples sent for pathology) is the gold standard for diagnosis. Can also be found in Crohns disease.

- **IMAGE**: http://www.naspghan.org/user-assets/Images/gallery_Endoscopic/S104.JPG

* CAMPYLOBACTER-LIKE ORGANISM TEST (aka CLO test or Rapid Urease Test): Just know that this can be used at the time of an EGD to help diagnose. It's faster and cheaper than sending a biopsy specimen to pathology, but it is not as specific as an EGD with biopsy.

* UREASE BREATH TEST: This is a noninvasive means to attempt diagnosis of Helicobacter pylori.

### NSAID-INDUCED DYSPEPSIA, ULCERS AND EROSIVE GASTRITIS

NSAID-induced dyspepsia, ulcers and erosive gastritis result from the inhibition of PGE and thus a decrease in the protection of the gastric mucosa.

**PEARL**: Especially consider this diagnosis in any patient with a chronic pain illness, such as Sickle Cell Anemia or Juvenile Rheumatoid Arthritis.

## EROSIVE GASTRITIS aka EROSIVE GASTROPATHY

Erosive gastritis (aka erosive gastropathy) may be caused by NSAIDS, exercise, *H. pylori* or the stress of surgery. NSAIDS is the most common cause. "Erosive" refers to mucosal injury. As mentioned above,

## NON-ULCER DYSPEPSIA

Non-ulcer dyspepsia is a chronic or recurrent epigastric discomfort associated with early satiety, bloating, belching, nausea and possibly reflux-like symptoms.

## NON-EROSIVE GASTRITIS

Non-erosive gastritis is associated with Crohns Disease, eosinophilic esophagitis, and of course, *H. pylori*. "Non-erosive" refers to the lack of mucosal injury (only erythema is noted on EGD).

## ZOLLINGER-ELLISON SYNDROME

Zollinger-Ellison Syndrome results in multiple GI ulcerations due to a gastrin tumor of the pancreas. Diagnose by checking for a high **fasting** gastrin level. If diagnosed, you will need to rule out Multiple Endocrine Neoplasia Type 1 (MEN Type 1 – associated with tumors of the PPP – Pituitary gland, Parathyroid hormones and Pancreas).

## INFANTILE GASTROESOPHAGEAL REFLUX (GERD)

Infantile gastroesophageal reflux (GERD) may be noted in up to half of all children 4-6 months of age. Usually resolves by 12 months of age. There is no need to treat if the child is healthy and doing well otherwise.

## IRRITABLE BOWEL SYNDROME (IBS)

Irritable bowel syndrome (IBS) is a crampy abdominal pain associated with diarrhea **or** constipation. May alternate. This is a diagnosis of EXCLUSION. Treat with fiber.

**PEARLS:** There must be some type of poop issue! There's often an emotional component as well. Do not choose this answer unless at least some type of workup has been done already. If no workup has been done, start with noninvasive tests such as a CBC, ESR, anti-TTG and stool guaiac. Do not choose an invasive test unless other tests are negative and the patient failed a FIBER trial. Non-invasive testing -> Fiber trial -> EGD and/or Colonoscopy

## INFLAMMATORY BOWEL DISEASE (IBD) – CROHNS AND ULCERATIVE COLITIS

Know the similarities & differences between the different inflammatory bowel diseases (Crohns Diseae and Ulcerative Colitis). Both are associated with HLA B27 and toxic megacolon. Both also have similar treatments.

* **ULCERATIVE COLITIS** (UC): Typically presents in a TEEN of an **Ashkenazi (European) Jewish** descent. Look for a history of chronic, crampy lower abdominal pain that may or may not be associated with bloody stool. If there is a severe colitis, fever may be present. Lab findings may include hypoalbuminemia and anemia. Diagnosis is by colonoscopy and biopsy. First-line treatment includes is 5-ASA (aka Mesalazine or 5-Aminosalicylic Acid). Second-line therapy options include steroids, Metronidazole, azathioprine, cyclosporine, methotrexate and tacrolimus.

- **PEARLS:** An acute ulcerative colitis flare can make the colon very fragile and susceptible to PERFORATION. If "barium enema," is an option, cross it out. Also, remember that this disease refers to a COLITIS and is primarily a LARGE BOWEL disease.

* **CROHNS DISEASE**: May present simply as short stature and weight loss prior to the onset of any of the usual GI symptom of diarrhea. GI findings may include transmural ulcers in a "skip lesion" pattern, noncaseating granulomas of the upper GI tract and perianal fistulas. Other manifestations include hepatic disease, erythema nodosum, pyoderma gangrenosum and uveitis. Supportive laboratory data may include elevated inflammatory markers (ESR or CRP) and anti-Saccharomyces antibodies. Treatment options include 5-ASA, steroids, Metronidazole and immunomodulators/immunosuppressives.

- **PEARL**: Unlike UC, Crohns lesions can occur anywhere from the mouth to the anus. Also, a future conversion to cancer is the RULE for ulcerative colitis patients, but that is not the case for Crohns patients.
- **MNEMONIC**: A positive "anti-sa**CROHN**amyces" antibody can help make the diagnosis.

## APPENDICITIS

Consider a diagnosis of appendicitis in any child **> 2 years of age** with abdominal pain. Look for a child that is not hungry, and has periumbilical abdominal pain that migrates to the right side of the abdomen (RLQ at McBurney's Point). May be associated with diarrhea. A psoas sign might be noted on exam and a "sentinel loop" of bowel with absence of air in the RLQ may be noted on x-ray. Inflammation may be seen on CT of the abdomen. An ultrasound can also be used, but all you truly need to make the diagnosis is a good history. Treatment is surgery (if the story fits, take the child to surgery). If the appendix has already ruptured, the patient may suddenly be PAIN FREE. In that case, an acceptable choice may be to give IV antibiotics with plans for a delayed appendectomy (weeks later!)

**PEARL**: The psoas sign is the finding of **abdominal pain** elicited by passively extending the thigh of a patient lying on his side with knees extended, or asking the patient to actively flex his thigh at the hip.

**IMAGE**: (video) http://www.youtube.com/watch?v=n0a0PCwsVQ4

## PANCREATITIS

Pancreatitis is an inflammation of the pancreas resulting in epigastric pain that can be associated with rebound, guarding, nausea, vomiting or fever. Pain often radiates to the back. Possible etiologies include trauma, cystic fibrosis, hereditary pancreatitis and pancreatic duct obstruction, but it's most often idiopathic. Diagnose with abdominal ULTRASOUND (extremely specific) or CT of the abdomen. Supportive labs include elevated biomarkers, such as ISOamylase, amylase and lipase. An ERCP may be obtained if the patient has RECURRENT bouts of pancreatitis.

**PEARL**: Many GI diseases result in hyperactive bowel sounds. With pancreatitis, bowel sounds can be INCREASED.

## INTUSSUSCEPTION

Intussusception is a telescoping of the bowel into an adjacent segment of bowel, often in the ileocecal area. Can result in **intermittent** episodes of abdominal pain, currant jelly stools, bilious emesis, a palpable mass and even a septic clinical picture without fever. Usually occurs in kids 3 months – 6 years of age. Treatment options include either an air contrast enema with a small amount of barium, or barium enema.

**PEARLS:** Buzz words include intermittent abdominal pain and currant jelly stools. Patient may not have ANY abdominal pain on exam. If given an option between air contrast enema and barium contrast enema, choose AIR contrast.

**MNEMONIC**: **intusSIXception** should help to remind you that this condition is seen in kids up to **SIX** years of age.

**NAME ALERT**: Current jelly SPUTUM is a buzz word for KLEBSIELLA pneumonia

### (DOUBLE TAKE) GIARDIA

Giardia presents with intermittent, watery diarrhea that has been going on for weeks. May be accompanied by abdominal distension, and may eventually cause malabsorption. Diagnose with a STRING TEST + ELISA, and **then** treat with Metronidazole.

**PEARLS:** History may include a camping trip or a child in daycare.

### ABDOMINAL PAIN PEARL

Keep in mind other causes of abdominal pain and discomfort, such as pneumonia, testicular torsion, torsion appendix and inguinal hernias.

## ***DIARRHEA***

### CHRONIC NONSPECIFIC DIARRHEA

Chronic nonspecific diarrhea is very common. Presents with malodorous stool that is often **intermittently** loose and may contain particle of food. Usually seen in children with a diet that is low in fat and high in carbohydrates. Treat by increasing fat intake and decreasing carbohydrates.

### (DOUBLE TAKE) LACTOSE INTOLERANCE (aka LACTASE DEFICIENCY)

It is **not** common for kids < 5 years old to have lactose intolerance (aka lactase deficiency). So for the pediatric boards, if the child is less than 5 years of age, suspect a different diagnosis!

* SYMPTOMS: Diarrhea +/- abdominal pain.

* **HYDROGEN BREATH TEST**: Can be used to help diagnose a lactase deficiency (as well as bacterial overgrowth). Patient takes a carbohydrate load, if unable to digest because of a lack of lactase, the bacteria will digest the carbs and release hydrogen (measured in the breath).

* TREATMENT: SOY milk. It does not contain lactose. It contains sucrose.

* **PEARLS:** Stool is NOT malodorous and does NOT have food particles. These patients do NOT vomit and do NOT have an associated rash.

* **MNEMONICS:** Most of you probably know of at least one friend, relative or patient that had lactose intolerance. Did they ever have a rash or vomiting after eating desert? Never! I'm guessing they had LACTAID in their fridge (my residency roommate certainly did). "**LACT**ose comes from the **LACT**ating breasts of women & cows. NOT soy beans."

## BACTERIAL OVERGROWTH

Bacterial overgrowth is often seen in children with short bowel syndrome. Will result in malabsorption. As mentioned above, an elevated fasting breath HYDROGEN test can make the diagnosis. Diagnosis is also supported by an elevated D-lactic acid level (not L-lactic acid level).

## CELIAC DISEASE (aka CELIAC SPRUE)

Classic symptoms of celiac disease (aka celiac sprue) include malodorous stool that is bulky and frothy. Patient may also have a distended abdomen and signs of malabsorption. Patients have a GLUTEN sensitivity. Malabsorption may be evidenced by noting reducing substances in the stool (caused by **patchy villous atrophy** of the GI mucosa leading to lactase deficiency), or noting split fats in the stool (indicates the pancreas is working and there is adequate lipase, but malabsorption is present). Supportive blood tests include the presence of antiendomysial antibodies and anti-Tissue TransGlutaminase (anti-TTG is much more specific). Gold standard for diagnosis is colonoscopy and biopsy showing villous atrophy. Treat with a gluten-free diet.

**PEARLS:** Often associated with Type 1 Diabetes Mellitus. If presented with a DM patient that is having GI symptoms, or is losing weight, consider CELIAC DISEASE. If there is a high suspicion for celiac disease, may empirically treat with a gluten-free diet and monitor for improvement rather than performing a traumatic procedure (colonoscopy).

## INFECTIOUS DIARRHEAL ILLNESSES

SEE ID SECTION FOR AN EXTENSIVE LIST OF INFECTIOUS DIARRHEAL ILLNESSES

# *CONSTIPATION*

## FUNCTIONAL CONSTIPATION

Look for functional constipation in a child that has, or had, difficult with toilet training and has had problems with frequent soiling or stool incontinence. Child may have a tendency to withhold stool.

**PEARL**: Diagnosis of exclusion. Look for a child that has a history of soiling himself who has **stool in the rectum** on rectal exam. Differentiated from IBS by the **absence** of diarrhea, bloating, etc.

## IRRITABLE BOWEL SYNDROME (IBS)

See ABDOMINAL DISCOMFORT section.

## CONGENITAL HYPOTHYROIDISM

Look for constipation + delayed anterior fontanelle closure, a hoarse cry, poor growth, or an umbilical hernia to indicate congenital hypothyroidism.

## CYSTIC FIBROSIS (CF)

Cystic fibrosis (CF) should be top on your list for any newborn that does not produce stool within 48 hrs!

## HIRSCHSPRUNG DISEASE

Hirschsprung Disease results in constipation **early** in infancy and tends to present prior to 2 years of age. No problems with soiling. Can be associated with poor oral intake, abdominal distension, occasional diarrhea and bilious emesis. May present as FTT. Boys are more often affected. Patient have an aganglionic (lack of parasympathetic innervation) segment of bowel that is **narrow or contracted, and can eventually result in megacolon proximal to that segment**. Strong associations with DOWN SYNDROME and CYSTIC FIBROSIS. Diagnosis is by biopsy.

**PEARLS:** Look for early history of constipation, no problems with soiling and the ABSENCE of stool in the rectum on rectal exam. Also keep in mind the associations with Trisomy 21 (constipation + syndromic features) and CF patients (constipation + foul smelling stools). If you are presented with images, remember that the **narrow** segment of the bowel (next to normal or dilated bowel) is the affected/aganglionic segment.

**IMAGE**: http://www.beltina.org/pics/hirschsprungs_disease.jpg

**IMAGE**: http://alturl.com/enzfv

## MECONIUM ILEUS

Meconium ileus results in abdominal distension and vomiting after birth. It's due to thickened meconium causing an obstruction in the ileum. May find palpable bowel cords on exam. X-ray may show ground glass or "soap and bubble" stool or air-fluid levels. Contrast enema may show a microcolon (small from the splenic flexure to the anus). There may be dry meconium pellets in the small intestine.

**PEARLS:** Meconium ileus is often the presenting symptom of **CYSTIC FIBROSIS**.

# *VOMITING*

## GASTROESOPHAGEAL REFLUX DISEASE (GERD)

Compared to pyloric stenosis, patients with gastroesophageal reflux disease (GERD) have vomiting that seems effortless. If the child is healthy, no need to treat. Will likely to resolve by 2 years of age. Always consider overfeeding as a possible etiology. If there is apnea, signs of esophagitis (posturing) or poor weight gain, start a workup and also treat. GI pH probe may help diagnose. Biopsy is unlikely to be an option, but choose GERD if eosinophils are noted on biopsy. May treat with an H2 blocker (cimetidine, nizatidine or ranitidine) or with a proton pump inhibitor (PPI), such as omeprazole.

**PEARL**: Metoclopramide and sitting upright during feeds have not been shown to decrease reflux.

## PYLORIC STENOSIS

Pyloric stenosis results from a gastric outlet obstruction due to a thickening or elongation of the pylorus. Look for **NON**-bilious, projectile emesis in a HUNGRY child. Labs may reveal a hypochloremic **hypOkalemic** metabolic alkalosis and possibly an elevated indirect bilirubin. An upper GI series may show the "string sign" or "railroad track "or "double track" sign. The railroad track sign is due to 2 lines of contrast created by thick muscle, with a connection due to contrast in ruggae. Diagnosis is made by ultrasound showing a pylorus that is **> 14 mm long** or **> 4 mm thick**.

* **SIDE NOTE**: Alkalosis is initially from vomiting out HCl. As the patient become dehydrated, there is a superimposed contraction alkalosis. Additionally, hypokalemia results in renal wasting of H+ ions in an effort to hold on to K+ ions. This results in even more alkalosis.

* **PEARLS:** If you see a normal potassium level in a patient with pyloric stenosis, know that the total body potassium is still low. If the serum pH is normal or acidotic, it is NOT pyloric stenosis. Occurs in boys > girls.

* **IMAGE**: (Railroad Track) http://alturl.com/4jtvv

* **IMAGE**: (String Sign) http://radiology.rsna.org/content/242/2/632/F3.large.jpg

* **MNEMONIC**: **4**yloric stenosis, 1**4** mm and **4** mm. Remembering the diagnostic criteria can be tough. Use "**4**yloric stenosis" to help you.

## ANTRAL WEB

An antral web is a membrane in the antrum of the stomach that can cause gastric outlet obstruction. Usually formed before birth. Can present as polyhydramnios in utero, or non-bilious emesis in an infant less than 6 months of age. Imaging with barium may reveal a filling defect in prepyloric region.

## ESOPHAGEAL WEB

An esophageal web can cause reflux-like symptoms, esophageal impaction and chest pain. It results from the failure of the esophagus to re-canalize in utero. The web then acts as an obstruction to the passage of a food bolus. Liquids, however, pass through more easily. Treatment requires dilation of the esophageal web.

**IMAGES**: http://bit.ly/ngX7zN and http://bit.ly/oF2ryU

**PEARL**: The "jet phenomenon" refers to the thin area of barium seen when looking at a barium swallow. It starts at initial point of constriction. When that area is tortuous (http://bit.ly/oF2ryU), it can resemble a TE fistula. When it's linear it does not (http://www.ajronline.org/content/129/4/747.full.pdf - page 1 - see it and move on!).

## ACHALASIA

Achalasia is caused by a dysmotility problem, or a lack of relaxation at the lower esophageal sphincter, which results in forceful emesis. There is eventual esophageal dilatation and loss of peristalsis ability. Look for forceful vomiting, difficulty swallowing (dysphagia) and weight loss or FTT.

## VOLVULUS

A volvulus can cause bilious emesis, abdominal distension and possibly even bloody stools (from ischemia and necrosis of bowel). Occurs due to a rotational defect during embryology resulting in poorly fixed bowel (rotational defect is call MALROTATION). Leads to bowel wrapping around the superior mesenteric artery (SMA) and causing bowel ischemia, and therefore requires emergent surgery. A double bubble may be seen on imaging. **PEARLS:** If given the option, confirm the diagnosis with a BARIUM **ENEMA** over an upper GI series because it can lead to aspiration! When looking for malrotation on imaging, the cecum will be in the wrong place (abnormally high). An abdominal X-ray may be shown with a "double bubble" sign (more on that below). Lastly, if a barium swallow (upper GI series) is done, there may be a "corkscrew" appearance of the duodenum.

**IMAGE**: http://www.hawaii.edu/medicine/pediatrics/pemxray/v3c17b.jpg (corkscrew)

**IMAGE**: http://www.hawaii.edu/medicine/pediatrics/pemxray/v3c17c.jpg (corkscrew)

**IMAGE**: http://www.learningradiology.com/caseofweek/caseoftheweekpix2008/cow308lg.jpg

## ANNULAR PANCREAS

Look for a history of polyhydramnios and then vomiting in a neonate. The annular pancreas forms a ring around the intestine. This causes poor swallowing in utero, resulting in polyhydramnios, and then vomiting in the neonatal period.

**IMAGE**: http://alturl.com/6eudk

## CYCLIC VOMITING

Cyclic vomiting is associated with intermittent episodes of repeated vomiting with periods of complete normalcy. There is likely to be an emotional component to the question, either in the patient or the family, and there may also be a history of migraines or IBS. Treatment may include hydration and/or prophylactic medications similar to those used in migraine patients: Amitriptyline (or similar tricyclic antidepressant/TCA), Cyproheptadine, or Propranolol.

**PEARL**: This is diagnosis of exclusion, so make sure somewhat of a workup has been done before choosing this answer.

## RUMINATION

Rumination is when a child chews something over and over again. Occurs in patients with mental retardation and in some children that are emotionally disturbed.

## BILIOUS EMESIS IN A NEWBORN

Bilious emesis is a surgical emergency in a newborn! Look for evidence of duodenal atresia or malrotation. In older children, bilious emesis can be less severe/emergent.

* Duodenal Atresia = bilious emesis on **1st DOL**!, double bubble on KUB. Could have jaundice due to low enterohepatic circulation. If there is "complete" atresia, there will be NO SECOND BUBBLE… NO AIR BEYOND ATRESIA. Remember, 1st DOL… NOT at 2 months (if a 2 month old baby has bilious emesis, consider pyloric stenosis, although that's typically NON-bilious!!!).

## DOUBLE BUBBLE

A "double bubble" refers to the radiologic sign noted when there is a duodenal obstruction. There will be a large "bubble" and a small "bubble." These represent a dilated stomach and a dilated duodenum, respectively. Associated conditions include **volvulus due to malrotation, duodenal atresia and antral webs.**

**IMAGE**: http://alturl.com/sx8my

## VOMITING PEARLS

* One episode of vomiting in an otherwise healthy child probably warrants reassurance.
* Always keep in mind infections (pneumonia, urinary tract infections, gastroenteritis, rotavirus, etc.) as possible causes of emesis.

* Inborn errors of metabolism and Diabetic Ketoacidosis (DKA) are a couple of metabolic causes of emesis.

# *GI BLEEDING*

## GI BLEEDING PEARL

**STEP 1 is to NG LAVAGE!** This is done to evaluate for an upper GI bleed. A brisk upper GI bleed can result in what looks like lower GI bleed because the blood acts like a laxative.

## LOWER GI BLEEDING (LGIB)

*** DIFFERENTIAL FOR LOWER GI BLEEDING (LGIB) IN THE NEONATAL PERIOD:**

- Maternal Blood: Perform an Apt test. Indirectly checks for fetal versus adult hemoglobin.
- Malrotation with Volvulus
- Necrotizing Enterocolitis (NEC): Particularly in premature neonates

*** DIFFERENTIAL FOR LOWER GI BLEEDING (LGIB) AT 1-2 YEARS OF AGE**

- Anal Fissure: Usually secondary to constipation. Most commonly located anteriorly.
- Intussusception
    - **MNEMONIC**: intus**SIX**eption) = 3 months - **6 yo**
- Juvenile Polyp: Painless bleeding. There may be a history of an intermittently seen mass protruding from the rectum.
    - **PEARL**: There is no increase in the risk of cancer with juvenile polyps.

**DIFFERENTIAL FOR LOWER GI BLEEDING (LGIB) AT 2-5 YEARS OF AGE**

* Meckel's Diverticulum: Painless. CAN be melanic. Technetium-99m study

*Juvenile Polyp: Painless. Not assoc with inc cancer risk

**DIFFERENTIAL FOR LOWER GI BLEEDING (LGIB) IN SCHOOL AGED KIDS**

* Meckel's Diverticulum: Painless. CAN be melanic. Technetium-99m study

* Juvenile Polyp: Painless. Not assoc with inc cancer risk

* Familial Adenomatous Polyposis: 100% chance of future malignancy.

* Ulcerative Colitis

* Crohns Disease

## PAINLESS RECTAL BLEEDING

Painless rectal bleeding can be due to anal fissures, polyps, hemorrhoids and Meckels Diverticulum

**PEARLS:** Polyps, fissures and hemorrhoids have small volume bleeding. Hemorrhoids usually do not result in streaks of blood on the stool Meckel's Diverticulum usually results in large volume bleeding.

## MECKEL'S DIVERTICULUM (aka MECKELS)

Meckels diverticulum is a true diverticulum of the small intestine containing all 3 layers of bowel wall. It is present at birth and can produce LARGE volumes of PAINLESS rectal bleeding. Diagnose with a Meckel's scan.

**PEARLS:** Bleeding is usually red, but CAN be melenic. Painless rectal bleeding due to Meckel's diverticulum is MUCH more common than bleeding due to polyps.

**MNEMONIC**: The "Rule of 2's" refers to the fact that 2% of the population have a Meckel's Diverticulum, most of them are located 2 feet from the ileocecal valve and most are 2 inches in length.

**FYI**: The Meckel's scan is a technetium-99 scan that looks for ectopic gastric or pancreatic cells in the small bowel. These ectopic cells are present in 50% of Meckel's diverticuli. Technetium scans can also be used to help diagnose a small bowel obstruction and intussusception.

### FAMILIAL ADENOMATOUS POLYPOSIS (FAP)

Juvenile polyps are extremely rare in children less than 10 years of age unless you have familial adenomatous polyposis (FAP). If a child younger than 10 presents with polyps, obtain GENETIC TESTING for APC mutation (Adenomatous Polyposis Coli mutation). These children have a 100% of ending up with cancer. For APC+ patients, screen for colonic adenomas every year starting at 10 years of age. In order to prevent cancer, the colon is resected once an adenoma is noted on colonoscopy, or once the patient turns 25.

## *MISCELLANEOUS GI CONDITIONS & TERMINOLOGY*

### NASOGASTRIC TUBE FEEDINGS (NG TUBE FEEDINGS)

Mild diarrhea is common with nasogastric tube feedings (NG tube feedings). Bolus feeds are recommended for children struggling with oral motor coordination in order to "help them learn." For all other patients, give CONTINUOUS NG tube feeds.

### ESOPHAGEAL PERFORATION

There is a strong association between esophageal perforation and Marfans Syndrome, Ehlers Danlos and Epidermolysis bullosa

### IMPERFORATE ANUS (aka ANAL ATRESIA)

Imperforate anus (aka anal atresia) presents with abdominal distension within the first 48 hours of life. Anus might be more anteriorly located. Some patients may have a fistula leading to the vaginal or urinary tract.

**IMAGE**: http://www.babymed.com/sites/default/files/imperforate_anus_newborn.png

* **(DOUBLE TAKE) VACTER-L (aka VACTERL or VATER) SYNDROME**: VACTER-L (aka VACTERL or VATER) syndrome is an acronym. VACTER**L** is now used instead of VATER because it stand for **V**ertebral anomalies, **A**nal atresia/imperforate anus, **C**ardiac defects (especially VSD), **T**racheoesophageal fistula, **R**adial hypoplasia and **R**enal anomalies, and **L**imb abnormalities. These children have a normal IQ. When associated with hydrocephalus, this can be an X-Linked disorder.

**PEARL**: May present with single umbilical artery.

**MNEMONIC**: Imagine Darth VACTER cutting off his own son's ARM (radial hypoplasia and limb abnormalities) and then using the ARM as a light saber to create ANAL ATRESIA and a TE Fistula."

**IMAGE:** (go to 5:30 in the video) http://www.youtube.com/watch?v=C-Del3ohVbY

**IMAGE**: http://www.youtube.com/watch?v=uXQxstTrhTI

## PERSISTENT CLOACA

Persistent cloaca refers to the presence of a single channel consisting of the rectum, vagina and the urinary tract (did not separate in utero). Requires immediate surgery to prevent severe urinary tract complications. Should be suspected clinically in any female child with an imperforate anus.

## RECTAL PROLAPSE

Rectal prolapses are usually due to **CONSTIPATION** or diarrhea (or a polyp), but **screen for cystic fibrosis**. Also caused by "TRICK YOu're A$$ **out**… or WHIP you're A$$ **out**!" = Tricuris = Whipworm. Also by Shigella!

## TYPHLITIS

If you are presented with a clinical description in an older child with leukopenia or neutropenia that sounds like NEC, choose typhlitis as the answer.

* **(DOUBLE TAKE) NECROTIZING ENTEROCOLITIS**: Necrotizing enterocolitis is often located at the ileocecal junction. Look for thrombocytopenia, bloody stool, an erythematous abdomen, poor feeding, PNEUMATOSIS INTESTINALIS +/- air in biliary tree. Associated with infection and hypoxic injury (apnea, asphyxia, RDS, etc). Do not feed these kids for **3 weeks**.

- **MNEMONIC**: NECKrotizing enterocolitis usually occurs at the NECK of the colon (ileocecal junction).

## OMPHALOCELE

An omphalocele is a herniation of bowel +/- organs through umbilicus. The herniated material is SEALED in by overlying membranes. Associated Beckwith-Wiedemann Syndrome and various chromosomal defects.

## GASTROSCHISIS

Gastroschisis occurs when there is herniation of UNCOVERED bowel NEAR (not through) the umbilicus. The organs remain in the abdomen. Treat by placing a nasogastric tube for decompression and keeping the bowel moist until surgical repair.

# PHARMACOLOGY & DRUG PEARLS

## *MISCELLANEOUS DRUGS*

### MISOPROSTOL

Misoprostol is a synthetic PROSTAGLANDIN. In patients with long-term NSAID use, it can help prevent GI ulcers.

**PEARL**: When referring to NSAIDS, the ABP might use terminology such as, "PROSTAGLANDING SYNTHASE (aka SYNTHETASE) inhibitor."

### SUCRALFATE (ALUMINUM HYDROXIDE COMPLEX)

Avoid use of sucralfate (aluminum hydroxide complex) in renal patients.

### MAGNESIUM SULFATE

When used as a tocolytic, magnesium sulfate can cause decreased fetal respiration, hypotonia and hypocalcemia.

**PEARLS**: **(DOUBLE TAKE) Magnesium Sulfate Infusions** (for t**ocolysis or** preeclampsia) can result in severe hypocalcemia and possible **hypo**magnesemia. Mechanism is complex/unknown. Treat with calcium gluconate, but keep in mind that they may not respond as quickly as would normally be expected. So if you are given a vignette about a hypocalcemic baby not responding to calcium gluconate, consider magnesium tocolysis as the etiology.

### TERBUTALINE

Terbutaline is a beta-adrenergic drug that can cause maternal hypERglycemia. This causes fetal hyperinsulinism and subsequent hypOglycemia in the neonate.

### ACE INHIBITORS

ACE inhibitors are contraindicated in pregnancy because of teratogenicity. Can cause oligohydramnios, neonatal anuria and cardiac malformations (especially a PDA).

### DIAZEPAM

Diazepam is contraindicated in breastfeeding.

### METOCLOPRAMIDE & PROMETHAZINE

Metoclopramide and promethazine can cause acute dystonic reactions. If presented with a child that suddenly has a fixed and uncomfortable looking "pose," or a fixed gaze, treat with Diphenhydramine!

### BLEOMYCIN

Bleomycin can cause Pulmonary Fibrosis.

**MNEMONIC**: "BLOW-mycin"

## VINCRISTINE & VINBLASTINE

Both vincristine and vinblastine can cause neurologic side effects (neuropathies, palsies, seizures, etc.) and SIADH.

**MNEMONIC**: I don't know about you, but wine gives me severe neurologic side effects. Did you know "VIN" is French for wine? As in "vin rose urine" see in iron chelation? If that doesn't work, how about vi**N**cristine and vi**N**blastine, where the N reminds you of neurologic side effects?

## DOXORUBICIN & DAUNOMYCIN

Doxorubicin and daunomycin are cardiotoxic **anthracyclines**.

**MNEMONIC**: doxoRUBIcin seems to help me remember the heart. RUBI RED HEART.

## CYCLOPHOSPHAMIDE

Hemorrhagic cystitis is the key side effect to remember for cyclophosphamide.

## ASPARAGINASE

Pancreatitis is the key side effect to remember forasparaginase.

**MNEMONIC IMAGE**: Note the ASPARAGUS in place of the PANCREAS.

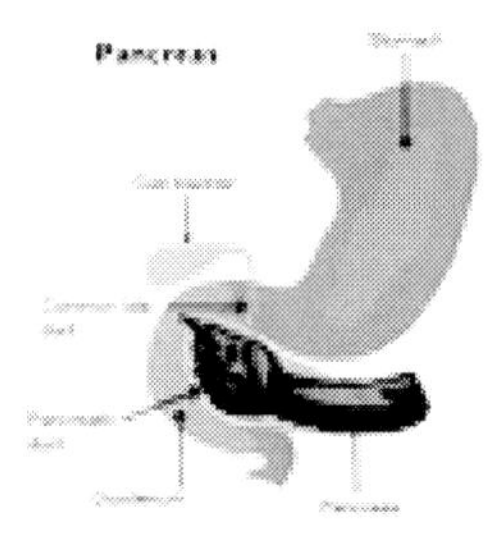


## METHOTREXATE (aka MTX)

Methotrexate (aka MTX) can cause ulcers along the GI tract.

**MNEMONIC**: mt**X** - Imagine the "**X**" being a pair of sharp scissors being swallowed and causing ulcers.

## MALIGNANT HYPERTHERMIA

Malignant hyperthermia should be suspected, in anyone who gets anesthesia and then has tachypnea + muscle rigidity + increased pCO2 + acidosis. Treat with DANTROLENE + Cooling + O2 + Hyperventilation beyond existing tachypnea. DO NOT give bromocriptime because it's a dopa agonist. Also look for high CK, arrhythmias, myoglobinuria. Kids with muscular dystrophy and other neuromuscular disorders are at risk for malignant hyperthermia.

**MNEMONIC**: Malignant Hyperthermia = "Caused by those MALICIOUS ANESTHESIA DOCS who sit there all day putting a *DENT* in every stool in the hospital"

## HEPATIC INDUCERS

Hepatic inducers are medications that work to increase the metabolism of certain medications and thus LOWER their effectiveness. Examples include Carbamazepine, Phenobarbital, Phenytoin and Rifampin.

**PEARL**: OCPs effectiveness is DECREASED with these medications so recommend a backup method of contraception.

## HEPATIC INHIBITORS

Fluconazole (and other –azole), Isoniazid, Sulfonamides and H2 Blockers (especially Cimetidine) are hepatic inhibitors. Focus on remembering these P450 INHIBITORS more than the inducers.

**MNEMONIC**: (It's PG-13) Imagine a racist trucker sitting in a diner. He sees a THAI DINEr sitting in a booth talking very loudly on her SUL-FONE. He gets pissed! He gets so angry that he turns as cold as ICE cold. He then climbs up onto a BLOCK of wood, bends over and starts shooting cubes/BLOCKS of ICE out of his AZOLE. His aim is pretty good and he completely knocks the SUL-FONE out of the THAI DINEr's hand!

* **KEY**:

- THAI DINEr should help your think of "ti-dine," as in cimeTIDINE and raniTIDINE.
- SUL-FONE represents a cell phone in the mnemonic, but should remind you of SULFONamides.
- BLOCK might help you remember that this mnemonic deals with BLOCKING of Cytochrome P450.
- ICE should help you remember ICE-oniazid/isoniazid.
- AZOLE represents #$$#*!% in the mnemonic, but should help you remember fluconAZOLE and itraconAZOLE.

## ALTERNATIVE MEDICATIONS

* ST. JOHN'S WORT: INDUCES P450. This alternative medication can pose a problem for patients on OCPs, digoxin, HIV medications or Warfarin.

* GINSENG: Interacts with numerous drugs, including aspirin, warfarin, diuretics and steroids.

- **MNEMONIC**: gin-SING interacts with just about everyTHING! If you see it in the history, it's probably related to your answer.

# *INTRAUTERINE DRUG EXPOSURES*

## COCAINE EXPOSURE

Premature birth is common with prenatal cocaine exposure. Surprisingly there are **minimal** withdrawal symptoms.

**MNEMONIC**: Somewhat of a "reverse mnemonic," but given that the withdrawal symptoms are minimal, if you ever hear a NICU nurse refer to a screaming and shivering newborn as a "crack baby," be sure to educate her (or him) on the fact that the term is a misnomer.

## HEROIN EXPOSURE

There is often intrauterine fetal demise (IUFD) from heroin exposure. If not, look for seizures, tremors, irritability, etc.

**PEARL**: Heroin induces lung maturity, so if the baby survives, s/he is less likely to have lung disease.

## METHADONE EXPOSURE

Newborn babies with a history of methadone exposure can have seizures. Symptoms may not present for up 4-6 weeks!

## (DOUBLE TAKE) MAGNESIUM SULFATE INFUSION

Magnesium sulfate infusions (for t**ocolysis or** preeclampsia) can result in severe hypocalcemia and possible **hypo**magnesemia. Mechanism is complex/unknown. Treat with calcium gluconate, but keep in mind that they may not respond as quickly as would normally be expected. So if you are given a vignette about a hypocalcemic baby not responding to calcium gluconate, consider magnesium tocolysis as the etiology.

## WARFARIN EXPOSURE

Warfarin exposure can result in FETAL WARFARIN SYNDROME. Look for a flat nasal bridge, a hypoplastic nose (possible upper airway obstruction), hypoplastic distal phalanges and "stippling of the vertebrae."

## ANTISEIZURE MEDICATIONS (aka ANTI-SEIZURE MEDICATIONS)

As a general rule for antiseizure medications (aka anti-seizure medications), look for cardiac defects, especially Tetralogy of Fallot, Aortic Stenosis and Pulmonary Stenosis. Certain medications can also cause MICROCEPHALY, anticonvulsant facies (broad bridge), intrauterine growth retardation (IUGR) and fingernail hypoplasia.

**MNEMONIC**: To remember the possible cardiac defects, think about what a seizure sounds like. TAP TAP TAP!!! As the patient seizes back and forth on a wooden floor, you notice bright red blood on the floor in the shape of a red HEART. KEY: TAP = TOF, AS and PS.

## PHENYTOIN EXPOSURE

Phenytoin exposure can cause FETAL HYDANTOIN SYNDROME (aka PHENYTOIN SYNDROME). Look for finger and nail hypoplasia, a wide anterior fontanelle, prominent metopic ridge (bony ridge secondary to abnormal closure of fontanelle), wide-spaced eyes, broad/depressed nasal bridge, short upturned nose and/or cleft lip/palate.

**IMAGE**: http://www.nlm.nih.gov/medlineplus/ency/images/ency/fullsize/17254.jpg

**MNEMONIC**: Imagine a baby sucking on "PHUNNY-TOES" and fingernails that fit very comfortably into a baby's CLEFT LIP/PALATE.

## VALPROIC ACID EXPOSURE

Valproic acid exposure can lead to a small mouth, a small chin, epicanthal folds, midface hypoplasia and SPINA BIFIDA.

**MNEMONIC**: Imagine what types of trophies a serial killer would have on his "**VAL**L" of prized possessions. Definitely a BIFURCATED SPINE. Possibly a skull with a small chin and small mouth?

## CARBAMAZEPINE EXPOSURE

Similar to VALproic acid. Spinal and facial deformities. Spinal deformities from carbamazepine exposure include spina bifida.

**MNEMONIC**: Car-Bam-azepine. Imagine a CAR that goes **BAM!** as it slams into a serial killer's **VAL**L!

## LITHIUM EXPOSURE

Lithium exposure can cause Ebstein's anomaly.

**MNEMONIC**: (VIDEO) Instead of lithium and Ebstein, if you can somehow associated lithium and EINSTEIN (since they sound alike), you'll be set. Now imagine **EINSTEIN** putting **LITHIUM** fuel in the flux capacitor of the Back to the Future car. For you junkies, yes... I know it's actually plutonium. Here's a great clip. In it, the doc **looks** like Einstein, and the dog's name **IS** Einstein) - http://www.hulu.com/watch/22646/back-to-the-future-the-delorean

## ETHANOL EXPOSURE

Ethanol exposure causes FETAL ALCOHOL SYNDROME (FAS). Problems include small for gestational age (SGA), midface hypoplasia, thin upper lip, a smooth philtrum, a small 5th fingernail, ASD and VSD.

* **PEARL**: Many of the symptoms overlap with those of anticonvulsant exposure. If you see SGA, ASD or VSD, consider choosing FAS over antiseizure medication exposure.

* **MNEMONIC**: Ever hear of WINE PINKY?

- **MNEMONIC**: (image) http://www.bit.ly/pzW64m
- **MNEMONIC**: (image) http://cdni.condenast.co.uk/642x390/d_f/Dr-Evil.jpg = Dr. Evil + Pinky + Philtrum + Lips. If you also want to also remember SGA, here you go (Mini-Me + Dr. Evil + Pinkies) - http://op-for.com/dr%20evil%20mini%20me.jpg.

* **MNEMONIC**: Imagine your favorite, or the most memorable, bottle of alcohol. Now imagine a child is born with two, mini, flight-sized bottles sticking out of his or her chest and pointing towards either flank. One bottle is in an ASD, and the other in a VSD.

* **MNEMONIC**: FAS = 3 letters... just like ASD and VSD.

* **IMAGES**: http://www.aafp.org/afp/2005/0715/p279.html (GREAT IMAGES. Don't read! Just look and move on!)

## VITAMIN A (aka RETINOL) EXPOSUE

Supplementation of Vitamin A during pregnancy is associated with Transposition of the Great Vessels (aka Trasnposition of the Great Arteries, or TGA, or TOGA).

**MNEMONIC**: Ever heard of Retin-**A**®? AKA tretinoin? Now try to associate this WHITE acne cream with a WHITE **TOGA** (TGA).

## ISOTRETINOIN EXPOSUE

Isotretinoin exposure IS associated with microcephaly, microphthalmia, hypoplasia of the ears, hypoplasia of the thymus, and truncus arteriosus.

**MNEMONIC**: ISOTRETINOIN not only SHRINKS PIMPLES, but it also shrinks the HEAD, EARS, EYES, and THYMUS!

# OPHTHALMOLOGY

## HORDEOLUM (aka STYE)

A hordeolum (aka stye) is a **red** and **painful** eye lesion noted at the rim of the eye lid (near the eyelashes). Prescribe warm compresses and possibly topical antibiotics as well.

**PEARL**: Do NOT prescribe oral antibiotics

**MNEMONIC**: ho-**RED**-e-**OWE**-lum should remind you that this is RED and PAINFUL eye lesion.

**IMAGE**: http://upload.wikimedia.org/wikipedia/commons/4/4a/Stye02.jpg

## CHELAZION

A chelazion is much slower growing, painless lesion that results from inflammation of the meibomian gland. There can be erythema as well, but it is much less painful. It resolves on its own and warm compresses do not help.

**PEARL**: Ophthalmology should be involved if it is chronic or interferes with a patient's vision.

**MNEMONIC**: "che-LAZY-on" the eye. This is the slow growing, LAZY eye lesion.

**IMAGE**: http://www.healthhype.com/wp-content/uploads/chalazion.jpg

## CORNEAL ABRASIONS

Use fluorescein staining to look for corneal abrasions. If present, order **topical antibiotics.**

**PEARL**: Do not order an eye patch.

## HYPHEMA

Hyphema is a condition where blood is present in the anterior chamber. Look for blood pooling at the bottom of the iris. If small, and due to trauma, may have the patient follow up with ophthalmology. If it is large, consider monitoring in the hospital and keeping the patient's bed at a 45' angle. There is a danger of developing glaucoma (increased intraocular pressure).

**IMAGE**: http://media.summitmedicalgroup.com/media/db/relayhealth-images/hyphema.jpg

**IMAGE**: http://www.kellogg.umich.edu/theeyeshaveit/trauma/images/hyphema.jpg

## PAPILLEDEMA

Papilledema is swelling of the optic disc. The patient's vision is usually not affected until there has been significant progression, so do NOT rule this out if the patient's vision is 20/20. On exam, they will have loss of venous pulsations, blurring of the optic disc margins and may have a bigger than usual blind spot.

**PEARL**: Papilledema is due to increased intracranial pressure, and is therefore a bilateral phenomenon. If you are given a patient with unilateral abnormalities, consider a different choice.

**IMAGES**: (three great images in the middle of the page) -
http://eyewiki.aao.org/Papilledema#Clinical_diagnosis

## PAPILLITIS

While papilledema is SWELLING of the optic disc, papillitis refers to **INFLAMMATION**. May look similar on exam, but will be associated with photophobia, pain, sensitivity to light pressure applied to the eye and may be noted following a recent viral illness.

**PEARL**: Choose this answer over papilledema if the finding is unilateral or associated with a loss of visual acuity or visual fields.

## CATARACTS

Cataracts opacification of the lens. For the pediatric boards, a cataract is going to be due to **GALACTOSEMIA or RUBELLA**. Last choice would be Neurofibromatosis Type 2.

## MYOPIA

Myopia means a child is nearsighted. **Neonates are initially myopic.**

## PRESBYOPIA

Presbyopia means a child is farsighted.

## VISUAL ACUITY BY AGE

* FIVE YEAR OLD VISION: Refer to an ophthalmologist if a child's visual acuity is 20/50 or worse at 5 years of age.

* SIX YEAR OLD VISION: Refer to an ophthalmologist if a child's visual acuity is 20/40 or worse at 6 years of age.

* **MNEMONIC**: At **5** years of age, kids can see 20/**5**0. At 6 years of age, kids can see 20/**4**0. So it goes age **5**, 6 = **5**0, 40. Also at age **7**, 20/**2**0 vision should be noted (7 looks like a 2).

## VISION SYMMETRY

Symmetry in visual acuity on exam is great! If a child's visual acuity is asymmetric (one eye has great acuity while the other does not), refer to ophthalmology for an evaluation for amblyopia.

## STRABISMUS

Strabismus basically means there is improper alignment of the eyes. It could be associated with an esotropia (eye looks more inward, medially) or an exotropia (eye looks more outward/laterally).

**IMAGE**: http://bit.ly/oGldjK - A left esotropia

## PSEUDOSTRABISMUS

Pseudostrabismus is the appearance of strabismus, or being "cross-eyed," even though the alignment of both eyes is appropriate. Often seen in patients with prominent, or uneven, epicanthal folds.

**IMAGE**: http://bit.ly/nBxjJM (2 images)

## AMBLYOPIA

Amblyopia refers to a structurally normal eye with poor vision. Often occurs when one eye can't see very well. Patients with strabismus are at risk for developing this.

**MNEMONIC**: The weaker eye "shuts down" as the dominant eye takes over.

## ESOTROPIA

Esotropia is an inward deviation of the eye. This is **common up until 3 months of age**, but it's abnormal after that. Esotropia can be unilateral or bilateral. If the right eye is staring at you and has a centered red light reflex while the left eye is deviated more inward, the left eye has the esotropia and is likely to have a light reflex that seems laterally displaced in comparison to the center of the pupil. The left eye is at risk for developing amblyopia. Treat by applying a patch to the "strong" right eye.

**IMAGE**: http://media.summitmedicalgroup.com/media/db/relayhealth-images/strabism.jpg

## EXOTROPIA

Exotropia is an outward deviation of the eye. May be uni- or bilateral. In infants, it is common (unilateral or bilateral) up until the age of 3 months. If the right eye is staring at you and has a centered red light reflex while the left eye is deviated outward, the left eye has the exotropia and is likely to have a light reflex that seems medially displaced in comparison to the center of the pupil. The left eye is at risk for developing amblyopia. Treat by applying a patch to the "strong" right eye.

**IMAGE**: http://media.summitmedicalgroup.com/media/db/relayhealth-images/strabism.jpg

## NYSTAGMUS

Nystagmus is an abnormal finding regardless of age.

## COLOR VISION

Color vision develops at 3 months of age. Binocular vision with convergence develops at 3 months of age. At 4 months, babies prefer to look at patterns and faces .

## CORNEAL LIGHT REFLEX TEST

The corneal reflex test  is the one you do with your flashlight to look for strabismus, esotropia, exotropia, pseudostrabismus, etc. If the refection is symmetrical and located at the center of the pupils, there's no problem.

**NOTE:** In case you're confused about the name, the "blink test" with a cotton swab to the eye has a similar name (Corneal Reflex Test).

# GENETICS & INHERITED DISEASES

**NOTE**: THE GENETICS SECTION IS EXTREMELY IMPORTANT! KNOWING IT COLD IS ESSENTIAL IF YOU WANT TO PASS THE PEDIATRIC EXAM. It contains a tremendous amount of information. Rote memorization is difficult! Please focus on the many images, mnemonics and pearls provided to help you quickly master the topics. If some of the links provided for images lead you to pages with words on then, try to strictly focus on the images and ignore the words! Also, some disorders may only be presented by name and mode of inheritance since they fit better elsewhere in the study guide. An effort has been made to at least provide a quick summary for virtually every disorder. Consider using the search function on our website if you need a full recap of certain disorders.

## *AUTOSOMAL DOMINANT DISORDERS*

### AUTOSOMAL DOMINANT DISORDERS

**An autosomal dominant disorder may be passed on from a mother to her male OR female child, and from a father to his male OR female child.**

### AUTOSOMAL DOMINANT MNEMONIC

FOR THIS AUTOSOMAL DOMINANT MNEMONIC, PLEASE EMPHASIZE THE CAPITALIZED WORDS TO YOURSELF AS YOU GO! IT'S A LONG MNEMONIC, BUT IT'S INVALUABLE ON THE EXAM!

Imagine a WAR between two tribes separated by a DAM OF MINTS (the mints are white and shaped like the MOON). The first tribe's name is MAN**N**TRA, and the second tribesman call themselves the WHITE HAIRS. You decide to learn more about the WHITE HAIRS. They live on an APE-alachian PEAR shaped mountain in eastern North America, and are led by three unique generals. The first is a black man named A-KON, who is very short and mini-me version of the similarly named singer/rapper named of AKON. He walks around bobbing his head and singing his mantra of, "down with the MAN**N**TRA!" The second general is a white man with prominent lips named MCJAGGER. The last general is a Mexican GARDNER with EXTRA TEETH. So, that's one black man who looks like a mini-me version of a famous rapper, one white person who looks a prominent lipped rock star (Mick Jagger), and one Mexican person with extra teeth. The WHITE HAIRS and the three generals go HUNT**ING** for a weak spot in their enemy's DAM OF MOON-SHAPED MINTS. Their HUNT**ING** weapon of choice is a TUBULAR bazooka that BLASTS out ASHY LEAVES which have small DANCING TICKS on them. As you can imagine, they probably are not going to win this WAR.

* AKON (Rapper): http://www.youtube.com/watch?v=rSOzN0eihsE

* A-KON (General) - http://img.medscape.com/pi/emed/ckb/pediatrics_genetics/941088-943343-388.jpg.

* MCJAGGER (General): http://www.cartoonstock.com/newscartoons/cartoonists/lal/lowres/laln210l.jpg

* Mexican GARDNER: http://emptybookshelf.com/wp-content/uploads/2006/07/kimmel.jpg

or http://www.moneymuseum.com/imgs/ximages/image/2010/5/I_DE_4974213_3.jpg

* The mnemonic above is designed to help you very quickly remember the names of the highest-yield AUTOSOMAL DOMINANT disorders. Please go over the story several times until you feel you have

memorized it. This WILL save you tons of time and WILL be worth **many** points on the exam! Before going over the mnemonic again, please review the key below:

- **DAM OF MINTS:** Represents DOMINANT
- **WAR:** Represents WAARdenburg Syndrome. Patient's have a forelock of WHITE HAIR
- **APE-alachian PEAR shaped mountains:** Represent APErt syndrome, in which the child's head may look somewhat like a pear.
- **MANNTRA:** Represents another mnemonic which reinforce some of the already mentioned autosomal dominant disorders, and also brings forth 3 more (Marfan, Noonan and Nail-patella).
    - M represents **M**arfan syndrome
    - A and APE-alachian represent **APE**rt syndrome
    - N represents Nail-patella syndrome
    - **N** and the "moon shaped mints" represent **N**oonan syndrome
    - T and TUBULAR both represent **TUBE**ROUS SCLEROSIS. Also, the ASHY LEAVES are meant to remind you of the associated Ash Leaf spots, and the DANCING TICKS are meant to remind you of hypsarrhythmia (infantile spasms associated with tuberous sclerosis)
    - R and BLAST represent **R**etino**BLAST**oma
    - A and A-KON both represent **ACHON**DROPLASIA. If you don't know the rapper, please substitute with an African American Munchkin from the Wizard of Oz.
- **MCJAGGER** = Peutz-Jegher syndrome
- **GARDNER** = Gardner syndrome (patients have extra teeth)
- **HUNTING** = HUNTINGTON'S disease (aka Huntingtons disease), **NOT** Hunter syndrome. Hunter syndrome is X-linked. Consider associating this mnemonic with HUNT**ING** at a the most DOMINANT online bank around, **ING** (because the HUNTER mnemonic also has to do with someone who hunts).

## WAARDENBURG SYNDROME

Patients with Waardenburg syndrome may present with albinism, a white forelock of hair, ocular albinism, different colored eyes (heterochromia) and/or deafness.

**PEARL**: The symptoms above are not a MUST. If you see a white forelock of hair in a patient with two blue eyes, you should still **strongly** consider this diagnosis.

**MNEMONIC:** Refer to "WAR" from the intro mnemonic.

**MNEMONIC:** Imagine a scary WAARDEN with bleach WHITE HAIR walking around the jail grounds. He wears an odd pair of multicolored sunglasses. One lens is GREEN and the other is BLUE. He always seems to be yelling angrily, but it's probably just because he's partially DEAF.

**IMAGE**: http://dermatology.cdlib.org/123/case_presentations/waardenburg/1.jpg

**IMAGE**: http://www.ghorayeb.com/files/WaardenbergHeterochromia_cropped.jpg

**IMAGE**: http://i10.photobucket.com/albums/a113/dorlandt/DeafBluebrowneye3H.jpg

**IMAGE**: http://download.imaging.consult.com/ic/images/S1933033208902378/gr30a-midi.jpg

## APERT SYNDROME (aka APERTS SYNDROME)

Patients with Apert syndrome (aka Apert's syndrome) have an early cranial suture closure along with bilateral syndactyly (fused digits). May also have choanal atresia and a cleft palate.

* **(DOUBLE TAKE) NAME ALERT/MNEMONICS:** Apert Syndrome, Alpers Syndrome and Alport's Syndrome can be confused because of their names. Within the autosomal dominant mnemonic, "Ape-alachian" mountains are used. The Appalachian Mountains are in North America so keep that setting in mind as you go through the mnemonic again. For Alpers Syndrome, just remember that the Swiss ALPS are nowhere close to here! Alport Syndrome can be a little confusing. It is more commonly an X-linked Dominant disease, but can be autosomal dominant as well. We recommend memorizing it as an X-linked disease. The mnemonic, "At Al's PORT, X marks the spot where the boats have to dock," can help you remember that Alport Syndrome is X-linked. If you are able to imagine that the X is made out of MINTS, it may help you remember that this is the disease that can be X-linked or autosomal dominant.

**IMAGES**: http://emedicine.medscape.com/article/941723-overview

**IMAGES**: (Try to focus on the images and move one)

http://dermatology.cdlib.org/111/case_reports/apert/verma.html

**MNEMONIC**: APE-alachian mountains in the intro mnemonic

**MNEMONIC**: "*A PEAR Syndrome,*" in which the head is shaped like a PEAR due to early suture closure. Now imagine the pear-shaped head being held by hands that look like a duck's BILATERAL webbed hands.

**MNEMONIC**: (image)

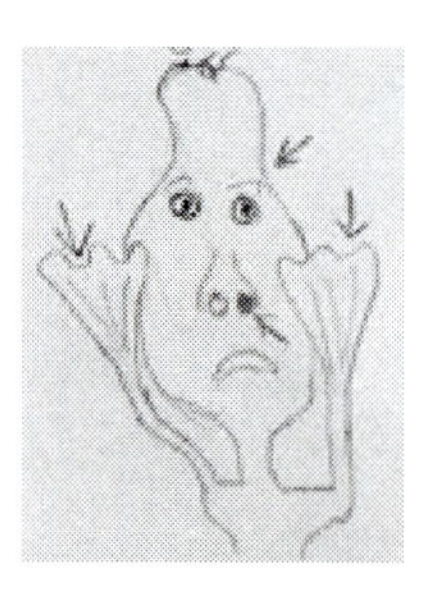

## NAIL PATELLA SYNDROME

Obviously, look for abnormal nails and patellae for nail patella syndrome. Thumbnails are most often affected, then fingernails and then toenails. Nail deformities may mild or severe (pitting, ridging, splitting, discolored or even absent). Patellae deformities also vary (small, irregular or absent). There are also sometimes elbow or iliac deformities (low yield).

**IMAGE**: http://www.ncbi.nlm.nih.gov/books/NBK1132/figure/nail-ps.F1/?report=objectonly

**IMAGE**: http://www.pedsradiology.com/Uploadimg/rtknee0243343470240.jpg

**MNEMONIC**: maNntra

## NOONAN SYNDROME (aka NOONANS SYNDROME)

Noonan syndrome (or Noonan's syndrome) patients have pulmonic stenosis, pectus excavatum, a webbed neck, low set ears and hypertrophic obstructive cardiomyopathy (aka HCM or HOCM). The webbed neck and low set ears are features common to Turner's syndrome, so look for mention of a normal karyotype to help you zone in on a diagnosis of Noonan syndrome. Also, remember that this disorder affects both females and males.

* **IMAGE**: http://reidhosp.adam.com/graphics/images/en/2927.jpg

* **IMAGE**: http://dermimages.med.jhmi.edu/images/noonan_syndrome_7_040212.jpg

* **MNEMONIC:** manNtra from the intro mnemonic

* **MNEMONIC:** Pul-NOOON-ic Stenosis

* **MNEMONIC:** Imagine pulling a Moonman/NOONman BY HIS LOW SET EARS. His skin stretches and causes WEBBING OF THE NECK. The NOONman eventually FALLS ON HIS CHEST creating a PECTUS DEFORMITY that pushes inward on his pulmonary artery and causes PUL-NOON-IC STENOSIS. He finally stands up, **HOC**s a loogie into his PALM and slaps you in the face!

- **HOC**s: Represents hypertrophic obstructive cardiomyopathy.
- PALM: Represents PUL-NOON-IC stenosis.

## ACHONDROPLASIA (aka DWARFISM)

While 15-20% of achondroplasia (aka dwarfism) cases are due to autosomal dominant inheritance, the rest are due to sporadic mutations. Patients are short, have macrocephaly, mild hypotonia, short arms/legs, a relatively normal torso and "trident-like" fingers. They also have a GENU VARUM deformity. They have a normal IQ. Cause of sudden death is COMPRESSION of the CERVICOMEDULLARY JUNCTION. Reproductive fitness is severely affected. Cases due to sporadic mutations are much more common than cases due to inheritance.

**IMAGE**: http://images.paraorkut.com/img/health/images/a/achondroplasia-108.jpg

**IMAGE**: http://images.paraorkut.com/img/health/images/a/achondroplasia-109.jpg

**IMAGE**: http://www.medindia.net/patients/patientinfo/images/shortman1.gif

**IMAGE**: http://4photos.net/photosv2/images_achondroplasia_1274964118.jpg

**IMAGE**: http://peterswordoftheday.files.wordpress.com/2008/09/mini_me.gif

**IMAGE**: http://3432-stoollala.voxcdn.com/wp-content/uploads/2010/09/weeman.jpg

**MNEMONIC**: Imagine a "mini-me" version of Akon walking around with bowed legs, chanting/rapping his mantra of, "down with the MAN**N**TRA!" You could also imagine Mini Me or Weeman being the naughty people/characters they are and running around with a devilish TRIDENT!

## PEUTZ-JEGHERS SYNDROME (aka HEREDITARY INTESTINAL POLYPOSIS)

Criteria for Peutz-Jeghers syndrome (aka hereditary intestinal polyposis) include family history, hamartomatous GI polyps and benign, hyperpigmented macules on the lips or oral mucosa. Diagnosed by noting two out of three criteria. Polyps can lead to obstruction or intussusception, so they should be removed. Lifetime risk of multiple types of cancer is increased, so frequent screenings are done (pap smears, colonoscopy, testicular exam, etc.).

**IMAGE**: http://www.dermis.net:80/bilder/CD006/550px/img0040.jpg

**IMAGE**: http://img.medscape.com/pi/emed/ckb/gastroenterology/169972-1340275-182006-1663047.jpg

**IMAGE**: http://img.medscape.com/pi/emed/ckb/gastroenterology/169972-1340275-182006-1663018tn.jpg

**MNEMONIC**: Imagine MCJAGGER making "kissy lips" to remind you of lip findings.

## GARDNER SYNDROME (aka GARDNERS SYNDROME)

Primary finding for Gardner syndrome (aka Gardner's syndrome) is the premalignant polyps throughout the intestines. Other findings include supranumery teeth (extra teeth) and tumors in other parts of the body (especially benign bone tumors, called osteomas).

**MNEMONIC**: Imagine a psychotic MEXICAN GARDNER using a MACHETE to harvest COLONIC POLYPS that look like red onions from GARDEN of death (premalignant), along with BONES and TEETH. (MACHETE represents need for surgical resection of premalignant polyps).

## (DOUBLE TAKE) RETINOBLASTOMA

The primary finding for retinoblastoma is LEUKOKORIA. There is increased risk of osteogenic sarcoma (aka osteosarcoma) of a long bone in the future. Inheritance pattern is actually a little confusing, with there being a strong genetic component, but also a high rate of spontaneous mutations.

**PEARLS:** Avoid completely discounting just because there is no family history. Retinoblastoma is a prime example of a disorder that often occurs due to spontaneous mutations. If you are given information about **chromosome 13** and an eye problem, or you are given the X-ray of a long bone (such as a femur, tibia, humerus, radius or ulna), consider this diagnosis.

**IMAGE**: http://rbne.files.wordpress.com/2008/08/joli-leukoria1.jpg?w=213&h=192

**MNEMONIC:** "BLAST" from the AD intro mnemonic.

**MNEMONIC**: RAT-inoblastoma. Imagine having a birthday cake made in the shape of your dad wearing a tuxedo. You go into the kitchen to get it, and you find two WHITE RATs knowing at the cake. One is nibbling on the right EYE (retinoblastoma), and the other is nibbling at the TIBIA (a long bone). Just as you get close, the tibia breaks and the entire thing collapses!

**REMINDER**: Leukocoria with loss of the red light reflex is seen in retinoblastoma as well as cataracts.

## OTHER AUTOSOMAL DOMINANT DISORDERS

* von Willebrand Factor Deficiency (aka vWF Deficiency): See Hematology

* HYPERTROPHIC OBSTRUCTIVE CARDIOMYOPATHY (HOCM): An autosomal dominant disorder like Noonan! See Cardiology

* ACUTE INTERMITTENT PORPHYRIA (AIP): A disorder of heme synthesis that causes symptoms when there is accumulation of a metabolite in the cytoplasm. Symptoms may include abdominal pain, confusion, weakness and headaches. Triggers may include infection, low carbohydrate intake and certain medications (seizure and sulfa medications). Treat with a D10 infusion and by stopping any offending agents.

# *AUTOSOMAL RECESSIVE*

**NOTE**: You are about to move on to the autosomal recessive section. The autosomal recessive disorder mnemonic has some overlap with diseases that start with the letters "T" and "A" from the autosomal dominant mnemonic. Please ensure you have solidified in your mind that Tubular Sclerosis, Achondroplasia and Apert Syndrome are autosomal dominant disorders before moving on.

## AUTOSOMAL RECESSIVE (AR) DISORDERS PEARLS

* Most inborn errors of metabolism (IEM) are due to autosomal recessive inheritance. Exceptions include Hunters ("X" marks the spot!) and Ornithine Transcarbamylase Deficiency, both of which are X-linked.

* An AR disorder is ruled out if vertical transmission is noted in 3 successive generations.

* You are much more likely to be asked to identify a dominant disorder than a recessive disorder. Hopefully the mnemonic below will help you rule out certain diseases when it is clear that the patient has a dominant disorder based on the mode of inheritance.

## AUTOSOMAL RECESSIVE MNEMONIC

This autosomal recessive mnemonic is strange, but easy to remember: "**PAT HAS WACK GAS** which made me **HURL** in the **BACK SEAT** of an **AUTO**mobile!" If you are unfamiliar with PAT, s/he is unique Saturday Night Live character. Possibly male... possibly female... definitely full of WACK GAS that would make you HURL in the BACK SEAT of an AUTOmobile!

* PAT: http://upload.wikimedia.org/wikipedia/en/9/99/Itispat.jpg
* P = PHENYLKETONURIA (aka PKU)
* A = ALPHA-1-ANTITRYPSIN DEFICIENCY
* T = TAY-SACHS DISEASE
* H & HURL = HURLER'S SYNDROME
* A = ATAXIA TELANGIECTASIA
* S = SICKLE CELL DISEASE and the THALASSEMIAS
* W = WILSONS DISEASE
* A = **ALP**ERS SYNDROME (AKA Progressive Sclerosing Poliodystrophy): This is a progressive neurologic disease. Patients do not meet their milestones and are noted to have ataxia, cognitive deficits and seizures. Also have liver disease. Die by 10 years of age.
    - **MNEMONIC**: The symptoms of ataxia and cognitive deficits are remarkably similar to how you would feel if you got ALTITUDE sickness in the Swiss **ALP**S!
* C = CYSTIC FIBROSIS
* K = KARTAGENER SYNDROME: = Immotile Celia & Sperm. Lung issues and infertile. "while playing KAR-TAG, atul's CELICA became immotile"
* G = GALACTOSEMIA
* A = see "A" disorders above
* S = see "S" disorder above
* BACK SEAT = May help you remember that this mnemonic deals with **recessive** disorders.
* AUTOmobile = AUTOsomal recessive.

## JOHANSON-BLIZZARD SYNDROME

Johanson-Blizzard syndrome is a multisystem congenital disorder. Findings include pancreatic insufficiency, hypoplasia (or aplasia) of the nostrils and mental retardation.

**MNEMONIC**: Imagine having a stuffy/TIGHT NOSE while being caught in a BLIZZARD.

**MNEMONIC**: (Image) Look! Scarlett Johansson is holding her PANCREAS! http://frillr.com/files/images/Scarlett%20Johansson.jpg

## X-LINKED DISORDERS

**PEARLS:**

* X-linked disorders are **RULED OUT** if you see male-to-male transmissions!

* A vignette about a male patient that discusses ANYTHING about a MATERNAL UNCLE should alert you to possible X-linked disorder.

* Most X-linked disorders are X-linked RECESSIVE. The dominant ones seem to select themselves out. These X-linked recessive disorders tend to be enzyme or protein deficiencies, thus several inborn errors of metabolism are X-linked recessive.

## X-LINKED DOMINANT DISORDERS

X-linked dominant disorders are extremely rare.

* **PEARL**: Because these are dominant disorders, the inheritance pattern will look very similar to an autosomal dominant pattern. A key difference is that the disorder will be present in **100%** of the daughters of an affected MALE.

### FAMILIAL HYPOPHOSPHATEMIC RICKETS

Familial Hypophosphatemic Rickets is also known as VITAMIN D RESISTANT RICKETS.

### AICARDI SYNDROME

In Aicardi syndrome, there is a missing corpus callosum and infantile spasms or epilepsy. Patients also have mental retardation.

### (DOUBLE TAKE) ALPORT SYNDROME (aka ALPORTS SYNDROME)

Alport syndrome (aka Alport's syndrome) is primarily an X-linked dominant disorder. The findings are more severe in males, and include RENAL disease (starts with hematuria and progresses to end-stage renal disease), bilateral sensorineural hearing loss and eye/vision problems. Females may be noted to have hematuria that an eventually progress to ESRD. To recap, boys have hearing, vision AND kidney problems. Girls only have renal issues.

**MNEMONIC**: At **Al's PORT**, **X** marks the spot where the boats have to dock. Consider adding a story about poor AL and the KIDNEY-SHAPED boat he crashed into his PORT because he had POOR VISION and POOR HEARING. It was so bad that he couldn't see or hear everyone on the dock yelling at him to slow down!

# X-LINKED RECESSIVE DISORDERS

**PEARLS**

* Any female who inherits an X chromosome with an X-linked recessive disorder should be a **CARRIER**, and should **NOT BE AFFECTED**. Whether or not this is true is up for debate, but it's a good rule of thumb for the exam. Just assume that there is always X-inactivation if it becomes a Barr body.

* If mom is a carrier and dad is affected, there is a 50% of a SON being affected. Work it out:

| | $X^r$ | Y |
|---|---|---|
| $X^r$ | - Girl | – Boy |
| X | - Girl | – Boy |

**MNEMONIC**: If you squint your eyes, an **X** can look like a tu**X**edo. If you don't like the mnemonics below, make your own by attaching a tuX to a related story.

## (DOUBLE TAKE) CHRONIC GRANULOMATOUS DISEASE (CGD) = SERRATIA

CHRONIC GRANULOMATOUS DISEASE (CGD) is an x-linked recessive disorder. If you see a boy on the boards diagnosed with Serratia, he's probably got CGD. CGD is the result of an enzyme deficiency within neutrophils that is needed to destroy cell walls and aid in phagocytosis. Chemotaxis is intact. This is not a cell line deficiency so there is **NO LYMPHOpenia and NO NEUTROpenia. This is a functional/oxidative burst issue.** The immune system is unable to destroy certain bacteria and fungi (**Aspergillus, Candida, E. coli, Serratia and Staphylococcus**) because they are all catalase positive organisms. The organisms/infections are usually considered to be **low-grade**. Patients may have **liver abscesses, osteomyelitis, recurrent lymphadenitis and recurrent skin infections**. Can also have GI infections and **granulomas of the skin, GI or GU tract**.

* DIAGNOSIS: **Nitroblue Tetrazolium (NBT) dye test or Dihydrorhodamine (DHR) Fluorescence**

* TREATMENT: Aggressive antibiotics at times of infections

* **PEARL**: Like the name says, it's a "CHRONIC" condition. Children usually present by **5 years of age.**

* **MNEMONICS**:

- CGD = This is X-linked disorder, ChroniX GranulomatuX Disease. Put an X where it makes sense for you to remember this key fact.
- NEUTROphil disorder = "**NITROphil** disorder = **NITROBLUE** tetrazolium test which tests for the NITROglycerin oxidative BURST." ChroniX GranulomatuX Disease seems to be all about colors!... BLUE, SERRATIA (RED)!!!

## (DOUBLE TAKE) DUCHENNE MUSCULAR DYSTROPHY

Duchenne muscular dystrophy is an X-Linked disorder (located at Xp21 – low yield) that causes a deficiency in the muscles' DYSTROPHIN protein. Children are usually wheelchair-bound by 7 or 8 years of age and are unable to walk by their 13th birthday. They can have severe scoliosis and the major cause of morbidity (and mortality) is the **involvement of respiratory muscles.** Children have a poor cough and frequent pneumonias. The heart can also be involved (cardiomyopathy). Other symptoms include toe walking, a waddling gait, use of the **Gower maneuver** to stand, **pseudohypertrophy of calf muscles** from fatty/fibrous tissue deposition, and poor head control. Start your workup with a **CK level (like Dermatomyositis)** since it's ALWAYS elevated. If high, refer to neurologist for further workup, including and electromyogram (EMG) and **muscle biopsy**.

**PEARLS:** Though it's an X-linked disorder there is a very **HIGH rate of spontaneous mutations**, so the mom does not have to be a carrier, and there does NOT have to be a positive family history. If the mom **is** a carrier, though, she **can** have elevated CK levels even though she has no symptoms.

**MNEMONIC**: C'mon. Have you **ever** seen a picture of a female patient doing the Gower maneuver? NO! So of course this is an X-linked disease. Now imaging a child going the maneuver while wearing a tu**X**.

**MNEMONIC VIDEO:** Imagine this kid in a tu**X**: http://www.youtube.com/watch?v=bI6utCce_3g

## (DOUBLE TAKE) GLUCOSE-6-PHOSPHATE DEHYDROGENASE DEFICIENCY (G6PD DEFICIENCY)

Glucose-6-Phosphate Dehydrogenase Deficiency (G6PD Deficiency) is an X-linked recessive disorder (so look for a male patient!) resulting in jaundice, dark urine and anemia due to hemolysis from oxidative injury. In newborns, this will present with jaundice in a MALE baby AFTER 24 HOURS. In other children, can result in hemolysis after ingestion of fava beans, malaria medications, trimethoprim-sulfamethoxazole, ciprofloxacin or nitrofurantoin. Look for HEINZ BODIES (purple granules noted in the red cells on microscopy). Associated with African-American and Mediterranean descent.

**PEARL**: Do not test for the deficiency during the acute hemolytic phase (wait a few weeks) because a false negative result can occur due to the build up of G6PD in reticulocytes.

**MNEMONIC**: Imagine a bottle of HEINZ BODIES ketchup that has an **X** located on the exact spot that you have to hit it to make the broken RBCs come out of the bottle.

## (DOUBLE TAKE) HEMOPHILIA A and HEMOPHILIA B (aka FACTOR VIII and FACTOR IX DEFICIENCY)

Hemophilia A and B refer to factor VIII and factor IX deficiencies, respectively. For both hemophilias, the **PTT** is increased. Since this is a factor deficiency, look for deep bleeds (muscle, joints, etc.)

**MNEMONIC**: Have you ever encountered a female hemophiliac? Hmm… You could also imagine the X as a pair of sharp blades going through blood vessels causing bleeding. That could remind you that hemophiliacs have an X-linked disorder.

## HUNTER SYNDROME

For more information on Hunter syndrome, see the Inborn Errors of Metabolism section.

**MNEMONIC**: **X** marks the spot!. Imagine a HUNTER taking aim at a target with an **X**! What do you think he's wearing as he practices his skills? A tu**X**!

## NEPHROGENIC DIABETES INSIPIDUS

Nephrogenic diabetes insipidus is a Foxa1 transcription factor deficiency leading to polyuria from a lack of ADH sensitivity. See Nephrology section.

## ORNITHINE TRANSCARBAMYLASE

Ornithine transcarbamylase is the most common urea cycle defect, and the only one that is X-linked. See the IEM section.

**MNEMONIC**: ornithine transcarbamyla**X**

## RETINITIS PIGMENTOSA

Retinitis pigmentosa is a retinal dystrophy that eventually leas to blindness.

**MNEMONICS:** PIG-men-tu**X**a. Or, imagine a blind, PIG-like man wearing sunglasses, a tu**X** and using a walking stick.

## TESTICULAR FEMINIZATION (aka ANDROGEN INSENSITIVITY)

Testicular feminization (aka androgen insensitivity) results from an androgen receptor deficiency. See the Endocrinology section.

**MNEMONICS:** te**X**ticular feminization. Also, you see "testicular," that should remind you that the patient has testicles and has an XY genotype, and the patient is a "male" with an X-linked disorder.

## (DOUBLE TAKE) WISKOTT-ALDRICH SYNDROME

For Wiskott-Aldrich syndrome, look for a patient with a history consistent with the triad of SMALL **T**hrombocytes, frequent **I**nfections and **E**czema. Platelets can be small in **size and/or number**. This is an X-linked so it will only be found in MALES. Patients develop both T and B cell problems as they ages. IgM is usually LOW, while IgA may be HIGH. A history may include findings such as petechiae, bloody diarrhea, bloody circumcision, eczema and/or otitis media. There is an increased risk of lymphoma in the 3rd decade of life. BMT is curative.

* **MNEMONIC**: Wiskott Aldrich = **wiX**ot**T ostrich** = Imagine a WISe OSTRICH wearing some nerdy black glasses, a tux and a TIE with the word "T-I-E" spelled down the front.

* **MNEMONIC**: (image)

T I E ↓WiXott IgM↓ ↑AIdrich IgA↑

- **KEY: X** Linked, small T = Represents small platelets (low MPV) and Thrombocytopenia, **I** = Infections, **E** = Eczema, Mirrored W = M and arrows show that Ig**M** is low, Duplicate A refers to Ig**A** and arrows show that IgA levels are increased.

* **PEARL**: In Idiopathic Thrombocytopenia Purpura (ITP), platelets are BIGGER

## *TRISOMY DISORDERS*

**PEARLS:** The risk **ALL** trisomy disorders increases with age. Women who are 35 years of age and older are considered high risk. There is an increased risk of having a **VSD**, especially with Trisomy 21. Know that Trisomy 21 is referred to as Down Syndrome, but you are **not** likely to be asked to recognize that Trisomy 18 is Edwards Syndrome, or that Trisomy 13 is Patau Syndrome.

## DOWN SYNDROME (aka DOWNS SYNDROME)

The term Down Syndrome (aka Downs Syndrome) specifically refers to having three Chromosome #21s, but does NOT differentiate between someone having a TRISOMY 21 Downs versus a Translocation Downs (due to a nondisjunction during gametogenesis). Down Syndrome is the most common cause of mental retardation. Fragile X is the most common **inherited** cause of mental retardation.

* **CLASSIC FINDINGS include a** rounded face, almond-shaped eyes with upslanting palpebral fissures, flat nasal bridge, macroglossia, single tansverse palmar crease (aka "simian crease"), poor muscle tone (hypotonia) and a wide space between the first and second toes (aka "sandal toe").

- **IMAGE**: http://newborns.stanford.edu/PhotoGallery/Downs1.html
- **IMAGE**: http://www.lucinafoundation.org/assets/downsyndrome1.jpg

* **OTHER POSSIBLE FINDINGS**

AV canal defect, VSD (**AV canal** >> VSD), atlantoaxial instability, small phallus (similar to Prader Willi), excess nuchal skin, clinodactyly (incurved pinkie finger), obstructive sleep apnea with right sided heart failure and mild mental retardation. Patients can usually live independently. Males are typically infertile while females are not. All patients have an increased risk of leukemia.

- **IMAGE**: (Brushfield Spots) http://www.milesresearch.com/eerf/images/brushfieldspots-600.jpg
- **PEARL**: **Knowing the etiology of a patient's Downs is extremely important for the purposes of counseling**. Since this is such a confusing topic, if all else fails, just choose to do a karyotype on the baby or the parents and MOVE ON TO THE NEXT QUESTION.

* **TRISOMY 21 DOWNS**: This is the most common form (95%) and is due to nondisjunction of chromosome 21. A karyotype of Mom and Dad's chromosomes is NOT needed because they would show normal chromosomes. The risk for recurrence of TRISOMY DOWNS in future pregnancies is equal to the mother's age-related risk + 1%.

* **TRANSLOCATION DOWNS**: Very confusing. Parental karyotyping IS needed. Please start by looking at the shortcut below. If that is enough for you, MOVE ON. If you need more info to make it stick in your mind, start by clicking on the link below and just reading the **first four** paragraphs (then stop!). So essentially, if the child's karyotype shows a translocation, get karyotyping of both parents' chromosomes. The carrier parent will have either the main part of chromosome 21 attached to a different numbered chromosome (usually chromosome 14, in which case this is a partial translocation), or the main part of one chromosome 21 will be attached to the main part of **another** chromosome 21. The second possibility may be shown as t(21q;21q) and carries a 100% risk of future Translocation Down Syndrome babies because that parent only has that chromosome (21 attached to 21) to donate. The first possibility (in which 21 is attached to 14) carries about a 10-15% chance of future Downs pregnancies if the MOM is the carrier, and about a 5% if the DAD is the carrier. If you take the time to work out the actual inheritance pattern, it's theoretically 33%. But nature doesn't seem to care. In any translocation carrier, that person's siblings should also be given the opportunity to be screened.

* **SHORTCUT:**

Trisomy 21: Age related risk + 1% is the risk of future Downs pregnancies.

Translocation in which the parent is carrying a 21 attached to 21: 100% Risk (all future children will have Downs).

Anything else: < 15% Risk

LINK: http://www.chkd.org/HealthLibrary/Content.aspx?pageid=P02153

SUMMARY TABLE:

| KARYOTYPE | | | |
|---|---|---|---|
| | **DAD** | **MOM** | **Risk of Recurrence** |
| Trisomy 21 | Normal | Normal | 1% + Age-related risk |
| 21/14 | Normal | Carrier | 10-15% |
| | Carrier | Normal | 5% |
| 21/21 | Normal | Carrier | 100% |
| | Carrier | Normal | 100% |

*** ATLANTOAXIAL INSTABILITY**

Patient may have a history of a gait disturbance and may have brisk reflexes on exam. The American Academy of Pediatrics (AAP) says to screen with X-rays between 3 to 5 years of age, and for anyone going to Special Olympics. This can be a tricky diagnosis to make because by definition, there is instability between C1 and C2 so the X-ray findings can change over time from normal and abnormal, and vice versa. So if you suspect the diagnosis, don't be shy about repeating an X-ray, or even getting an MRI (required if there are clear symptoms).

## TRISOMY 18 (aka EDWARDS SYNDROME)

Trisomy 18 (aka Edwards syndrome) findings include rocker-bottom feet (rounded, rocking-chair like plantar surface), clenched fist with overlapping fingers, underdeveloped/hypoplastic nails, horseshoe kidneys, crossed legs, small jaw (micrognathia), prominent occiput and pectus excavatum. Can also have cardiac defects such as an ASD or VSD. Horrible prognosis. Most die in utero or by the age of 1.

**IMAGE**: http://www.lucinafoundation.org/assets/trisomy18a.jpg

**IMAGES**: http://emedicine.medscape.com/article/943463-overview#showall

**MNEMONIC**: Imagine an 18 year old, named Ed, who is dressed in a red, white and blue suit for his birthday party. He's feeling so patriotic because he can finally vote! He jumps onto and stands on a ROCKING chair that has legs which look like HORSESHOE KIDNEYS. He gives a thank you speech. At the end he throws his right hand up in the air like a ROCKER with a CLENCHED FIST AND OVERLAPPING FINGERS three times as he says, "Thank you! Thank you! Thank you!" It's a little shaky up there on the ROCKING chair and he falls. As he falls, he hits his CHEST on an arm of the chair, and as he tries to brace himself his CLENCHED FIST get run over by the HORSESHOE KIDNEY shaped legs and CRUSH HIS FINGER NAILS.

## TRISOMY 13 (aka PATAU SYNDROME)

Trisomy 13 (aka Patau syndrome) findings include cardiac defects, cleft lip +/- palate, punched out scalp lesions, polydactyly, low set ears, cystic kidneys, microcephaly and microphthalmia. Poor prognosis. Most die by 6 months of age.

**IMAGE**: http://www.lucinafoundation.org/assets/trisomy13.jpg

**PEARL**: Likely a low-yield disorder. Move on if you find yourself spending too much time here.

**MNEMONIC**: Imagine JASON from FRIDAY THE **13**$^{TH}$ wearing his hockey mask to cover up his CLEFT LIP. Note that it has PUNCHED OUT LESIONS on it. In one hand he has a sword, and the other hand has "EXTRA FINGERS." See the image below for this visual pun - http://popdose.com/wp-content/uploads/jason01.jpg

## *MISCELLANEOUS GENETIC FINDINGS & DISORDERS*

**NOTE**: This section includes disorders without an established inheritance pattern or genetic etiology.

### TERMINOLOGY

* SYNDACTYLY: Fusion of two or more fingers. May also include fusion of bones.

* CLINODACTYLY: Permanent deviation of one or more fingers.

- **PEARL**: It's usually the 5th digit. For the boards, ALWAYS look at the pinkie finger when a picture of a hand is shown.
- **IMAGE**: http://www.aafp.org/afp/2005/0715/afp20050715p279-f5.jpg
- **IMAGE**: http://upload.wikimedia.org/wikipedia/commons/3/3f/Sampleclinodactyly.jpg

### CLEFT DISORDERS

Cleft disorders are multifactorial conditions, which means that there is a significant chance of recurrence in future pregnancies (about 5%). Multifactorial conditions are frequently associated with other anomalies.

**IMAGE**: www.thechildrenshospital.org/imgs/KidsHealth/image/ial/images/1322/1322_image.gif

* **CLEFT PALATE**: A child born with a CLEFT PALATE is much more likely to have other anomalies than a child born with a cleft lip. These patients are also more likely to have an underlying chromosomal anomaly than a cleft lip patient OR even a cleft palate with a cleft lip. So, get a karyotype! There is a high risk of recurrence, HIGH risk of associated anomalies, and HIGH risk of an associated chromosomal defect so GET A KARYOTYPE.

* **CLEFT LIP**: High risk of recurrence, low risk of associated anomalies, low risk of an associated chromosomal defect.

* **CLEFT PALATE & CLEFT LIP**: High risk of recurrence, low risk of associated anomalies, low risk of an associated chromosomal defect.

* **DISEASE ASSOCIATIONS**: Cleft lips **and/or** palate are associated with various disorders including Treacher Collins, DiGeorge Syndrome, Crouzon Syndrome, Apert Syndrome, Pierre-Robin Syndrome and Trisomy 13

- **MNEMONIC**: Imagine that a PINK PLATE represents any sort of cleft deformity. This has been chosen because lips and palates are pink, and "plate" looks/sounds like palate. Here's the story: Your favorite TREACHER loves animals and the color pink. She's wearing a pink dress as she walks out of her classroom with a PINK PLATE filled with CROUTONS and PEARS. As soon as she pus it down, a ROBIN flies over and a CURIOUS MONKEY (named Curious George) shuffles over to eat the yummy food on the PINK PLATE.
- **KEY**: TREACHER Collins, Cruzon (CROUTONS), Apert (PEARS), Pierre-Robin (ROBIN) and DiGeorge (CURIOUS GEORGE).

* **SUBMUCOUS CLEFT PALATE**

These are difficult to see because they are covered by a layer of normal looking mucosa, so look for a BIFID UVULA. Patients can have hearing deficits due to problems with tympanic membrane mobility so obtain tympanometry. If the patient has difficulty forming certain words after the age of 5, surgically repair.

- **IMAGE**: http://upload.wikimedia.org/wikipedia/commons/b/bf/Bifid_uvula.JPG

## WILLIAMS SYNDROME (aka WILLIAM'S SYNDROME)

In Williams Syndrome (aka William's Syndrome), while patients have mild mental retardation, they also have a fun, "cocktail party" personality, so they get along great with other people. Facial features include an elf-like face ("ELFIN" FACE), a low and flat nasal bridge, an upturned nose, a smooth philtrum and wide-spaced teeth. Other problems include SUPRAvalvular aortic stenosis (not involving the valve, but the aorta itself), peripheral pulmonic stenosis and hypercalcemia. Can occasionally also have hypothyroidism.

**PEARL**: If you see the word SUPRAvalvular stenosis, have a HIGH suspicion for this diagnosis.

**IMAGE**: http://en.wikipedia.org/wiki/File:Poor_little_birdie_teased_by_Richard_Doyle.jpg

**IMAGE**: http://www.vaheart.com/wp-content/uploads/2010/11/Aortic-Stenosis.jpg

**MNEMONIC**: Think of your favorite (or least favorite) celeb with the last name Williams for this mnemonic. Serena Williams? Robbie Williams? Robin Williams? Vanessa Williams?. Now, imagine __________ WILLIAMS at a PARTY riding a COW around the PERIPHERY of the group. The cow has an ELFIN FACE and WIDELY SPACED TEETH. WILLIAMS is having fun, but ISN'T TOO SMART. S/he tries to make the cow go fast and then JUMP OVER A HURDLE. The COW suddenly stops and __________ WILLIAMS goes flying off the COW. S/he flies OVER THE HURDLE, and as s/he lands his/her PHILTRUM hits a rock. WILLIAMS' PHILTRUM is now LONG and the nose is UPTURNED. Now WILLIAMS is in the middle with the group on the PERIPHERY taking care of him/her.

* **KEY:** PARTY represents the personality, COW represents calcium for dairy products, PERIPHERY represents peripheral pulmonic stenosis, ISN'T TOO SMART represents the mild mental retardation and JUMP OVER THE HURDLE represents **supra**valvular stenosis (if you remember nothing else, remember THIS finding).

## HOLT ORAM SYNDROME

Holt Oram Syndrome findings include upper limb and cardiac defects. Look for an ABSENT or HYPOPLASTIC radius and thumb, TRIPLE-jointed thumb and/or septal defects (usually a VSD).

**MNEMONIC**: Repeat the following phrase over and over again - "HOLD an ARM by a TRIPLE-JOINTED THUMB." Now imagine that the first joint of the thumb says V, the next says S and the last says D (V-S-D).

* **KEY:** HOLD an ARM represents Holt Oram, the ARM represents upper limb deformities and the V-S-D written on the thumb joints represents the most common septal defect.

## CRI-DU-CHAT SYNDROME (aka 5p- or 5p SHORT ARM DELETION

Cri-Du-Chat Syndrome (aka 5p- or 5p short arm deletion) findings include a high pitched cry, developmental delay, microcephaly, wide-set eyes (hypertelorism), a high palate and possibly a VSD. The name means "cry of a cat." Cries sound like the meow of a cat.

**IMAGE**: http://upload.wikimedia.org/wikipedia/commons/c/c9/Criduchat.jpg

## CROUZON SYNDROME (aka CRANIOFACIAL DYSOSTOSIS)

Crouzon syndrome (aka craniofacial dysostosis) findings include craniosynostosis (early suture closure), with a prominent forehead, proptosis, a cleft lip/palate and the **lack of syndactyly**.

**PEARL**: Early suture closure also happens in Apert Syndrome ("A PEAR"), but there is no fusion of digits (syndactyly) in Crouzon.

**IMAGE**: http://geneticpeople.com/wp-content/uploads/2010/02/pla_crouz2.jpg

**IMAGE**: http://giemsanotserology.cascadiat.net/wp-content/uploads/2010/06/crouzon.jpg

**MNEMONIC:** Remember the pink plate (palate) with CROUTONS on it from the mnemonic earlier?

**MNEMONIC:** Sorry for any weak stomachs. Imagine a person who is missing eyes and a cleft palate. You put two SUTURES in the FOREHEAD of the patient above the eyebrows. You leave some extra SUTURE material and then SUTURE through two CROUTONS as well and let them hang in front of the empty eye sockets (CROUTON EYES). You then take some crushed CROUTONS and shove them into the CLEFT PALATE to make it nice and even.

* **KEY:** SUTURES represent craniosynostosis, FOREHEAD represents a prominent forehead, CROUTON EYES represent proptosis and the filling of the cleft with croutons should help solidify the finding of cleft lip/palate in patients with Crouzons.

## FRAGILE X SYNDROME

Fragile X syndrome findings include mental retardation, delayed milestones, socially awkward, a long face with a prominent jaw, large ears and macro-orchidism (oversized testicles).

**PEARLS:** This is the most common **inherited** cause of mental retardation. The most common cause of MR is Downs. Though both boys and girls carry the defective gene, it's not usually symptomatic in females. So, look for a MALE patient on the pediatric exam. Assume females don't express it because the bad X becomes inactivated.

**IMAGES**: http://medgen.genetics.utah.edu/photographs/pages/fragile_x.htm

**MNEMONIC**: (and mnemonic image) Note the X's below. The top makes a large V to remind you of the LONG face and CHIN. The smaller X's represent LARGE EARS. Note how they are fairly large compared to the rest of the face. You could also imagine that they represent X-shaped earrings. Lastly, the OO represent macro-orchidism.

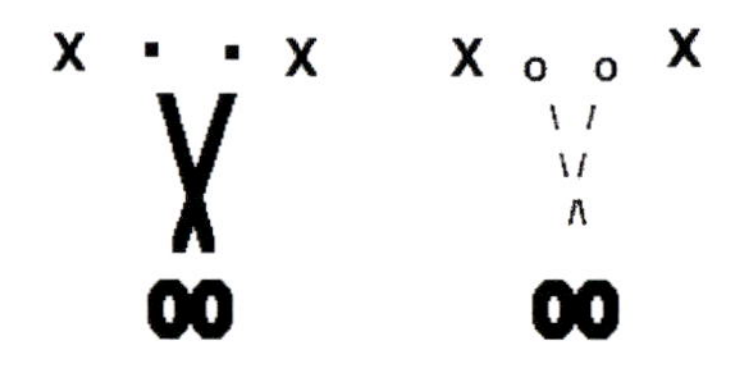


## ANGELMAN SYNDROME (aka ANGELMANS SYNDROME)

Angelman syndrome (aka Angelman's syndrome) patients can be MALE or FEMALE so don't let the name fool you! These patients tend to be "angelic" and laugh frequently. They are easily excited, can have seizures and have what is referred to as a "puppet gait." The condition occurs due to PATERNAL DISOMY, in which certain MATERNAL genes are deleted and there is imprinting of the PATERNAL genes.

**PEARL**: The "puppet gait" terminology may have fallen out favor, but refers to an unstable and jerky gait like that of a puppet. The concept of unilateral parental disomy can be confusing and is great exam question for the pediatric boards. Hopefully the mnemonic below helps.

**MNEMONIC:** Rename the disease AngelWOMAN to remind you that it can occur in females.

**MNEMONIC:** Imagine a FEMALE ANGEL with wings. She has a SMILE on her face and is wearing a t-shirt that says "DADDY'S LITTLE ANGEL." Even though she has wings, the man upstairs decides he is going to pull her to the clouds with some PUPPET STRINGS.

* **KEY:** FEMALE ANGEL represents that this is a male OR female disorder, SMILE represents frequent laughter, DADDY'S LITTLE ANGEL represents paternal disomy in which she is carrying two sets of Dad's genes and the PUPPET STRINGS is to remind you of the "puppet gait."

## PRADER-WILLI SYNDROME (aka PRADER WILLI SYNDROME)

Prader-Willi syndrome (aka Prader Willi syndrome) patients can have hypotonia (floppy baby), mild retardation, almond shaped eyes (often with mild strabismus), small hands, a HUGE appetite, obesity and small testicles/penis in boys.

* **PEARLS:** Like Angelman's, it can also be found in BOYS or GIRLS. This disorder occurs due to MATERNAL DISOMY with MATERNAL IMPRINTING. That means Dad's genes have been deleted and the patient has two of mom's copies of this particular gene. Symptoms are much milder in females. The gene location is probably not important for the test, but the fact that it is KNOWN (15q11-13) means that this disorder can be diagnosed by FISH.

* **IMAGE**: http://www.aafp.org/afp/2005/0901/afp20050901p827-f1.jpg

* **IMAGE**: http://www.nature.com/ng/journal/v40/n6/images/ng.158-F1.jpg

* **MNEMONIC:** Imagine a FAT Will Smith with TINY HANDS shoving tons of ALMONDS in his mouth. He's wearing a t-shirt that says, "MOMMY'S LITTLE FATTY."

- **KEY:** ALMONDS represent the shape of the eyes and the t-shirt represents maternal imprinting.

* **MNEMONIC:** Sorry, this is a good mnemonic, but not politically correct. FISH + WILLY = _______ WILLY! Imagine a HUGE, OBESE and DUMB FISH/WHALE named WILLY with such a SMALL PENIS that you can hardly see it. It's so DUMB that it tried to jump over a dock, but ended up landing on it instead. Now he's stuck on the dock, HUNGRY and just FLOPPING around.

* **MNEMONIC IMAGE**: http://www.imdb.com/media/rm2549784064/tt0106965

- **KEY:** FISH represents the mode of diagnosis, HUGE/OBESE represents obesity, DUMB represents mental retardation, HUNGRY represents the insatiable appetite and FLOPPING represents the hypotonia. In this mnemonic, you could also make Willy's eyes almond shaped and imagine that he has small hands/fins.

## LAURENCE MOON BIEDL SYNDROME

Laurence Moon Biedl syndrome is a condition similar to Prader Willi, but it is PROGRESSIVE. It involves the eyes (retinitis pigmentosa), the brain and the endocrine system. In males, there is gynecomastia.

**MNEMONIC**: Rename it "insurance mOoOn." The two capital "O's" look like EYES and bOObs. The lower case "o" represents a small brain to help you remember that there are PROGRESSIVE CNS issues. INSURANCE kind of sounds like Laurence, and may help you think of PROGRESSIVE INSURANCE!

**(DOUBLE TAKE) MNEMONIC:** If you pee all over the floor because of hypospadias, you will likely be called a **SLOB**." **S**ilver Russell Syndrome, **L**aurence-Moon-Biedle Syndrome, **O**pitz Syndrome and **B**eckwith-Wiedemann Syndrome.

## BECKWITH-WIEDEMANN SYNDROME

Beckwith-Wiedemann syndrome findings include MACROglossia, MICROcephaly, MACROsomia (large body), midline abdominal wall defects and evidence of **hemihipertrophy (aka hemihyperplasia)**. The midline defect may be an omphalocele (bowel through the umbilicus) or umbilical hernia. Babies will often have neonatal hypoglycemia. Children are at increased risk for malignancies, especially Wilms Tumor. Look for a picture of a patient with a big body, big tongue and some type of overgrowth symptom (unilateral overgrowth of a leg, face, arm, etc.).

**PEARL**: It's the most common overgrowth syndrome in infancy and it would be VERY, VERY worthwhile to learn about it. .

**IMAGE**: http://graphics8.nytimes.com/images/2007/08/01/health/adam/17076.jpg

**IMAGE**: http://www.endotext.org/pediatrics/pediatrics1b/figures1b/figure6.jpg

**IMAGE**: http://www.sonoworld.com/fetus/page.aspx?id=2257 (DON'T READ, JUST LOOK!)

**MNEMONIC**: Replace the "W's" with "V's". Rename it BeckVvith-VvIDE mouth. The different sized V's are to remind you of the hemihypertrophy (often seen on the ABP exam). The "Vvide mouth" is needed to accommodate the large tongue in this very small head!

## (DOUBLE TAKE) KLIPPEL-TRENAUNAY SYNDROME

Klippel-Trenaunay syndrome is associated with AV fistulae causing skeletal or limb OVERGROWTH (hemihypertrophy). Patients with Klippel-Trenaunay have Port Wine Stains and overgrowth of tissue, bones and soft tissue. Look for unilateral limb overgrowth and CHF.

* **IMAGES**: http://geneticpeople.com/wp-content/uploads/2011/03/bobbyreport001xb.jpg

* **PEARL**: Hemihypertrophy images on the pediatric exam should very quickly clue you in to a few disorders. Highest on your differential should be Beckwith-Wiedemann Syndrome, then Klippel-Trenaunay, then Russell-Silver Syndrome and then possibly Proteus Syndrome.

* **MNEMONIC**: From now on, say CRIPPLE-**T**. Think of these patients as having a CRIPPLING disorder in which they have one HUGE leg that prevents them from getting around.

* **NAME ALERT:** KLIPPEL-FEIL SYNDROME. Completely different disorder. Look for a Torticollis-like photograph (due to fused cervical vertebrae).

- **IMAGES**: http://emedicine.medscape.com/article/1264848-media

## PROTEUS SYNDROME

Proteus syndrome is a rare condition that is highly variable, but can also cause hemihypertrophy. Probably low-yield for the pediatric boards, but as mnemonic, let's rename it "HEMIHYProteus" to remind us of the HEMIHYPertrophy that may be seen.

## PIERRE-ROBIN SYNDROME (aka PIERRE-ROBIN SEQUENCE)

In Pierre-Robin syndrome (aka Pierre-Robin sequence), there is a triad of cleft palate, micrognathia (small chin) and glossoptosis (tongue sticks out because of their relatively small chin, not because of macroglossia). May also have extremity (syndactyly or clinodactyly), spine and CNS issues (delay or seizures). Patients are at risk for difficulty breathing and developing cor pulmonale (right sided heart failure) within the first year of life.

**IMAGE**: http://img.medscape.com/pi/emed/ckb/pediatrics_surgery/993855-995706-1587.jpg

**IMAGE**: http://img.medscape.com/pi/emed/ckb/otolaryngology/834279-844143-28.jpg

**MNEMONIC**: Remember the ROBIN from the PINK PALATE (cleft deformities) mnemonic? With the PEAR and CROUTONS? Now imagine that ROBIN has such a SMALL JAW that the CURIOUS MONKEY eats all the food while the ROBIN struggles to get any food at all.

## CHARGE SYNDROME

CHARGE syndrome is an acronym, which stands for Coloboma of the eye, Heart defects, Atresia (coanal), Retarded **growth**, Genitourinary abnormalities and Ear anomalies. If you see a few of these, make the diagnosis.

**MNEMONIC**: Imagine a CHARGE card machine that is shaped like a HEART. It has a single EYE, a single NOSTRIL, and a single EAR.

* **KEY:** HEART represents cardiac defects, SINGLE EYE represents coloboma, SINGLE NOSTRIL represents coanal atresia, SINGLE EAR represents ear deformities and the $-earring is to remind you that this mnemonic is for the CHARGE CARD SYNDROME!

## COCKAYNE SYNDROME

Cockayne syndrome is characterized by PREMATURE AGING, photosensitive skin, growth failure, hearing loss and vision problems.

**MNEMONIC**: Imagine a COCKY OLD MAN who lives in the woods as a hermit. The woods are being torn down to build new homes. When the bulldozers are close his home, he FINALLY HEARS something and goes outside. He is BLINDED by the sun and quickly gets SUNBURNED. Since he's such a COCKY OLD MAN, he raises his cane at the bulldozers and refuses to leave.

* **KEY:** COCKY = name of disease, OLD MAN = premature aging, FINALLY HEARS = hearing problems, BLINDED = vision problems and SUNBURNED = photosensitivity.

## AUTISM

Children with autism have delayed language, serious social impairment, compulsive and repetitive behaviors. This is a CLINICAL diagnosis and is based on observations of the behavior of a child. At least one of the mentioned impairments must have been prior to the age of 3.

**PEARL**: For autism, or any clinical suspicion of a language/mental delay, start your workup with a HEARING TEST.

**IMAGE**: http://en.wikipedia.org/wiki/File:Autism-stacking-cans_2nd_edit.jpg

## ASPERGER SYNDROME (aka ASPERGERS SYNDROME)

Asperger syndrome (aka Asperger's syndrome) is similar to autism because patients have impaired social interactions and repetitive stereotypical behavior, but there is **NO delay in language or cognitive development**

## RETT SYNDROME (RETTS SYNDROME)

Rett syndrome (Rett's syndrome) results in a regression of developmental milestones and deceleration of growth (look for MICROCEPHALY) after about a year. They often have breath holding spells and seizure. This is an **X-linked** disorder, but it tends to affect GIRLS more than boys.

**MNEMONIC**: we REgre**X** to inform you that your child, RE**X**, is REgre**XX**ing and might be REtarded. *PBR's REgre**X** for the non-PC mnemonic, but hopefully it helps.*

**MNEMONIC VIDEO:** http://www.youtube.com/watch?v=wPLb7LT9Rq0 This is a long video so only watch it if you like "dance," or if you struggle to remember that this is the disease with regreXXion of motor skills that tends to affect girls > boys. *PBR's REgre**X** if this makes you cry.*

## (DOUBLE TAKE) KLINEFELTER SYNDROME (aka KLINEFELTERS)

Klinefelter Syndrome (aka Klinefelters) = XXY. Gynecomastia, small testicles/infertile and normal intelligence to MILD MR. May have a mild motor or speech delay. Tall stature with long arms/legs. Low upper to lower segment ratio. Workup = REFER FOR CHROMOSOMAL ANALYSIS!

**MNEMONIC:** Gynecomastia = Kalvin Kline FELT HER BREASTS

**PEARL**: Mild mental retardation may be described as a patient who is "awkward," "below average in school" or even just "shy." Generally, though, these patients have some learning disabilities but their IQ can be normal.

**PEARL**: Someone who is tall with long arms/legs could also be described as having a LOW "upper to lower segment ratio."

## (DOUBLE TAKE) MARFAN'S SYNDROME (aka MARFANS SYNDROME)

Classic features of Marfan's Syndrome (aka Marfans Syndrome) include tall stature with long and thin upper extremities, long fingers, a pectus deformity, joint flexibility/hypermobility and possible cardiac problems. Cardiac problems may include mitral **valve prolapse (MVP), aortic dissection** and mitral or aortic regurgitation. Patients may have a high arched palate and a speech disorder, but do NOT have cognitive deficits. Patient's are also at risk for esophageal perforation.

* **PEARLS:** Patients can have subluxation of the lens, which may also be seen in Ehlers Danlos and homocysteinuria. If they mention SUPERIOR subluxation of lens, pick Marfans. Any patient with Marfans should not be cleared for sports participation until they have had an echocardiogram and an evaluation by a cardiologist. If they mention "arm span greater than height," you're done.

* **IMAGES**: Please do not get distracted by the reading. Look at the images and move on.

- http://trialx.com/curebyte/2011/05/21/marfan-syndrome-photos/
- http://www.hughston.com/hha/b_12_2_4a.jpg

* **MNEMONIC**: http://www.clevelandleader.com/node/6446 Michael Phelps won several gold medals. Isn't that just like winning the Most Valuable Plater (MVP = Mitral Valve Prolapse)?

## EHLERS DANLOS SYNDROME

Ehlers Danlos syndrome has many similarities to Marfans Syndrome, including joint hypermobility, possible mitral valve prolapse (MVP) and aortic dissection. **Skin laxity (loose skin)** is a key differentiating feature. Other unique features include contractures, skin nodules, **easy bruisability** and poor wound healing. Also, after getting wounded patients form "cigarette paper" scars.

**IMAGE**: http://www.gurumed.org/wp-content/uploads/2010/11/EhlersDanlos.jpg

**IMAGE**: http://www.netterimages.com/images/vpv/000/000/012/12481-0550x0475.jpg

**IMAGE**: http://geneticpeople.com/wp-content/uploads/2009/09/untitled1wk3.jpg

**IMAGE**: http://dermatlas.med.jhmi.edu/data/images/ehlers_danlos_1_050124.jpg

**MNEMONICS:** Rename it as "That feller's *daaamn* loose!" = "That fella is damn loose," said with a southern drawl/accent). Also, if you get confused between Marfans and this one, remind yourself that Michael Phelps doesn't have any skin findings.

## (DOUBLE TAKE) HOMOCYSTEINURIA

Homocysteinuria is an aminoacidopathy with physical features that are Marfanoid, but this includes COGNITIVE DEFICITS and neurologic deficits. Also, like Marfan Syndrome, there can be displacement of the lens. However, the displacement is DOWNWARDS or POSTERIORLY. These patients are hypercoagulable so there may be a history of DVT. It can be diagnosed by noting high levels of homocysteine in the urine. Treat with pyridoxine or diet high in cysteine and low in methionine.

**MNEMONIC**: (Image of Homoerectus) http://bit.ly/roDnZq

**MNEMONIC:** Imagine a proud and TALL HOMOerectus next to a fire. He suddenly develops CLOTS in his BRAIN that cause him to FALL DOWN, face forward so that he injures his EYES. He now has NEUROLOGIC DEFICITS because of the CLOTS in the brain, and his ocular lens is DOWNWARDLY displaced.

**MNEMONIC:** This sounds intuitive, but how do you treat someone that is loosing an amino acid in their urine? Give them more of it! Treat with **cysteine.**

## (DOUBLE TAKE) TURNER SYNDROME (aka TURNERS)

Turner Syndrome (aka Turners) has a high association with Aortic Coarctation. Look for a WEBBED NECK (also called CYSTIC HYGROMA). May have a history of having only breast buds, but they do NOT have breasts. Short stature, pedal edema, wide spaced nipples, short 4th and 5th metacarpals and may have either scant or no pubic hair. Primary amenorrhea occurs because of streak ovaries and ovarian failure. They **DO** have a uterus. High FSH always means ovarian failure (top 2 reasons are Turners and autoimmune ovarian failure) = HYPERGONADOTROPIC HYPOGONADISM. Renal anomalies including a possible HORSESHOE kidney.

**IMAGE**: http://bioh.wikispaces.com/file/view/turner.jpg/33628275/427x427/turner.jpg

**IMAGE**: http://www.endocrineonline.org/gif%20box/Ullrich.gif

**IMAGE**: http://newborns.stanford.edu/images/footedemaDAClark.jpg

**MNEMONIC**: Association between XO and coarctation of the Aorta - "XO kind of looks like AO (abbreviation for aorta), doesn't it?"

**MNEMONIC**: Coarctation of the Aorta and Horseshoe Kidney = In case you didn't know, Tina TURNER was married to a man who was reportedly abusive. His name was Ike. Now, imagine Tina TURNER'S hand thrusting through Ike's chest and squeezing his AO... He TURNS and hits her in the head with a HORSESHOE!" KEY: Squeezing his AO = Coarctation of the Aorta. Horseshoe trauma = Horseshoe Kidney.

## RUSSELL-SILVER SYNDROME (aka SILVER RUSSELL SYNDROME)

Russell-Silver syndrome (aka Silver Russell syndrome) is a type of dwarfism. Characteristics include growth retardation, clinodactyly, a small chin that gives a TRIANGULAR FACE and hypospadias. Patients can also have hemihypertrophy of the head, trunk, arms or legs.

**IMAGES**: http://drugster.info/img/ail/1961_1974_3.jpg

**IMAGE**: http://molconditions.files.wordpress.com/2011/04/silverrussellfig2.jpg?w=155&h=378

**MNEMONIC**: Imagine a "RESSELLER" (wrestler) with a face that is a lop-sided TRIANGLE (small chin with possible facial hemihypertrophy). The face is made out of SILVER to make it easy for him to head butt people. Also, every time he RESSELLS, his yellow PEE gets all over the white RESSELLING matt (hypospadias). With tricks like that, of course he's going to win a silver medal!

## POTTERS SYNDROME

Potters syndrome is characterized by poor production of urine in utero. Due to the lack of fluid (oligohydramnios) babies are born with multiple problems, including pulmonary hypoplasia and club feet. Other findings may include excess/baggy skin that looks like a loose fitting glove and fetal membranes with yellowish nodules.

**IMAGE**: http://bit.ly/U1simV

**MNEMONIC**: Imagine that you are a POTTER. You forget a clay POT on the porch. You find it weeks later with only has a few drops of water (oligohydramnios) at the bottom of it, and it now has nasty YELLOW NODULES along the inside wall. You're so disgusted that you can BARELY BREATHE (pulmonary hypoplasia). You put on a couple of GLOVES (glove-like skin), grab two CLUBS (club feet) and smash the nasty thing to pieces!

## (DOUBLE TAKE) VACTER-L (aka VACTERL or VATER) SYNDROME

VACTER-L (aka VACTERL or VATER) syndrome is an acronym. VACTER**L** is now used instead of VATER because it stands for **V**ertebral anomalies, **A**nal atresia/imperforate anus, **C**ardiac defects (especially VSD), **T**racheoesophageal fistula, **R**adial hypoplasia and **R**enal anomalies, and **L**imb abnormalities. These children have a normal IQ. When associated with hydrocephalus, this can be an X-Linked disorder.

**PEARL**: May present with single umbilical artery.

**MNEMONIC**: Imagine Darth VACTER cutting off his own son's ARM (radial hypoplasia and limb abnormalities) and then using the ARM as a light saber to create ANAL ATRESIA and a TE Fistula."

**IMAGE**: (go to 5:30 in the video) http://www.youtube.com/watch?v=C-Del3ohVbY

**IMAGE**: http://www.youtube.com/watch?v=uXQxstTrhTI

## (DOUBLE TAKE) PRUNE BELLY SYNDROME

Prune belly syndrome results in weak abdominal musculature (due to deficient abdominal muscles), cryptorchidism (**missing** testicles), undescended testicles and **urinary tract anomalies.** The urinary tract anomalies include vesico**ureter**al reflux (VUR), posterior urethral valves (PUV) and renal dysplasia. The renal anomalies often result in **hydronephrosis, oligo**hydramnios and pulmonary hypoplasia. Children have a poor prognosis and most unfortunately die within a few months.

**PEARL**: Look for a description of a **newborn** with weak abdominal muscles, "undescended testes" (may actually be missing) and a midline abdominal mass (distended bladder or hydronephrosis).

## GENETIC TESTING

Genetic testing is **NOT** recommended when dealing with a disease of **adult onset** because the child has no choice in the matter. Recommend waiting until s/he's old enough to make his/her own decision about testing.

## MISCELLANEOUS ABNORMALITIES OF FINGERS AND TOES

* **CLINODACTYLY**: Most commonly the pinkie finger (5th digit). Associated with Down Syndrome, Fetal Alcohol Syndrome, Russell Silver Syndrome and Pierre-Robin Syndrome.

- **MNEMONIC**: downs = down5.
- **MNEMONIC**: If you're having a hard time remembering the definition, think of what it means to CLING on to something. You have to "CURVE" your fingers!

* **SANDAL TOE DEFORMITY**: Down Syndrome

* **2-3 SYNDACTYLY**: Fused 2nd and 3rd digits are found in Smith-Lemli-Opitz

- **IMAGE**: http://newborns.stanford.edu/images/syndactylyDAClark.jpg
- **MNEMONIC**: smYth lemlY opYtz since they look like a "Y"

* **SMALL 5th FINGERNAIL**: Fetal Alcohol Syndrome

* **SHORT 4th AND 5TH FINGER**: Turners Syndrome

* **TRIPHALANGEAL THUMBS**: Holt Oram SYNDROME and Diamond-Blackfan Anemia

* **HYPOPLASTIC DISTAL PHALANGES**: Due to Warfarin exposure in utero

# HEMATOLOGY & ONCOLOGY

**NOTE**: The ABP content specifications have consistently said that about 2.5% of the exam will cover hematologic disorders. In general, there tends to be a much greater focus on non-malignant processes.

## *PEDIATRIC LEUKEMIAS*

**(DOUBLE TAKE) DEFINITIONS/PEARLS:** Leukemias and lymphomas are both LYMPHOblastic disorders of T and B cells.

* **LEUKEMIA**: Bone marrow biopsy shows **> 25% blasts**

* **LYMPHOMA**: Bone marrow biopsy shows < 25% blasts, and NONTENDER lymphadenopathy is present somewhere (neck, mediastinum, groin, etc.)

* **CELL LINES IMAGE**: http://updates.clltopics.org/wp-content/uploads/2010/03/myeloid-and-lymphoid-lines.gif

### LEUKEMIA PEARL

Look for increased **leuko**cytes (WBCs) in the blood and marrow. The primary leukemia in kids is ALL. It's doubtful that you would be tested on anything else.

### ACUTE LYMPHOCYTIC LEUKEMIA (ALL)

Remember, LYMPHoid cells are T and B cells. Acute Lymphocytic Leukemia (ALL) is usually a B-cell (aka plasma cell) issue. It may be due to a translocation, t(4;11). ALL is the most common pediatric malignancy and has a great cure rate if caught in time. As the name suggests, it's presents **ACUTELY**. Symptoms may include fever, organomegaly (liver or spleen) or joint pain. The leukocyte count (WBC) will either be **very** high (>50) with a high **LYMPHO**CYTE predominance. Alternative, the child may present with **pancytopenia** and a low WBC. The pancytopenic pictures occurs due have marrow failure if the child presents later in the disease process. Patient go into remission quickly after chemo but continue to get intrathecal chemotherapy to prevent a relapse. If they relapse, they will need a bone marrow transplant.

**PEARL**: If given the results of a bone marrow biopsy, look for > 25% blasts to diagnose a leukemia.

**PEARL**: If the patient has a pancytopenia, they might only give you a low WBC in the labs section. You will have to figure out the rest by circling words like "pallor" and "easy bruising." Also, the pancytopenic picture could look like aplastic anemia, so look for organomegaly to help you diagnose ALL.

**MNEMONIC**: ALL kind of looks like 4;LL, which should help you think of t(4;11).

### ACUTE MYELOID LEUKEMIA (AML)

Acute myeloid leukemia (AML) may present with similar symptoms to ALL. WBC will be high, but RBCs and platelets may be low due to marrow infiltration. Infections can occur because the leukemic WBCs are defective. You may be shown a myeloblast. If so, look for AUER RODS.

**IMAGE**: http://en.wikipedia.org/wiki/File:Auer_rods.PNG

## CHRONIC MYELOGENOUS LEUKEMIA (CML) & CHRONIC LYMPHOCYTIC LEUKEMIA CLL

Chronic Myelogenous Leukemia (CML or Chronic Myeloid Leukemia) and Chronic Lymphocytic Leukemia (CLL). If by some rare chance they refer to a Philadelphia chromosome, choose CML. The Philadelphia chromosome is a t(9;22) translocation.

**MNEMONIC**: **P**hiladelphia = t(**P**;22)

**PEARL**: CML and CLL tend to be adult diagnoses so these are low yield topics, EXCEPT for the fact that they can probably be excluded from any answer choices.

# *PEDIATRIC LYMPHOMAS*

**NOTE**: Break the lymphomas up into Hodgkin's versus non-Hodgkins lymphomas. Hodkin's = GOOD PROGNOSIS. Non-Hodkin's = NON-good prognosis!

**(DOUBLE TAKE) DEFINITIONS/PEARLS:** Leukemias and lymphomas are both LYMPHOblastic disorders of T and B cells.

* **LEUKEMIA**: Bone marrow biopsy shows **> 25% blasts**

* **LYMPHOMA**: Bone marrow biopsy shows < 25% blasts, and NONTENDER lymphadenopathy is present somewhere (neck, mediastinum, groin, etc.)

* **CELL LINES IMAGE**: http://updates.clltopics.org/wp-content/uploads/2010/03/myeloid-and-lymphoid-lines.gif

## HODGKINS LYMPHOMA

Hodgkins lymphoma accounts for almost HALF of lymphoma cases and ~~has~~ a GOOD prognosis with GOOD cure rates. Signs and symptoms progress in a stepwise fashion. Look for a **TEENAGE** patient with **NONTENDER** CERVICAL or SUPRACLAVICULAR lymphadenopathy. It always starts this way! That will then be accompanied by fevers, fatigue, night sweats and weight loss. Later in the course, there is also generalized lymphadenopathy. On labs, patient may nave a normal or high LEUKOcyte count (WBC), but must have either **high or low LYMPHOcytes**. DIAGNOSE with an excisional node biopsy and noting **REED STERNBERG CELLS**, which are usually of B-cell origin.

**PEARL**: Your differential may include TB or Atypical Mycobacteria. For a NEGATIVE PPD, choose Hodgkins. For a slightly positive PPD (5-10 mm) choose Atypical. For a PPD that is > 10 mm, choose MTB.

**IMAGE**: http://hematopathology.stanford.edu/images/HP-Home-Image.jpg = RS Cells

**MNEMONIC**: **REED** Sternberg cells kind of look like OWLS' EYES, so imagine an OWL wearing glasses and **REED**ing a book on a farm. The farmer comes outside and takes a shot at the OWL. He misses the OWL but hits the book. The OWL flies away SAFELY (high cure rate). Whew… that owl really "HODGED" a bullet.

## NON-HODGKIN LYMPHOMA (NHL)

A preadolescent child with a **nontender mass** in the **HEAD, NECK** or abdomen will likely be presented for Non-Hodgkin lymphoma (aka NHL or non-Hodgkins or non-Hodgkin's lymphoma). The mass is usually **rapidly growing**. **LEUKO**cyte count may be normal, while other cell lines may be low due to marrow infiltration.

* **BURKITT'S B-CELL LYMPHOMA**: This is a B-cell, non-Hodgkin's lymphoma that is associated with the Ebstein Bar Virus (EBV) and results in a translocation, t(8,14).

- **MNEMONIC**: **B**urkitt's = **B** cell = e**B**v = t(**B**,14)

# BONE TUMORS

## LONG BONE TUMORS PEARL

**PEARL**: The definition of a long bone is one that is **longer than it is wide**. Therefore, the phalanges count. For the pediatric exam, you will likely be given images of only the following bones for this section: Humerus, Radius, Ulna, Femur, Tibia and Fibula.

## OSTEOGENIC SARCOMA & EWINGS SARCOMA (aka EWING SARCOMA)

Osteogenic sarcoma and Ewings sarcoma (aka Ewing sarcoma) are both malignancies of **LONG BONES**. They **can** metastasize (look especially for lung metastases). Treatment requires chemotherapy, radiation and surgery.

* **MNEMONIC**: Imagine a child with a large bony tumor of the humerus extending out of his arm. He's competing in a Special Olympics race. He starts off well, but with each step closer to the finish line his bony arm tumor is punching HOLES IN HIS LUNGS and limiting his breathing (mets to lungs).

* **OSTEOGENIC SARCOMA (aka OSTEOSARCOMA)**: Usually at **proximal or distal** aspect of the long bones. Look for a long bone with a SUNBURST PATTERN/LESION.

- **PEARLS:** May present as a teen going through a GROWTH SPURT and having **UNILATERAL** bone pain (**GROWING PAINS** should be **bilateral**, at **night,** better with NSAIDS and without edema). May also present as post-traumatic pain that actually gets **worse** instead of getting better.
- **IMAGE**: http://medicalimages.allrefer.com/large/bone-tumor.jpg
- **IMAGE**: http://bit.ly/N3Vt7d

* **EWING SARCOMA**: Look for a long bone with **lamellated "ONION SKIN" lesions**. These lesions look like a very thin extra layer of bone, or like a thin onion peel laying over the bone. Can also have a SUNBURST pattern similar to osteogenic sarcomas. May affect the pelvic bones and **soft tissues** as well. Lesions are extremely **painful**.

- **IMAGE**: http://www.learningradiology.com/caseofweek/caseoftheweekpix2007-1/cow279arr.jpg - White arrows point to ONION SKIN lesions, red circle shows a SUNBURST pattern and the blue circle shows an osteolytic lesion.
- **IMAGE**: http://www.nlm.nih.gov/medlineplus/ency/imagepages/1233.htm
- **IMAGE**: Proximal Humerus - http://bit.ly/M1iVjJ
- **IMAGE**: Onion skin lesion. http://bit.ly/UqATEn
- **MNEMONIC**: U-WING's sarcoma. U looks like two LONG BONES next to each other.

## OSTEOCHONROMA

Osteochonroma is a very common and BENIGN bony tumor that is usually located near GROWTH PLATES. The bony tumor has a cartilaginous cap.

**IMAGE**: http://upload.wikimedia.org/wikipedia/commons/4/45/Osteochondroma_X-ray.jpg

**IMAGE**: http://www.ojrd.com/content/figures/1750-1172-3-3-1-l.jpg

### OSTEOID OSTEOMA

Osteoid osteoma descriptions sounds a lot like "growing pains." **BENIGN,** worse at **night** and **RELIEVED BY ASA**. The distinguishing feature is that it is **UNILATERAL**. X-ray will show a **central radiolucency with thickened bone around it.**

**IMAGES**: http://bit.ly/qpkbL9 and http://bit.ly/qUdZty

**PEARL/SHORTCUT:** Nighttime pain is all g**OO**d! It's either a benign osteoid osteoma (if unilateral), or it's growing pains (bilateral). **OO** = Osteoid Osteoma.

## *OTHER MALIGNANCIES, TUMORS & SYNDROMES*

### WILMS TUMOR

A Wilms tumor is also known as a NEPHROblastoma (kidney tumor). It's the most common abdominal malignancy in kids! Look for an otherwise asymptomatic unilateral flank mass in a child older than 6. Can be associated with **hemihypertrophy,** aniridia (no iris), and possible hypertension or gross hematuria. Most classic association is with **WAGR syndrome**(*Wilms tumor, Anirida, GU anomalies & Retardation*).

**PEARL**: All children with HEMIHYPERTROPHY are at increased risk for having Wilms tumor and Hepatoblastomas.

**(DOUBLE TAKE) PEARL:** Hemihypertrophy images on the pediatric exam should very quickly clue you in to a few disorders. Highest on your differential should be Beckwith-Wiedemann Syndrome, then Klippel-Trenaunay, then Russell-Silver Syndrome and then possibly Proteus Syndrome.

### (DOUBLE TAKE) RETINOBLASTOMA

The primary finding for retinoblastoma is LEUKOKORIA. There is increased risk of osteogenic sarcoma (aka osteosarcoma) of a long bone in the future. Inheritance pattern is actually a little confusing, with there being a strong genetic component, but also a high rate of spontaneous mutations.

**PEARLS:** Avoid completely discounting just because there is no family history. Retinoblastoma is a prime example of a disorder that often occurs due to spontaneous mutations. If you are given information about **chromosome 13** and an eye problem, or you are given the X-ray of a long bone (such as a femur, tibia, humerus, radius or ulna), consider this diagnosis.

**IMAGE**: http://rbne.files.wordpress.com/2008/08/joli-leukoria1.jpg?w=213&h=192

**MNEMONIC:** "BLAST" from the AD intro mnemonic.

**MNEMONIC:** RAT-inoblastoma. Imagine having a birthday cake made in the shape of your dad wearing a tuxedo. You go into the kitchen to get it, and you find two WHITE RATs knowing at the cake. One is nibbling on the right EYE (retinoblastoma), and the other is nibbling at the TIBIA (a long bone). Just as you get close, the tibia breaks and the entire thing collapses!

**REMINDER:** Leukocoria with loss of the red light reflex is seen in retinoblastoma as well as cataracts.

## NEUROBLASTOMA

Neuroblastoma is the most common **EXTRACRANIAL solid tumor** of childhood. Most commonly presents in the **abdomen** of a young child (< 5 yo). Occurs in males more than females. Unique features include a possible **PURPLE HUE** TO EYELIDS and possible echymoses in the periorbital area. This results from **lymphoid tissue involvement**, not child abuse! May also have **opsoclonus-myoclonus (aka "Dancing Eyes" and myoclonic jerking)**. Other findings may include abdominal pain, an elevated ferritin or an elevated LDH. Prognosis is based on age. **Good if < 1 yo, bad if > 1 yo**. Urine VMA can be elevated.

**IMAGE**: (VIDEO) http://www.youtube.com/watch?v=UCiAz8YA0iY

## BRAIN TUMORS

In general, brain tumors are the most common **solid tumors** found in pediatrics. Look for a child with a headache that is worse in the morning. Most people with regular headaches like to lie down. These kids feel better when they stand up! May also feel better after vomiting.

* **CRANIOPHARYNGIOMA**: Will cause problems with the vision, the pituitary gland or the hypothalamus. Look for calcifications near the pituitary if given a skull X-ray.

* **POSTERIOR FOSSA TUMORS**: Common after age 2. **Medulloblastoma** is the most common.

## (DOUBLE TAKE) LANGERHANS CELL HISTIOCYTOSIS (LCH) = HISTIOCYTOSIS X

Langerhans Cell Histiocytosis (LCH), aka Histiocytosis X, is a **PAPULAR** rash that is sometimes associated with petechiae. The rash is located **in the folds** (inguinal folds, supra-pubic folds, perianal area). It can resemble eczema, but the petechiae or PAPULES should guide you towards this diagnosis. LCH is a type of **cancer**. You may be shown lytic bone lesion (possibly of the skull). Diagnose by skin biopsy. Can also be associated with DIABETES INSIPIDUS. Treat by removing the lesion and giving steroids, +/- chemotherapy.

**PEARLS:** Do not confuse this with Wiskott-Aldrich (WiXotT-Aldrich, X-linked, low IgM, high IgA, TIE = Thrombocytopenia, small platelets, Infections and Eczema). Also, if they describe an eczema or seborrheic dermatitis type of rash in a patient with high urine output, LCH is your diagnosis.

**IMAGE**: http://bit.ly/pJIQSN

**IMAGE**: http://bit.ly/pc5dEG

## RHABDOMYOSARCOMA

Rhabdomyosarcoma is the most common **SOFT TISSUE** sarcoma in kids. May present as unusual or worsening post-traumatic pain, a head/neck tumor, a rectal mass, a **vaginal mass (grapelike)** or another soft tissue mass. Diagnose with biopsy.

## TUMOR LYSIS SYNDROME

Tumor lysis syndrome is an **ONCOLOGIC EMERGENCY**. Occurs due to lysis of large tumors or "blood tumors" (lymphoma). Potassium, phosphorus and LDH levels are elevated due to release from lysed cells. Treat with allopurinol, aggressive hydration and **alkalinization with sodium bicarb** (helps increase clearance of phosphorus and uric acid).

## CORD COMPRESSION

Treat cord compression with IV steroids +/- radiation therapy Marfan

## ANTERIOR MEDIASTINAL MASS

An anterior mediastinal mass is another ONCOLOGIC EMERGENCY! Can cause respiratory failure or superior vena cava syndrome (SVC syndrome). Mass may be a lymphoma, ectopic thyroid tissue or a thymoma (most common tumor found in the anterior mediastinum).

* RESPIRATORY FAILURE: May result from tracheal **compression BELOW the vocal cords** making it difficult to intubate. Do NOT give sedation to attempt intubation because you WILL loose the ability to maintain oxygenation in the patient.

* SVC SYNDROME: Compression of the SVC by an anterior mediastinal mass. Symptoms include facial swelling, dusky color due to venostasis, weight loss and night sweats due to the presence of a malignancy. Most commonly due to a lymphomatous mass. Treat with steroids + radiation + whatever the definitive treatment is for the anterior mediastinal mass.

# ***RBC BASICS & SOME HEMOGLOBIN FACTS***

## (DOUBLE TAKE) CELL LIFE SPANS

The red blood cell (RBC) has a life span of about 120 days (about 3 months). Platelets live for about 10 days.

**MNEMONIC**: Imagine a carton meant for EGGS (12) that has RBCs in it. Or, if you are worried that will make you think of 12 days, then how about imagining a STOOL (3) with the round top made out of a RED BLOOD CELL to remind you that it lives for 3 months?

**MNEMONIC**: Imagine going BOWLING (10) at a place that uses 10 PLATES instead of pins.

**MNEMONIC**: Bigger cell (RBC) = Bigger lifespan!

## FETAL & ADULT HEMOGLOBIN STRUCTURE

* FETAL HEMOGLOBIN: Structure is made of 2 alpha chains and 2 gamma chains, each with 2 alleles. Fetal hemoglobin is referred to as Hemoglobin F (Hgb F).

- NEWBORNS: 50-80% of the circulating hemoglobin is fetal hemoglobin (Hgb F)
- 6 MONTHS: 10% of the circulating hemoglobin is fetal hemoglobin (Hgb F)
- 12 MONTHS: < 2% of the circulating hemoglobin is fetal hemoglobin (Hgb F)

* ADULT HEMOGLOBIN: Structure is made of 2 alpha chains and 2 beta chains. The alpha chains have 4 alleles and the beta chains have 2 alleles = $a_{1234}a_{1234}b_{12}b_{12}$. Note that "a" = alpha, "b" = beta. Normal adult hemoglobin is referred to **A1**.

- THALASSEMIA: Alpha thalassemia refers to a mutation in an alpha chain allele. Beta thalassemia refers to a mutation in a beta chain allele.

## NEWBORN ANEMIA

**Anemia in a newborn is defined as a hemoglobin less than 13 (KNOW THIS)!** The table below provides you with some average hemoglobin values at different ages.

| AGE | HGB | HCT |
|---|---|---|
| 28 wks gestation | 14.5 | 45 |
| 32 wks gestation | 15 | 47 |
| Full Term | 16.5 | 51 |
| DOL 2-4 | 18.5 | 56 |
| DOL 15 | 16.6 | 53 |

## RBC MCV

* NEWBORNS: May have an RBC MCV up to **110**. That is considered NORMAL.

* 6 MONTHS – 2 YEARS: At this age, the normal range is **70-90**. Note that this is a LOWER range than older children and adults (80-100).

## POLYCYTHEMIA

Polycythemia is defined by a hematocrit **> 65%**, but treatment is not required unless is > 70% by a VENOUS blood sample. Patients are at risk for problems associated with hyperviscosity. Potential signs and symptoms include hypoglycemia, thrombocytopenia, joint pain, clots (deep vein thrombosis or pulmonary embolus), stroke, hemoptysis, lethargy and hypotonia. Risk factors include intrauterine growth retardation (IUGR), delayed clamping of the umbilical cord, twin to twin transfusion, infant of a diabetic mother (IDM), Down Syndrome and chronic hypoxia (refers more to an older child). Treat with partial volume exchange transfusion.

**PEARLS:** If the patient is hypoglycemic due polycythemia, the blood glucose may not correct as expected with the usual dextrose infusion.

**MNEMONIC**: Assume the RBC are sucking up all the glucose!

## PRBC TRANSFUSIONS

A standard PRBC (packed red blood cell) transfusion is 10-20 cc/kg.

**PEARLS:** Massive PRBC transfusions and exchange transfusions both have similar complications - **Hypo**calcemia, **hyper**kalcmia and thrombocytopenia.

**MNEMONICS:** Regarding the 10-20 cc/kg, think of it as a giving a "bolus" of PRBCs. A bolus of IVF is **20 cc/kg**. Regarding complications, think of calcium precipitating out as the 2 different bloods mix (actually it's a citrate toxicity). For the hyperkalemia, imagine RBCs hemolyzing as they get sheared through the IV. For thrombocytopenia, assume it's a hemodilutional effect!

# ***NORMOCYTIC ANEMIA***

## PHYSIOLOGIC ANEMIA

The phrase "physiologic anemia" refers to a normal process. Fetal hemoglobin has a high affinity for oxygen. It slowly gets replaced by adult hemoglobin (A1), which has a lower affinity for oxygen. Since A1 doesn't hold on to oxygen very well, it allows the release of O2 into the tissues. As more A1 is made and more O2 is released, there is a decreased need for erythropoietin. That results in decreased RBC production and a "physiologic anemia." The Hgb nadir (low point) occurs sometime between 6 and 16 weeks. **It should not be lower than a hemoglobin of 9 for term infants and 7 for premature infants**. That is about the level at which erythropoietin production increases again, resulting in a rise in A1 hemoglobin production. There is no iron deficiency or structural problem. Everything is normal, so it will be a NORMOCYTIC anemia.

## HEMOLYTIC ANEMIAS

Hemolytic anemias are normocytic anemias because normal sized red cells are being lysed, and the normal sized red cells that are not hemolyzed have a normal MCV. Look for JAUNDICE (as mentioned in the neonatology section). Other findings may include an elevated reticulocyte count (> 1%), elevated indirect/unconjugated bilirubin, hemosiderin/bilirubin in the urine, a low serum haptoglobin or a positive Coombs test if it is an autoimmune hemolytic anemia.

## COOMBS TEST PEARLS

The Coombs test helps to identify AUTOIMMUNE hemolytic anemias. In terms of DIRECT versus INDIRECT, it's always confusing and is probably LOW YIELD. If it causes you anxiety, here's the simplified version.

* **DIRECT COOMBS TEST**: The patient's **RBCs** are placed DIRECTLY into a solution containing "anti-antibody antibodies." If the RBCs taken from the patient are being destroyed by an autoimmune process, then the cells will be coated with antibodies and the "anti-antibody antibodies" will latch onto them.

- **MNEMONIC**: DIRECT the patient to the lab and then DIRECT the patient's RBCs DIRECTLY into the solution.

* **INDIRECT COOMBS TEST**: The patient's **serum** is used to indirectly figure out if the patient is possibly having an autoimmune hemolytic anemia. Someone else's RBCs that are similar to the patient's own RBC (Rh compatible & ABO compatible) are placed in the patient's serum. If they serum causes hemolysis of the alien RBCs, it's assumed that the same thing is happening in the patient.

## (DOUBLE TAKE) RHESUS DISEASE (aka RH DISEASE)

When checking for Rhesus Disease (aka RH Disease), look for an Rh- mom in her SECOND pregnancy: Maternal IgM antibodies are made during the FIRST pregnancy and are too large to cross over into the fetal circulation. During the SECOND pregnancy, IgG antibodies are present, which are small enough to cross (can cause ERYTHROBLASTOSIS FETALIS if they cross early in pregnancy). Rh- mom's are supposed to get RHOGAM at 20 wks, and then again after delivery if baby is found to be Rh+.

* KLEIHAUER BETKE TEST: Check MATERNAL blood to see if there are FETAL red cells present.

## (DOUBLE TAKE) ABO INCOMPATIBILITY

ABO incompatibility usually occurs in mothers with an "O" blood type. Naturally occurring "anti-A" or "anti-B" IgG antibodies may be present. This can result in hemolytic disease of the newborn in a FIRST TIME pregnancy. So for hemolysis in a G1P1 baby, consider ABO incompatibility as the etiology.

## (DOUBLE TAKE) GLUCOSE-6-PHOSPHATE DEHYDROGENASE DEFICIENCY (G6PD DEFICIENCY)

Glucose-6-Phosphate Dehydrogenase Deficiency (G6PD Deficiency) is an X-linked recessive disorder (so look for a male patient!) resulting in jaundice, dark urine and anemia due to hemolysis from oxidative injury. In newborns, this will present with jaundice in a MALE baby AFTER 24 HOURS. In other children, can result in hemolysis after ingestion of fava beans, malaria medications, trimethoprim-sulfamethoxazole, ciprofloxacin or nitrofurantoin. Look for HEINZ BODIES (purple granules noted in the red cells on microscopy). Associated with African-American and Mediterranean descent.

**PEARL**: Do not test for the deficiency during the acute hemolytic phase (wait a few weeks) because a false negative result can occur due to the build up of G6PD in reticulocytes.

**MNEMONIC**: Imagine a bottle of HEINZ BODIES ketchup that has an **X** located on the exact spot that you have to hit it to make the broken RBCs come out of the bottle.

## PYRUVATE KINASE DEFICIENCY

Pyruvate kinase deficiency is an enzyme deficiency resulting in energy issues we will not go into. The end result is an unstable RBC structure causing RBC deformities and hemolysis. The hemolytic anemia can cause hyperbilirubinemia and jaundice.

**PEARL**: One of the metabolites that builds up due to the PK deficiency is 2,3-diphosphoglycerol (DPG). DPG causes the oxygen dissociation curve to shift to the right, resulting in the hemoglobin have less O2 affinity. So the patient can have a surprisingly **good exercise tolerance even though s/he is anemic**.

**IMAGE**: http://www.bio.davidson.edu/Courses/anphys/1999/Dickens/labsk_hgbdiso2curve.gif

**MNEMONIC**: Some people know the DPG as Snoop's "Dog Pound Gang." Now imagine Snoop and his gang having surprisingly good exercise tolerance. Given how much they smoke, it truly would be impressive.

**MNEMONIC VIDEO:** http://www.youtube.com/watch?v=trzLD3b5QAQ&feature=channel_video_title

## HEREDITARY SPHEROCYTOSIS

Look for intermittent anemia and possibly a similar family history to diagnose hereditary spherocytosis. It's an inherited disorder resulting in spectrin deficiency (usually autosomal dominant). Cells as more fragile and as pieces of the RBC membrane are ripped off, spherical cells are left behind. The amount of hemoglobin remains the SAME in this slightly smaller, spherical cell. The results is an **ELEVATED MCHC**. Diagnose with **osmotic fragility** testing. Treat with folic acid and transfusions (when needed).

**PEARLS:** Parvovirus B19 can cause an aplastic crisis. The resulting constellation of lab values is extremely unique. LOW hemoglobin + **LOW reticulocyte count + HIGH MCHC**. If you see this combination on the pediatric boards, select HS.

**MNEMONICS:** HS = Hereditary SpherOcytOsis = Hereditary Spectrin deficiency. So **low retic + high MCHC = HS**. Spher**O**cyt**O**sis = **O**sm**O**tic fragility testing.

## (DOUBLE TAKE) ERYTHEMA INFECTIOUSUM

Erythema infectiousum IS an INFECTIOUS rash!!! Caused by Parvovirus B19. It is also called Fifth Disease Look for erythematous facial flushing of the cheeks (sometimes described as "slapped cheeks" appearance).

The extremities will have diffuse macular (or morbilliform) erythema (especially on the extensor surfaces) referred to as "lacy" or "reticular." Diagnose with Ig**M titers** (there is no culture or rapid antigen available).

**PEARLS:** The rash occurs AFTER the slapped cheeks rash (often a week later). Patients may also have knee or ankle pain. Parvovirus B19 infection can result in **APLASTIC CRISIS**. Intrauterine exposure can result in **hydrops fetalis**.

**MNEMONIC**: infectio**5**u**M** = FIFTH disease = "Fiver fingers." Imagine a cheek being SLAPPED with FIVE fingers covered by a white LACY/LACEY glove with a red M on the back of it (extensor surface). **M** = ig**M** titers

**MNEMONIC**: ParVoVirus B19: From now on, say/think "par**V**o**V**irus **V**19." **V = Roman numeral 5!**

## PAROXYSMAL NOCTURNAL HEMOGLOBINURIA (PNH)

Look for dark colored urine **in the mornings** are keywords that might indicate a diagnosis of paroxysmal nocturnal hemoglobinuria (PNH). This is a complement-mediated hemolysis, which means there are NO antibodies involved, which means that it is a COOMBS NEGATIVE autohemolytic anemia. Diagnose by using either **flow cytometry** analysis of CD59 expression on the cells, a DF assay or by doings a Ham's test (checks for fragility with exposure to mild ascetic acid).

**MNEMONIC**: pnH = pnHam. Or, consider having a Paroxysmal Nocturnal HAM EATING CONTEST. A ham-eating contest that happens every once in a while (paroxysmal) at night.

## SICKLE CELL ANEMIA

* **PRESENTATION**: The signs or symptoms of sickle cell anemia could include the following:

- APLASTIC CRISIS: Will be due to Parvovirus B19
- SEQUESTRATION CRISIS (aka SPLENIC SEQUESTRATION CRISIS): Look for abdominal pain, splenomegaly and signs of SHOCK. Cells get trapped in the organ (usually the spleen, sometimes the liver). The pooling of the blood results in a lower effective volume of circulating blood which results in SHOCK. Emergently transfuse!
  - **PEARL**: If given an option of IVF or PRBCs, go with what is FASTEST to bring up the blood pressure - IVF!
- ACUTE CHEST SYNDROME: Look for the triad of PAIN + INFILTRATE + HYPOXIA. If there's any doubt about the O2 sat, confirm with an ABG. Treatment is dictated by the **hemoglobin level**. **LOW** = TRANSFUSION. **HIGH** = EXCHANGE Transfusion.
- VASOOCCLUSIVE CRISIS: Acute pain! If severe, can cause ischemia, infarction of bone, stroke or dactylitis (swelling of the hands or feet). Initial therapy is IV fluids and pain control. For suspected STROKE, get an **MRI**. For an obvious or confirmed stroke, give PRBCs.
  - **IMAGE**: www.mdnotebook.com/wp-content/uploads/2011/05/dactylitis-scd.jpg
- OSTEOMYELITIS: There's an increased risk of Salmonella osteomyelitis.
  - **PEARLS:** What's the most common etiology of osteomyelitis in patients with sickle cell anemia? STAPHYLOCOCCUS AUREUS! DO NOT FALL INTO THE TRAP! Also, remember that sickle cell patients can present with sepsis due to encapsulated organisms.

  - **<u>(DOUBLE TAKE) MNEMONIC</u>**: A complete list of encapsulated organisms can be recalled by remembering that "Some Nasty Killers Have Some Capsule Protection." Streptococcus pneumoniase, Neisseria meningitidis, Klebsiella pneumoniae, Haemophilus influenzae, Salmonella typhi, Cryptococcus neoformans and Pseudomonas aeruginosa. Bruton's agammaglobulinemia and sickle cell patients are especially susceptible to encapsulated organisms.
- DILUTE URINE: Due to renal damage, the urine may be dilute as evidenced by a low specific gravity. This is also known as HYPOSTHENURIA.
  - **<u>MNEMONIC</u>**: HYPOsthenURIA

* **DIAGNOSIS** of sickle cell anemia is done by **HEMOGLOBIN ELECTROPHORESIS** will show hemoglobin F and S (no hemoglobin A). Might be shown as SF. The hemoglobin S develops due to a "Glu-Val" amino acid swap (**Glutamic acid substitution of Valine**) in the Beta chain. Consider these patients as being functionally asplenic. Give PCN prophylaxis through 5 years of age. Look for Howell-Jolly bodies on the smear.

* **<u>PEARL</u>**: These are normal sized RBCs with the Glu-Val swap that predisposes them to sickle, but they are otherwise of NORMAL size, so this is a NORMOCYTIC anemia. So, any macrocytosis or microcytosis has to be accounted for by an additional diagnosis! B12 or Folate deficiency? Iron deficiency? Concomitant thalassemia?

* **<u>PEARL</u>**: If the electrophoresis results come back with Hgb SA, that means the patient has sickle cell **trait**. Hemoglobin SC and S-beta thalassemia can result in a palpable spleen at the age of 6, but SF (aka SS) causes the spleen to shrivel up by then. So a palpable spleen in an 8 year old is NOT due to sickle cell anemia.

* **<u>IMAGE</u>**: http://www.wadsworth.org/chemheme/heme/glass/slide_005_howell-jolly_bod.htm

* **<u>MNEMONIC</u>**: Sickle cell anemia results in **HOW**ell-JOLLY bodies. Have you ever noticed **HOW JOLLY** these patients get once they are given morphine?

* **<u>MNEMONIC</u>**: Regarding the glu-val swap mutation, there's a mnemonic in there somewhere about a SICKLED CELL splattering over a BED (beta) on a WALL with GLUE on it. Thoughts?

## TRANSIENT ERYTHROBLASTOPENIA OF CHILDHOOD

Look for a healthy child with erythropenia. Typical age of onset for transient erythroblastopenia of childhood is around 18-24 months, but can happen earlier or later. The etiology is unknown.

**<u>PEARLS:</u>** Even if the hemoglobin is 4.2, the treatment is to OBSERVE unless there is hemodynamic compromise due to high output cardiac failure. Can look similar to Diamond Blackfan (just the red cell is affected), but the age of presentation is different. **Diamond Blackfan occurs around 3-12 months of age, and is associated with dysmorphic features. Also, for DB the anemia is MACROCYTIC**.

## ACUTE BLOOD LOSS ANEMIA

In cases of acuta blood loss anemia, there can be a normal MCV because there has been no time for the MCV to change.

## (DOUBLE TAKE) ANEMIA OF CHRONIC DISEASE

Anemia of a chronic disease is a chronic inflammation leads to a defect in the **production** of RBCs. There are plenty of building blocks so the MCV is usually NORMAL, but it can be LOW NORMAL or only BORDERLINE LOW. So this can be a normocytic or a slightly microcytic anemia. Also look for a low TIBC (Total Iron Binding Capacity). The TIBC measures the ability of transferrin to take on more iron. If the value is low, that means there's plenty of iron around.

## END STAGE RENAL DISEASE (aka ESRD or RENAL FAILURE)

In end-stage renal disease (aka ESRD or renal failure), the kidneys fail to produce erythropoietin. Again, it's a defect in the production of RBCs. There are plenty of building blocks so the MCV is normal.

## PEARLY REMINDERS:

* The combination of ANEMIA + JAUNDICE = **HEMOLYSIS**! May last for WEEKS in ABO incompatibilities and Rh disease as the maternal antibodies slowly clear. Treatment with phototherapy, transfusions and exchange transfusions (if severe). **Remember, an MCV of 70-90 is NORMAL up to 2 years of age.**

# ***MICROCYTIC ANEMIA***

## MICROCYTIC ANEMIA DEFINITION

An MCV < 70 IS DEFINITELY CONSIDERED MICROCYTIC ANEMIA. The 70-80 range is NOT microcytic for a child less than 2 years of age.

## IRON DEFICIENCY ANEMIA

Look for a LOW transferring saturation and LOW reticulocyte count to diagnose iron deficiency anemia. Also, if you are given ferritin level < 15, that **confirms** the diagnosis. The lower limit of iron varies by age, but for infants 30 is the cutoff. For children, 55 is the cutoff.

* **PEARLS:** Transferrin transports iron, therefore if there's no iron around, the transferrin saturation will be low. The RDW (RBC Distribution Width) can also help you if it is HIGH because this suggests that there are old normal-sized RBCs mixed in with new MICROCYTIC ones (only helpful in the initial phase). You will NOT be asked to calculate the Mentzer index (MCV/RBC), but if you can it may help. An index > 12 = Iron Deficiency Anemia. An index < 11 = Thalassemia.

* **PEARLS:** WHOLE COW MILK (not formula) has LOW iron and may be a cause of SEVERE anemia if started in a child before the age of 1. Choose nutritional deficiency as the etiology (not bleeding (unless of course the stool guaiac is positive).

* **PEARLS:** Sorry if these are obvious, but... Viruses can cause marrow suppression, and possibly a transient anemia. Don't treat with iron. Keep in mind that some of the above lab values have a very high POSITIVE predictive value, but a LOW negative predictive value. So if a ferritin level is < 15, it is VERY helpful (high PPV). If it is normal or high (can be elevated as an acute phase reactant), then the result is "negative" or "normal," and it doesn't help you at all (low NPV).

## (DOUBLE TAKE) ANEMIA OF CHRONIC DISEASE

Anemia of a chronic disease is a chronic inflammation leads to a defect in the **production** of RBCs. There are plenty of building blocks so the MCV is usually NORMAL, but it can be LOW NORMAL or only BORDERLINE LOW. So this can be a normocytic or a slightly microcytic anemia. Also look for a low TIBC (Total Iron Binding Capacity). The TIBC measures the ability of transferrin to take on more iron. If the value is low, that means there's plenty of iron around.

## THALASSEMIAS

The thalassemias are autosomal recessive disorders, so parents can have "traits." As a reminder, normal adult hemoglobin (**A1**) is made up of 2 alpha chains and 2 beta chains. The alpha chains have 4 alleles and the beta chains have 2 alleles = $a_{1234}a_{1234}b_{12}b_{12}$.

## ALPHA THALASSEMIA

Alpha thalassemia refers to a mutation in an alpha chain allele. Because of the defect, other types of hemoglobin persist. You will find **elevated** levels of fetal hemoglobin (hgb F) as well as elevated levels of "minor adult hemoglobin" (hgb A2). There will only be low levels of A1. Patients with alpha thalassemia do fairly well and only have a mild microcytosis.

**PEARL**: The only thing you probably need to know for the exam is that alpha thalassemia can cause a microcytic anemia and **cannot** be diagnosed by hemoglobin electrophoresis.

**MNEMONIC**: Be familiar with, but do not memorize this. It's doubtful that you will be asked to name the types of alpha thalassemia on the exam, but if you are, try this – "silent… trait (or minor)… barts… DEAD!" If one allele is mutated the patient is said to be a "silent carrier", if 2 = trait or minor, 3 = Hemoglobin H disease (or Barts) and 4 = Major (or hydrops fetalis)! Four defective alleles are not compatible with life since ALL types of hemoglobin have alpha chains. Hgb A1 = alpha-beta, Hgb S = alpha-beta (but with a defective beta due to glu-val), Hgb A2 = alpha-delta and Hgb F = alpha-gamma.

## BETA THALASSEMIA

Beta thalassemia is much more important. Refers to mutation in the beta chains. Like alpha thal, other types of hemoglobin persist. **Elevated** levels of Hgb F and Hgb A2. Low Hgb A1. There are 2 beta alleles, but can somehow have THREE forms of beta thal, trait/minor, intermedia and major/Cooley's.

* BETA THALASSEMIA TRAIT (aka BETA THALASSEMIA MINOR): Only one allele is defective. Patients are asymptomatic. May have a mild microcytosis.

* BETA THALASSEMIA INTERMEDIA: Both alleles are affected and there will be a very obvious microcytosis, but it's not as severe as MAJOR. Patients are able to maintain their hemoglobin in the 6-9 range and may only require an occasional transfusion after an illness.

* BETA THALASSEMIA MAJOR (aka COOLEY'S ANEMIA): Both alleles are affected and patients have a **SEVERE microcytic anemia** within the first year of life. They need frequent transfusions and are a setup for complications such related to hemochromatosis, organomegaly and gallstones. Patients can also have bone thickening, arterial and venous clots. **Blood smears will show microcytosis, target cells and basophilic stippling**. HEMOCHROMATOSIS (aka HEMOSIDEROSIS): The deposition of iron into various organs

causing problems such as heart failure, liver failure and renal disease. Can be hereditary or due to chronic transfusions for other problems, such as beta thalassemia major. Consider this diagnosis in any patient with symptoms of congestive heart failure. Look for other signs, such as polyuria. Treat with DESFEROXIME ("de-iron" – "de-ferrous").

* **PEARLS:** If tested on a thalassemia, it will likely be beta thalassemia major. If you are provided with an MCV that is low (60-70) and "out of proportion" with the fairly mild anemia, pick a thalassemia (but not beta thal major). Look for microcytosis and target cells on a blood smear.

* **IMAGE**: http://bit.ly/qiRGml (click Show All) **Don't read! Move on!**

* **IMAGE**: (Basophilic Stippling) http://bit.ly/nSGV7a

## (DOUBLE TAKE) LEAD TOXICITY

Lead toxicity often results because kids love its sweet taste. Sources of lead toxicity are paint in old homes and dust from home renovations. Because lead blocks iron from being incorporation into heme, lead toxicity can look very much like iron-deficiency anemia (MICROcytic). To differentiate this from iron deficiency, look for BASOPHILIC STIPPLING of RBC's, constipation, neurologic issues and LEAD LINES (obtain imaging of long bones). Check a lead level at 12 and 24 months of age. Even mild elevations (level of 10-20) can cause neurologic problems with learning and behavior. For mild elevations, make changes in the home and educate the family. For levels > 45, CHELATION therapy is needed for lead poisoning with **edetate disodium calcium (EDTA)**, dimercaprol or d-penicillamine. Always confirm an elevated capillary sample with a VENOUS sample because these are more reflective of RECENT exposure.

**PEARLS:** Iron levels should be normal and TIBC should be normal or low.

**IMAGE**: (Lead Lines) - http://www.alsehha.gov.sa/dpcc/images/lead_lines.gif

**IMAGE**: (Lead Lines) - http://en.wikipedia.org/wiki/File:Lead_PoisoningRadio.jpg

**IMAGE**: (Basophilic Stippling) - http://bit.ly/oeWPKR

**MNEMONIC**: The treatment for lead is "LEADetate."

## LAB REVIEWS - FERRITIN, TIBC, RDW, & TRANSFERRIN SATURATION

* **FERRITIN**:

- Extremely low (< 15) = Iron deficiency anemia.
- HIGH = Anemia of chronic disease or INFLAMMATION

* **TIBC**:

- High = Iron deficiency anemia
- LOW = Anemia of chronic disease, lead poisoning and just about anything else.

* **RDW**:

- Anemia + LOW RDW may mean thaLOWssemia.
- Anemia + HIGH RDW may mean early iron deficiency anemia

* **TRANSFERRIN SATURATION:**

- LOW = Low iron
- HIGH = Plenty of iron around to bind the empty sites

# *MACROCYTIC ANEMIA*

## MACROCYTIC ANEMIAS (aka MEGALOBLASTIC ANEMIA)

Look for an MCV > 100 to indicate macrocytic anemias, aka megaloblastic anemia. Keep in mind that an MCV up to 110 is NORMAL for a newborn.

## (DOUBLE TAKE) FOLATE (B9) DEFICIENCY

Folate (Vitamin B9) deficiency results in a **macrocytic** anemia and macroglossia. In pregnant women, deficiency can lead to neural tube defects (such as spina bifida) in the unborn fetus.

**PEARLS:** Look for a diet poor in veggies, or a history of **goat milk** ingestion.

**PEARL**: Both B12 and Folate deficiency can lead to hypersegmented neutrophils and neural tube defects. Both can be masked by treatment with the other. Use the symptoms, diet history and blood testing to get the diagnosis right the first time!

**MNEMONIC**: FOLATE deficiency form FOLIAGE (veggies) deficiency!

## (DOUBLE TAKE) B12 DEFICIENCY (aka CYANOCOBALAMIN DEFICIENCY)

Cyanocobalamin (Vitamin B12) deficiency can present as a macrocytic anemia in a patient with neurologic issues (paresthesias, ataxia and decreased reflexes). B12 deficiency can also result in neural tube defects (such as spina bifida) in the unborn fetus if a pregnant woman is deficient. Common causes of B12 deficiency include diet (VEGANS) and an inability to absorb B12. Look for INTRINSIC FACTOR DEFICIENCY (Pernicious Anemia), or an inability to absorb B12 due to gastric surgery (diagnosed with a Schilling test). B12 is absorbed in the terminal ileum, which means it may be lacking patients with Short Bowel Syndrome, bacterial overgrowth or Inflammatory Bowel Disease (IBD). Treat with IM B12 injections.

**PEARLS:** B12 will likely be used on the exam, but KNOW that B12 = CYANOCOBALAMIN. In a B12 deficient patient, if a macrocytic anemia is noted and mistakenly treated with folate supplements, the B12 deficiency will be MASKED as the megaloblastic anemia improves. This can eventually lead to **irreversible** neurologic damage.

## (DOUBLE TAKE) FANCONI ANEMIA

Fanconi **anemia** is an APLASTIC anemia that is MACROCYTIC. Patients are usually > 4 years of age. Look for multiple anomalies including **short stature, café au lait spots, renal abnormalities**, microcephaly, hypogonadism and upper limb/hand anomalies. Treatment is with a bone marrow transplant.

**MNEMONIC**: Imagine you are a private investigator. You enter a mass murderers home and see **A PLASTIC ceiling FAN**. It has four ORANGE PLASTIC (**aplastic**) blades that look like **CON**es. We'll call them **FAN CONES**. Each of these ceiling **FAN CONES** has body parts hidden inside them. You look up right as **HUGE BLOOD CELLS** start to drip out of them. You find a stepladder, look inside the FAN CONES and find a hidden GONAD, a SMALL HEAD, an ARM and a HAND.... Remember, this is a CEILING fan. The Blackfan mnemonic is a standing plastic black fan.

**NAME ALERT**: The name alert is for FANCONI **SYNDROME**.

**(DOUBLE TAKE) FANCONI SYNDROME**

For Fanconi **syndrome**, look for polydipsia, polyuria, failure to thrive (FTT) and a **ton** of abnormal labs. It all happens due to **cystine** deposits in the kidneys causing proximal tubule damage (called CYSTINOSIS). The renal problems result in losses of amino acids, proteins, electrolytes (sodium, potassium, bicarbonate and phosphorus) and glucose. The loss of bicarb results in a **GAP** metabolic acidosis. The loss of phosphorus can present as RICKETS. Diagnose by obtaining a leukocyte cystine level.

**PEARLS:** There are MANY possible disease and points of confusion with these diseases. In Fanconi Syndrome, there is loss of sodium and the serum sodium level is low to low-normal. In diabetes insipidus, there is a loss of WATER that can result in hypERnatremia if there's no access to free water. Renal Tubular Acidosis can also result in electrolyte abnormalities and failure to thrive (FTT), but Renal Tubular Acidosis results in a NONGAP acidosis! Lastly, between Fanconi Anemia, Fanconi Syndrome, Diamond-Blackfan and Shwachman-Diamond, there is tremendous room for confusion. Use mnemonics (ours or yours) to keep them all straight!

**NAME ALERT**: The name alert is for FANCONI **ANEMIA!**

## (DOUBLE TAKE) DIAMOND-BLACKFAN ANEMIA

DIAMOND-BLACKFAN ANEMIA is an aplastic anemia due to **pure red cell aplasia**. That means that unlike other aplastic anemias, ONLY THE RED CELL LINE IS AFFECTED. Red cells are MACROCYTIC, anemia is typically severe, patients present around 3 months of age and they may have TRIPHALANGEAL THUMBS and craniofacial anomalies. Look for low reticulocyte counts. If asked about treatment, use steroids chronically and transfuse for severe anemia.

* **MNEMONIC**: Imagine looking at a standing BLACK FAN with THREE GIANT PLASTIC BLADES. The blades are BLACK, but are SHAPED LIKE RED CELLS.

- **KEY**: BLACK FAN = Name, THREE = TRIphalangeal thumbs and 3 months old at presentation, GIANT = MACROCYTOSIS, PLASTIC = aPLASTIC anemia, SHAPE = Reminder that this is a pure RED CELL aplasia.

* **PEARL**: If you see a patient with anemia and icterus or hyperbilirubinemia suggesting hemolysis, it is NOT Diamond-Blackfan Anemia. Since there is a red cell aplasia, the anemia is from a lack of production, NOT from hemolysis.

* **NAME ALERT**: The name alert is for SHWACHMAN-**DIAMOND** SYNDROME.

**(DOUBLE TAKE) SHWACHMAN-DIAMOND SYNDROME**

SHWACHMAN-DIAMOND SYNDROME is a rare, autosomal recessive (AR) disorder associated with bone marrow failure (**pancytopenia/neutropenia), pancreatic insufficiency**, **bone/skeletal abnormalities** and possible **neutrophil defects** (migration and chemotaxis). Recurrent bacterial infections are common, including **recurrent pneumonias**, skin infections, otitis media and sinusitis. Patients may also have aphthous stomatitis.

* **PEARL:** Due to some overlapping symptoms with Cystic Fibrosis (CF), such as pancreatic insufficiency, you may have to choose between this disorder and CF. Remember, this one has **cell line issues** and the electrolytes and lungs are NORMAL.

* **MNEMONIC**: Imagine a DIAMOND in a SHWACH WATCH being used to make a CREASE in a black PAN with a white handle made of HOLLOW BONE."

- **KEY**: PAN/HOLLOW = **PAN**CYTOPENIA/NEUTROPENIA!!! CREASE = panCREASE, PAN = PANcytopenia, HOLLOW BONE = Aplasia resulting in PANcytopenia and neutropenia.

* **NAME ALERT**: The name alert is for **DIAMOND**-BLACKFAN ANEMIA.

## *APLASTIC ANEMIA*

* Be sure to also read the FANCONI ANEMIA and Diamond-Blackfan sections.

**PEARLS:** When an aplastic anemia is present in isolation, there is NO jaundice. Aplasia is just the lack of production of RBCs, NOT the destruction (there is NO jaundice). Aplastic anemias often have associated pancytopenia due to overall bone marrow suppression. Diamond Blackfan is unique in that is is a PURE RED CELL APLASIA, and there will be no other cell line affected! Aplastic anemias can be MACROcytic (diamond blackFAN and FAN**coni** anemia – both "a plastic fan" anemias) or normocytic (sickle cell anemia + Parvovirus B19, hereditary spherocytosis + parvo). Aplastic anemias an look similar to leukemia but there shouldn't be any lymphadenopathy, hepatomegaly, elevated LDH or Uric as would be found in leukemic patients. Aplastic Crisis is usually due to Parvovirus B19, but it can also be caused by CHLORAMPHENICOL

## *PLATELET DISORDERS*

### (DOUBLE TAKE) CELL LIFE SPANS

The red blood cell (RBC) has a life span of about 120 days (about 3 months). Platelets live for about 10 days.

**MNEMONIC**: Imagine a carton meant for EGGS (12) that has RBCs in it. Or, if you are worried that will make you think of 12 days, then how about imagining a STOOL (3) with the round top made out of a RED BLOOD CELL to remind you that it lives for 3 months?

**MNEMONIC**: Imagine going BOWLING (10) at a place that uses 10 PLATES instead of pins.

**MNEMONIC**: Bigger cell (RBC) = Bigger lifespan!

## THROMBOCYTOPENIA

**Thrombocytopenia is defined in children as < 250, and in adolescents/adults as < 150.**

## MATERNAL IDIOPATHIC THROMBOCYTOPENIA PURPURA (ITP)

Maternal Idiopathic Thrombocytopenia Purpura (ITP) can cause thrombocytopenia in neonates that lasts for **weeks**.

## NEONATAL SEPSIS-INDUCED THROMBOCYTOPENIA

**Neonatal sepsis m**ay result in thrombocytopenia, or platelet dysfunction, which can cause mucosal or GI bleeding, bruising, nosebleeds and petechiae.

- **PEARLS:** In general, epistaxis is commonly due to dry air and does not usually require a workup if the patient is otherwise healthy. If you are see a deep muscle bleed, that's more likely to be due to a coagulopathy issue rather than a platelet problem.

## THROMBOCYTOPENIA AND ABSENT RADIUS (aka TAR SYNDROME)

In Thrombocytopenia and Absent Radius syndrome (aka TAR syndrome) patient have thrombocytopenia and absent RADII. Hand or shoulder anomalies may also be seen. The thrombocytopenia is **AMEGAKARYOCYTIC** (low or absent number of megakaryocytes on bone marrow biopsy).

- **IMAGES**: http://emedicine.medscape.com/article/959262-overview#showall **Move on! PEARL**: This is usually a BILATERAL problem with both RADII missing. The thumbs are PRESENT.

## IDIOPATHIC THROMBOCYTOPENIA (ITP)

Look for LARGE platelets with a high Mean Platelet Volume (MPV) to indicate idiopathic thrombocytopenia (ITP). Often occurs after a viral syndrome. Treat with **IVIG** or steroids. If both options are present, choose IVIG over steroids. Rhogam (WinRho) can also be used if the patient is Rh+ to induce a mild hemolysis which then "distracts" the immune system. For kids under 10, it usually resolves after a few months. Transfuse if there are signs of bleeding. If given a platelet count of < 20, transfuse if BLEEDING.

**PEARLS:** Look for a low platelet count (thrombocytopenia), **but** BIG platelets (high MPV). The disorder can look like leukemia, so don't give steroids until LEUKEMIA HAS BEEN RULED OUT. Steroids given to a leukemic patient can induce a short **remission** resulting in delayed diagnosis and treatment.

## (DOUBLE TAKE) WISKOTT-ALDRICH SYNDROME

For Wiskott-Aldrich syndrome, look for a patient with a history consistent with the triad of SMALL **T**hrombocytes, frequent **I**nfections and **E**czema. Platelets can be small in **size and/or number**. This is an X-linked so it will only be found in MALES. Patients develop both T and B cell problems as they ages. IgM is usually LOW, while IgA may be HIGH. A history may include findings such as petechiae, bloody diarrhea, bloody circumcision, eczema and/or otitis media. There is an increased risk of lymphoma in the 3rd decade of life. BMT is curative.

* **MNEMONIC**: Wiskott Aldrich = **wiX**ot**T ostrich** = Imagine a WISe OSTRICH wearing some nerdy black glasses, a tux and a TIE with the word "T-I-E" spelled down the front.

* **MNEMONIC**: (image)

T I E ↓WiXott ↑Aldrich
IgM↓ IgA↑

- **KEY: X** Linked, small T = Represents small platelets (low MPV) and Thrombocytopenia, **I** = Infections, **E** = Eczema, Mirrored W = M and arrows show that Ig**M** is low, Duplicate A refers to Ig**A** and arrows show that IgA levels are increased.

* **PEARL**: In Idiopathic Thrombocytopenia Purpura (ITP), platelets are BIGGER

## (DOUBLE TAKE) KASABACH MERRITT SYNDROME

In Kasabach Merritt syndrome, there are large, congenital vascular tumors. They are not true hemangiomas but can cause a severe CONSUMPTIVE COAGULOPATHY (in the form of **thrombocytopenia** and the consumption of coagulation factors) and death.

**IMAGE**: http://img.medscape.com/pi/emed/ckb/hematology/197800-202455-5213tn.jpg

**IMAGE**: http://img.medscape.com/pi/emed/ckb/hematology/197800-202455-5214tn.jpg

**PEARL**: Look at the above images closely. Make sure you look closely at images so that you do not get this vascular tumor confused with hemihypertrophy.

**MNEMONIC**:

- >---< is used by many of us when recording CBC results.
- ↓---<ASSABACH = low platelets, risk of bleeding and death

## GLANZMANN THROMBASTHENIA

For Glanzmann thrombasthenia, look for **unexplained bleeding** in a patient with a normal platelet count, normal PTT and a normal INR. The bleeding is from **dysfunctional** platelets.

## BERNARD-SOULIER SYNDROME

In Bernard-Soulier syndrome, there is mild thrombocytopenia + LARGE platelets + Prolonged Bleeding Time.

**MNEMONIC**: Platelets are LARGE, like a Saint Bernard.

# *COAGULOPATHY*

**PEARL**: Platelet problems can result in bleeding **or** petechiae. Problem with a coagulation factor typically only result in bleeding.

## VITAMIN K DEPENDENT FACTORS

The Vitamin K dependent factors include II, VII, IX and X.

**MNEMONIC**: 2 + 7 = 9…10

## COAGULATION CASCADE

Factors III and VII are the main EXTRINSIC factors in the coagulation cascade. Factor X is part of the common pathway.

## VITAMIN K DEFICIENCY

Although Vitamin K deficiency CAN result in a high PTT, for the boards you will likely be shown a **high PT/INR** and asked to identify the vitamin deficiency.

## (DOUBLE TAKE) HEMOPHILIA A and HEMOPHILIA B (aka FACTOR VIII and FACTOR IX DEFICIENCY)

Hemophilia A and B refer to factor VIII and factor IX deficiencies, respectively. For both hemophilias, the **PTT** is increased. Since this is a factor deficiency, look for deep bleeds (muscle, joints, etc.)

**MNEMONIC**: Have you ever encountered a female hemophiliac? Hmm... You could also imagine the X as a pair of sharp blades going through blood vessels causing bleeding. That could remind you that hemophiliacs have an X-linked disorder.

## BLEEDING CIRCUMCISION

If there is significant bleeding after a circumcision, workup the patient for a factor deficiency and consider **maternal** medications as a possible etiology (antibiotics, seizure meds and warfarin are some common culprits).

## VON WILLEBRAND DISEASE (aka VON WILLEBRAND FACTOR DEFICIENCY)

von Willebrand disease (aka von Willebrand factor deficiency) is an autosomal dominant disorder. Both **factor** VIII and for platelets depend on the vW factor (vWF), so there is an elevated PTT **and** a prolonged **BLEEDING TIME**. Measure VIII and IX (to rule out hemophilia) and the vWF (to confirm that it is low). Keep in mind that factor VIIi can be slightly low. No treatment is needed for minor bleeding, but DDAVP can be used to increase plasma vWF and Factor VIII. For EXTREME bleeding, GIVE FACTOR VIII CONCENTRATE (aka HAMATE P). Cryoprecipitate can be used as a last resort (it has Fibrinogen, VIII, vWf). Aminocaproic acid is used sometimes for mucosal bleeds since it helps **inhibit** fibrinolysis.

## DISSEMINATED INTRAVASCULAR COAGULATION (DIC)

Risk factors for disseminated intravascular coagulation (DIC) include burns, malignancy, pregnancy and sepsis. Look for THROMBOCYTOPENIA, low fibrinogen, elevated fibrinogen split products (D-Dimer), elevated thrombin time, PT **and** PTT. Treat as follows:

* Thrombocytopenia = Platelets.

* Severe Anemia = PRBCs

*** Low Fibrinogen = CRYOPRECIPITATE** (includes Fibrinogen, factor VIII, vWf)

* **Low clotting factors = FRESH FROZEN PLASMA (FFP)**. It has MOST of the clotting factors ans is very expensive. FFP is also used to reverse Warfarin.

* FIXING THE UNDERLYING CONDITION!

# INFECTIOUS DISEASES

**NOTE/PEARL**: Know the infectious diseases section well for the pediatric boards! Board questions with a great deal of cross-subject testing is common. The American Board of Pediatrics (ABP) expects you to be able to differentiate infections from metabolic disorders.

## ANTIBIOTICS – A BRIEF REVIEW

### ANTIBIOTIC AGE PEARLS

AGE LIMITATIONS: When choosing an antibiotic (especially tetracycline, doxycycline, fluoroquinolones and macrolides), ALWAYS look at the age of the patient.

* TETRACYCLINE & DOXYCYCLINE: May be given **after the age of 8**.

- **PEARL**: If you diagnose a 5 year old child with Rocky Mountain Spotted Fever, go ahead and GIVE doxycycline. That is the first-line therapy regardless of age.

* FLUOROQUINOLONES: Give after **18 years** of age due to possible tendonitis and tendon rupture. Avoid choosing this as an answer for your adolescents with STDs.

* MACROLIDES (ERYTHROMYCIN): Avoid in children less than 6 years of age due to an association with pyloric stenosis. Give azithromycin instead.

### PENICILLIN

Yes! Penicillin is still the TREATMENT OF CHOICE for Streptococcal pharyngitis, dental infections and dental-related infections (including parotiditis). It's also used for Syphilis and CAN be used for other infections if the bacteria is sensitive to it.

* PENICILLIN ALLERGY: Most patients who claim to have this allergy do not have a true allergy. If skin testing has been done and confirms the allergy, do not give penicillins or cephalosporins. If no testing has been done, look for other options. If no other options are available, choose a cephalosporin since the likelihood of a concurrent cephalosporin allergy is less than 10%.

### CLINDAMYCIN

Clindamycin is active against most Gram positives. Does have some MRSA coverage.

**PEARL**: It is NOT active against Enterococcus.

### VANCOMYCIN, LINEZOLID & AMPICILLIN

Don't you DARE pick anything other than vancomycin, linezolid or ampicillin to cover for ENTEROCOCCUS! Ampicillin actually covers it **better** than vancomycin!

**PEARL**: Enterococcus is also covered by rifampin and quinolones, but hopefully that will not be tested. Know that it is NOT covered by cephalosporins.

## CEPHALOSPORINS

* FIRST GENERATION CEPHALOSPORINS: Cephalexin (Keflex - PO) and Cefazolin (Ancef - IV). Good activity against Staph and Strep. No activity against MRSA. Good activity against some Gram negatives (Proteus, E. coli, and Klebsiella).

- **MNEMONIC**: Coverage against PEcK.

* SECOND GENERATION CEPHALOSPORINS: Cefaclor, Cefuroxime, Cefotetan and Cefoxitin. No good for Gram positives. BETTER Gram negative coverage. Covers above organisms and Haemophilus influenza, Enterobacter and Neisseria.

- **MNEMONIC**: Coverage against HEN PEcK
- **MNEMONIC**: There are no "T's" near the "F" of these names, except for except for cefoTeTan. Imagine two big mountains (or lady parts) called the grand TeTons to help remind you that it is a 2nd generation.

* THIRD GENERATION CEPHALOSPORINS: Ceftriaxone, Cefotaxime, Ceftazidime and Cefpodoxime. Also not very good for Gram positive coverage. Even BETTER Gram negative coverage. Good for hospital acquired Gram negative infections. Great for CNS infections, which is why we give CEFTRIAXONE for meningitis. Ceftazidime is a great drug for Pseudomonas.

- (DOUBLE TAKE) **PEARL**: Ceftriaxone is still **great** for Streptococcus. When treating for meningitis, also give Vancomycin. The vancomycin is NOT being given for Staph aureus coverage. It's being given for possible Strep that is RESISTANT to ceftriaxone!
- **MNEMONIC**: Most **T**HREE generation drugs have a "**T**" within **T**HREE letters of the "F." cefpodo**X**ime is the **X**-cep**T**ion.
- **MNEMONIC**: cef**TAZ**adime is good for pseudo**MON**as. Think of TAZ-MON-ian devil
- **MNEMONIC IMAGE**: http://upload.wikimedia.org/wikipedia/en/c/c4/Taz-Looney_Tunes.svg

* FOURTH GENERATION CEPHALOSPORINS: Cefepime. Extended spectrum agent with Gram positive coverage that's just as good as the FIRST generation's. Good CNS penetration and broader Gram negative coverage. More resistant to beta-lactamases.

## MACROLIDES

Macrolides include azithromycin, clarithromycin, erythromycin… anything with a –THROMYCIN. Common side effect is increase GI motility.

**MNEMONIC**: Every use erythromycin for diabetic gastroparesis? Maybe in med school?

## CARBAPENEMS

Imipenem could be seen on the exam. Use carbapenems for extended-spectrum beta-lactamase (ESBL) producing organisms. EnteroBACTER (a Gram negative organism) is a common one.

## MEBENDIZOLE, ALBENDAZOLE & PYRANTEL PAMOATE

Mebendizole, albendazole and pyrantel pamoate are for WORM infections.

**PEARL**: For Enterobius (aka PINWORMS), it's a ONE TIME dose. Repeat after 2 weeks if it recurs. For all other worm infections, treatment requires multiple doses.

## METRONIDAZOLE

Metronidazole is active against parasites and anaerobic infections. Often used for intra-abdominal infections. Some of the commonly treated organism include Giardia, Entamoeba, Trichomonas, Bacteroides, Clostridium and Gardnerella.

**PEARL**: For Trichomonas, it's a one time dose. If the adolescent female you are treating has a child, stop breastfeeding for 24 hours.

**MNEMONIC**: Commonly treated organisms = **GET BCG**! There is a women's clothing line called BCG. There's also a consulting firm called BCG. Imagine a woman traveling on an underground METRO to go and **GET BCG** clothing/consulting! It's stuffy underground and it feels like there's not much air (anaerobes).

# *GRAM POSITIVE ORGANISMS*

**PEARL/NOTE**: I cannot stress enough how important it is to know the difference between infections caused by Streptococcus PYOGENES, Streptococcus pneumoniae and Staphylococcus aureus. Knowing Strep from Staph is not enough. Memorize EVERYTHING because this will probably be worth a TON on the exam.

## ENTEROCOCCUS FAECALIS

Enterococcus faecalis is a Gram positive diplococcus (can look similar to Strep on a Gram stain). Can cause sepsis and necrotizing enterocolitis. For treatment, don't you dare choose anything except VANCOMYCIN, LINEZOLID or AMPICILLIN to treat it!

**PEARL**: Enterococcus is also covered by rifampin and quinolones, but hopefully that will not be tested. Know that it is NOT covered by cephalosporins.

## LISTERIA MONOCYTOGENES

Listeria monocytogenes is a Gram positive DIPHTHEROID (rod). Look for a history of NODULES ON THE PLACENTA, or a mild maternal fever with virus-like symptoms. The treatment is the same as for suspected GBS, ampicillin.

**PEARL**: If the bug has not been clearly identified, treat with both ampicillin **and** gentamicin for neonatal sepsis. The ampicillin will cover GBS, Enterococcus and Listeria. The gentamicin will cover some Gram negatives.

## CLOSTRIDIUM TETANI (aka TETANUS)

Clostridium tetani (aka tetanus) is an anaerobic organism. Look for a history of a puncture wound. Causes uncontrollable muscle spasms and contractions. Look for an arched back or "lock jaw." Treat with debridement of infected tissue + tetanus immunoglobulin + metronidazole **OR** penicillin.

## (DOUBLE TAKE) CLOSTRIDIUM BOTULISM

Clostridium botulism is a Gram positive organism. Look for a child **less than 6 months** of age with **progressive descending** weakness. Can progress from **progressive ptosis** and a poor suck to urinary retention **within hours**. The botulism toxin **inhibits release of acetylcholine** into the neurosynaptic junction. Treatment involves either supportive care (intubate if needed) or the antitoxin if it's available.

**PEARLS:** For the exam, **progressive** ptosis in a baby probably means Botulism! Children are usually less than 6 months of age since they are too young at that age to prevent colonization in gut.

## (DOUBLE TAKE) CORYNEBACTERIUM DIPHTHERIAE

Corynebacterium diphtheriae is a Gram positive rod that can cause Diphtheria. It starts off with a low fever and URI symptoms, including a sore throat. Eventually a **pseudomembrane forms on the tonsils and pharynx**. The swelling can be so bad that intubation is needed. It can also cause motor and sensory problems along with a loss of reflexes. The D in DTaP is for Diphtheria, so it's no longer seen in the U.S. but could be seen in a child of immigrant status. Treat with metronidazole or erythromycin.

**IMAGE**: http://bit.ly/117OOkz

**IMAGE**: http://bit.ly/SoKGtL

**IMAGE**: http://www.dipnet.org/img/mem.jpg

# *STREPTOCOCCAL INFECTIONS*

**NOTE:** ALWAYS CHECK WHICH STREP YOU ARE DEALING WITH. The ones most commonly seen on the exam are Group A Streptococcus (pyogenes), Group B Streptococcus (GBS/agalactiae) and Streptococcus pneumoniae. The others are less commonly seen (viridans, mutans and bovis).

## STREPTOCOCCUS (aka STREP)

In cases of streptococcus (Strep), look for Gram positive cocci in PAIRS and CHAINS! Gram staining will show BLUE organisms.

## ALPHA HEMOLYTIC STREPTOCOCCUS (VIRIDANS & PNEUMONIAE)

Streptococcus *viridans* and Streptococcus *pneumoniae* (aka STREP PNEUMONIAE or PNEUMOCOCCUS) are alpha hemolytic strep species

- **MNEMONIC**: Imagine a **GREEN ALF** doing a **"STREP"** tease dance and showing you his **NEW MOVE TODAY** (or **NEW MOON TODAY**).
  - **KEY:** Green = Verde = Viridans. ALF = Alien Life Form = Alpha. New Move Today (or New Moon Today) = Pneumonia
- **MNEMONIC**: (VIDEO - watch the dance!) - http://www.youtube.com/watch?v=gjo_GoJ03ps&feature=related
- **MNEMONIC**: (Image - Green ALF) http://walyou.com/wp-content/uploads/2009/04/probo-robot-alf-1.jpg

## BETA HEMOLYTIC STREPTOCOCCUS (AGALACTIAE & PYOGENES)

Streptococcus *agalactiae* (a **Group B** Strep, aka **the GBS you need to know!**) and Streptococcus *pyogenes* (a **Group A** Strep) are beta hemolytic strep species.

- **MNEMONIC**: The names and groups are confusing, so go over this again and again until you get it. Or, create your own mnemonics.

- **aBpA!:** If you can remember that acronym, you're set! If you have trouble remembering the order of the letters, maybe you've heard of **A**cute **B**roncho**P**ulmonary **A**spergillosis? For ***aBpA***, the lower case letter represent the correct nomenclature of ***a**galactiase* and ***p**yogenes*! The upper case letters are always upper case when written – group **A** and group **B**! Write ***aBpA*** and circle the two groups of letters when you get confused. It helps!
  - ***aB*** = ***a**galactiae* of group **B** strep = G**B**S = g**B**s = **B**aby!
  - ***pA*** = pyogenes of group **A = "GAS"** = ga**S** = **S**kin!

- **<u>MNEMONIC</u>**: Inhaling **GAS** makes my THROAT HURT! This may help you remember that group **A** strep (***p**yogenes*) is the one that cause ***p***HARYNGITIS and oral abscesses! ***pYOGENes*** **is** ***p*YOGENic** in the skin and in the oral cavity!

## STREPTOCOCCAL PHARYNGITIS (aka STREP PHARYNGITIS or STREP THROAT)

To identify Streptococcal pharyngitis (aka Strep pharyngitis or Strep throat), look for fever, lymphadenopathy, sore throat, erythematous or exudative tonsils and the ABSENCE of coughing, sneezing and rhinorrhea (those are viral symptoms). The pharyngitis is due to Group A Streptococcus and it usually resolves in 2-5 days. Diagnose by CULTURE. If a rapid Strep is positive, TREAT. If negative, send for a CULTURE. The **only** reason Strep throat is such a big deal is because it can lead to **RHEUMATIC FEVER, which is PREVENTABLE!** Treat with **PENICILLIN or AMOXICILLIN**. If allergic, use erythromycin or clindamycin.

**<u>PEARLS</u>:** Antibiotics are NOT given to shorten the course of the illness. They are given to prevent RHEUMATIC FEVER! Even given **9 days after the onset of symptoms,** RHEUMATIC FEVER can be **prevented**! ALWAYS get a culture. Consider treating if you have a strong clinical case or suspect poor clinical follow up. Rheumatic fever is **NOT** due to GAS skin infections!

## (DOUBLE TAKE) POST STREPTOCOCCAL GLOMERULONEPHRITIS (PSGN, aka POST INFECTIOUS GLOMERULONEPHRITIS)

Post streptococcal glomerulonephritis (PSGN, aka post infectious glomerulonephritis) is a Group A Streptococcus (GAS) syndrome. Look for **HEMATURIA + PROTEINURIA +/- swelling +/- HTN**. There may be a history of a skin infection 1 month before, or a throat infection 1-2 weeks earlier. Labs will show **a low C3 and NORMAL C4 +/- renal impairment**. Biopsy will show LUMPY BUMPY IgG deposits (biopsy is not required to make the diagnosis). Treat with IVF and a loop diuretic if there's HTN. Generally has a GOOD prognosis, but may consider steroids or cyclophosphamide for cases that are not improving. C3 should be followed until it's back to normal.

**<u>PEARL</u>: Antibiotics given for a Group A Strep infection can prevent rheumatic fever, but they CANNOT prevent PSGN. Also, PSGN can occur from pharyngitis OR skin infections, but rheumatic fever only occurs after a PHARYNGITIS.**

**<u>PEARL</u>**: If the C3 level does not come back down to normal after 6 weeks, **consider a different diagnosis**, such as **MPGN (membranoproliferative glomerulonephritis has a low C3 and C4)** or lupus nephritis (+ANA, low C3 **and C4**). If the renal impairment is severe, obtain a biopsy to look for rapidly progressive glomerulonephritis (RPGN).

## PERITONSILLAR ABSCESS

A peritonsillar abscess presents with a deviated uvula due to unilateral tonsillar swelling, difficulty with opening mouth (trismus) and speaking ("hot-potato voice") and drooling because of pain with swallowing (odynophagia). Usually due to Group A Streptococcus (GAS). Treat with **IV antibiotics to also cover for anaerobes**. Clindamycin and ampicillin-sulbactam are great. Once the lesion has been drained and the patient can swallow, can treat with oral amoxicillin-clavulanate.

**IMAGE**:

http://www.meddean.luc.edu/lumen/MedEd/medicine/pulmonar/pdself/Acute_ottitis_media/Slide22.JPG

## RETROPHARYNGEAL ABSCESS

A retropharyngeal abscess will present with a young child (< 6 years old) with fever, lymphadenopathy, pain and difficulty with swallowing, drooling and **HYPEREXTENSION of the neck**. The child may also have refused to eat. Usually due to Group A Streptococcus (GAS) and can be diagnosed with a lateral X-ray of the neck showing widening of the retropharyngeal space in which the pus resides.

**PEARL**: Don't confuse hyperextension of the neck in a drooling child with **the LEANING forward posture and drooling of a chld with epiglottitis.**

**IMAGE**: http://img.medscape.com/pi/emed/ckb/emergency_medicine/756148-764421-110tn.jpg

## SCARLET FEVER

Scarlet fever is also due to Group A Strep (GAS). Look for a rash in areas with CREASES (axilla and groin). There is initially a "sandpaper" rash. It then becomes more erythematous and less sandpaper-like. There can also be pink or red lines in/near the creases called Pastia's lines. Other symptoms may include perioral pallor and a strawberry, or **white**, tongue. Though many GAS infectious hurt, this rash does NOT.

**IMAGES**: http://emboardreviewblog.blogspot.com/2011/06/scarlet-fever.html

**MNEMONIC**: The rash of SCARLET fever tends to go to the "naughty" and hidden areas of the groin and axilla. This only works if you are familiar with the story of The Scarlet Letter.

## OCCULT BACTEREMIA

Streptococcus pneumoniae (Pneumococcus) is the most common etiology of occult bacteremia (no obvious source). For Streptococcal bacteremia found incidentally, if there are NO symptoms, then you do NOT need to treat!

## PNEUMONIA

Overall, Streptococcus pneumoniae (Pneumococcus) is the most common etiology of pneumonia in children.

**PEARL**: If asked what the most common etiology of pneumonia is in a patient with cystic fibrosis, pick THIS!

## GROUP B STREPTOCOCCAL SEPSIS (GBS SEPSIS)

"Early onset" Group B streptococcal sepsis (GBS sepsis) refers to sepsis occurring within the first 3 days of life. It is usually due to a GBS pneumonia. "Late onset" GBS sepsis refers to after 3 days and up until the first

90 days of life. Infections tend to be more focal. The most concerning would be meningitis. Also look for cellulitis and osteomyelitis. GBS is pretty much susceptible to anything, so TREAT WITH PENICILLIN G.

**PEARL**: Differential of sepsis-like picture in the newborn includes Congenital Adrenal Hyperplasia, Inborn Errors of Metabolism and Heart Failure. If WBC given, look at clinical picture. Look at ratio of Bands:Neutrophils rather than the VERY UNRELIABLE WBC. If > 0.2 then suggestive of infection.

## GBS SCREENING & PROPHYLAXIS MADE EASY!

HERE ARE SOME KEY POINTS AND PEARLS ABOUT GBS SCREENING AND PROPHYLAXIS:

* GROUP B BETA HEMOLYTIC STREPTOCOCCUS (GBS) SCREEN: Occurs at 35-37 weeks.

* PROPHYLAXIS: If indicated, give penicillin at least 4 hours prior to delivery.

*** INTRAPARTUM GBS PROPHYLAXIS – TO GIVE OR NOT TO GIVE?**

- PREMATURE RUPTURE OF MEMBRANES (PROM) OVER 18 HOURS AGO: YES
- PRIOR PCN TREATMENT during this pregnancy for a positive screen: YES
- PRIOR GBS POSITIVITY in a **previous** pregnancy with a **recent negative screen**: NO
- STATUS UNKNOWN and complicated pregnancy or PREMATURE (< 37 weeks): YES
- STATUS UNKNOWN and **UN**complicated pregnancy and **TERM** (> 37 weeks): **NO**
- C-SECTION with artificial rupture of membranes (AROM) with +GBS screen: **NO**
    - **PEARL**: Any child born to a mother who is GBS+ should be watched in the hospital for at least 48 hours.

## STAPHYLOCOCCUS AUREUS & EPIDERMIDIS

* STAPHYLOCOCCUS AUREUS: Look for Gram positive in CLUSTERS on Gram staining. If Staph aureus is growing in the BLOOD, it requires prolonged IV antibiotics and and ECHOCARDIOGRAM to look for endocarditis! Pleare refer to the chart above for everything else you need to know about Staph aureus.

* COAGULASE NEGATIVE STAPHYLOCOCCUS (aka CONS): STAPHYLOCOCCUS EPIDERMIDIS is a CONS. Regardless of what you learned in residency, if a blood culture comes back with "Coagulase negative Staphylococcus", TREAT! And treat BIG, with **VANCOMYCIN.** Staph epi is extremely resistant to common antibiotics. Also, the final culture results may actually show MRSA!

**PEARL**: For SECONDARY BACTERIAL PERITONITIS in a patient on peritoneal dialysis, and for patients with a VENTRICULOPERITONEAL SHUNT (VPS) INFECTION, assume the organism causing the infection is methicillin-resistant Staphylococcus epidermidis. Treat with VANCOMYCIN until the culture results come back.

**MNEMONIC**: Don't let CONS "con" you into leaving it alone. TREAT!

## STAPHYLOCOCCUS AND STREPTOCOCCUS COMPARISON CHART

| STREPTOCOCCUS | STAPHYLOCOCCUS AUREUS |
|---|---|
| **GROUP A STREP (GAS = STREP *pyogenes*):**<br>* PHARYNGITIS<br>* PERITONSILLAR ABSCESSES: Use ampicillin-sulbactam, IV ampicillin-sulbactam, or clindamycin! These all help with anaerobes!<br>* SKIN INFECTIONS: Including cellulitis, necrotizing fasciitis, impetigo (less bullous than Staph impetigo, but **much more common**), and erysipelas (fever + cellulitis). Rashes are usually erythematous and painful!<br>* POST STREP GLOMERULONEPHRITIS (PSGN): Look for a low C3. PSGN is **NOT** preventable by early antibiotic treatment of the Strep infection!<br>* SCARLET FEVER: Covered in the Strep section.<br>* STREP TOXIC SHOCK SYNDROME: Can have concurrent *pyogenes* NECROTIZING FASCIITIS<br>* RHEUMATIC FEVER: Covered in the Cardiology section.<br><br>**GROUP B STREP (GBS):**<br>* NEONATAL SEPSIS early<br>- NEONATAL PNEUMONIA late<br><br>**STREP PNEUMONIAE:**<br>* PNEUMONIA<br>* MENINGITIS<br>* OTITIS MEDIA<br>* OCCULT BACTEREMIA:<br>* PERITONITIS<br>* PARANASAL SINUSITIS<br>* SEPTIC ARTHRITIS<br>* OSTEOMYELITIS: Staph is much more common.<br>* RARE: Cellulitis and brain abscesses<br><br>**STREP VIRIDANS, MUTANS & BOVIS:**<br>* ENDOCARDITIS: VIRIDANS >> mutans or bovis | * **PURULENT** SKIN INFECTIONS<br>* CARBUNCLES AND FURUNCLES: If < 5 cm, I&D only. No antibiotics indicated even if it's CA-MRSA. If > 5 cm, use clindamycin or trimethoprim-sulfamethoxazole for outpatient treatment.<br>* BULLOUS IMPETIGO: More bullous than strep. VERY thin blisters. Less common than GAS.<br>* FOLLICULITIS: Pustules with surrounding erythema<br>* STAPH TOXIC SHOCK SYNDROME: Associated with emesis, watery diarrhea, emesis, foreign bodies (tampon)<br>* **CHRONIC** SINUSITIS: Means >90 days. Staph is more common than the H-M-S bugs (Haemophilus, Moraxella, Strep pneumoniae)<br>* SEPTIC ARTHRITIS: Much more common than Pneumococcus<br>* OSTEOMYELITIS: Most common etiology.<br>* BREAST ABSCESSES: If found in a **neonate**, give **IV** antibiotics.<br><br>* **PEARL**: If MRSA is said to be susceptible to Clindamycin but resistant to Erythromycin, give something else!<br><br>* **PEARL**: For most skin and soft tissue infections, cephalexin, cefazolin, clindamycin and trimethoprim-sulfamethoxazole are very acceptable therapies. If the patient has known MRSA or risk factors, clindamycin and trimethoprim-sulfamethoxazole are good first line agents. Vancomycin is IV only. Linezolid is IV or PO, but is more expensive. |

# GRAM NEGATIVE ORGANISMS

## RICKETTSIA RICKETTSII and ROCKY MOUNTAIN SPOTTED FEVER (RMSF)

Rickettsia rickettsii causes ROCKY MOUNTAIN SPOTTED FEVER (RMSF). It is the most common tick-borne illness. Look for hematologic and neurologic signs, such as thrombocytopenia, petechiae, HEADACHE, lethargy and altered mentation. The spotted **rash starts at the palms, wrists and soles**, but then spreads to the trunk within **hours**. Diagnose first by clinical findings, but then confirmed with a biopsy of a lesion and direct immunofluorescence. Treat with **DOXYCYCLINE.**

**PEARLS:** Start treatment based your clinical suspicion BEFORE sending the biopsy. Do NOT delay treatment if there is NOT a history of a tick bite. Treat patients of ALL AGES with DOXYCYCLINE because the benefits outweigh the risks. Also, it's possible that an EKG may show conduction defects (low-yield).

**PEARLS:** EHRLICHIA CHAFFEENSIS can have a very similar presentation but would have associated liver dysfunction and possible leukopenia and/or lymphopenia. It's also known as Human "Monocytic" Ehrlichiosis even though there is no monocytosis. Fortunately, the treatment is the SAME.

**MNEMONIC**: To remember the direction of the rash, imagine ROCKY balboa initially running around the PERIPHERY of Philadelphia, and then running up a MOUNTAIN of stairs in the CENTER of the city.

**MNEMONIC**: (video) - http://www.youtube.com/watch?v=NubH5BDOaD8

## ENTEROBACTER

Enterobacter is usually hospital acquired, resulting in urinary or respiratory tract infections. Bad bug. Try treating with cephalosporins, though it's frequently an ESBL organism requiring a CARBAPENEM (imipenem).

## (DOUBLE TAKE) BARTONELLA HENSELAE

Bartonella henselae causes CAT SCRATCH DISEASE. Look for **DRAINING and TENDER** lymphadenopathy and possibly a history of cat (or dog) exposure. Resolves on it's own so there's no need to treat unless the patient is immunocompromised or has severe symptoms including hepatosplenomegaly. If so, treat with non-penicillin drugs, like macrolides, cephalosporins, doxycycline or trimethoprim-sulfamethoxazole.

**MNEMONIC**: Imagine BART simpson's cat being a TENDER kitty that's fine when it's LEFT ALONE, but it SCRATCHES you when you try to pet it. Solution is to LEAVE IT ALONE!

* **KEY:** TENDER = tender lymphadenopathy, LEAVE IT ALONE = no treatment needed.

## CITROBACTER FREUNDII

If you read about a BRAIN ABSCESS on the peds exam and see Citrobacter freundii in the answer choices. Pick it and move on! It can be a hospital acquired infection, even in neonates.

## (DOUBLE TAKE) CHLAMYDIA TRACHOMATIS

Chlamydia trachomatis can cause urethritis, conjunctivitis and pelvic inflammatory disease (PID). In neonates, it can cause a pneumonia associated with a **staccato** cough. It's the most common STD. PID can lead to ectopic pregnancies and infertility. Eye infections can lead to blindness. Conjunctivitis in a neonate (less than a month old) should raise concern for this as the etiology (vertical transmission). It's an obligate **intracellular** anaerobe. Getting cultures is difficult, so order PCR of CELLS, secretions or urine. Chlamydia can also cause lymphogranuloma venereum (LGV), which is an STD that initially starts with **small nontender** papules or shallow ulcers that resolve. **Then** a **TENDER** UNILATERAL INGUINAL lymph node appears that can rupture, relieve the pain, and then possibly drain for months.

* **SEXUALLY TRANSMITTED DISEASE (STD)**: Treat with **DOXYCYCLINE** for 7 days. For PELVIC INFLAMMATORY DISEASE, treat for 14 days. For STD recurrence, treatment failure or noncompliance concerns, treat with Azithromycin x 1. Other possible medications include erythromycin, levofloxacin or ofloxacin for multiple doses/day x 7 days.

* **CONJUNCTIVITIS**: Treat with **oral erythromycin** to eradicate nasopharyngeal colonization which can lead to pneumonia.

* **(DOUBLE TAKE) LYMPHOGRANULOMA VENEREUM SEROVAR**: Lymphogranuloma venereum serovar is an STD caused by Chlamydia trachomatis. Rare in the U.S. More common in tropical areas. Starts as small nontender papules or shallow ulcers which resolve. Eventually, a **TENDER UNILATERAL INGUINAL** lymph node appears. Pain is relieved when it ruptures. The node can **continue to drain** for months. Treat with DOXYCYCLINE or erythromycin.

* **PEARL**: In general, when you think the diagnosis is due to a Chlamydia species, choose doxycycline if the child is > 8 years of age, or choose a macrolide (usually erythromycin). Also, this is an intracellular organisms. Look for the phrase "**intracytoplasmic inclusions**."

## CHLAMYDIA PNEUMONIAE

Chlamydia pneumoniae causes an atypical pneumonia. Often noted in a child less than 2 months of age. The patient may have tachypnea, a staccato cough +/- eye discharge. For the boards, the patient may be AFEBRILE. Look for intracytoplasmic inclusion bodies and a Gram negative organism. Can also do a PCR. Treat with doxycycline or macrolides.

**PEARL**: If you see the phrase "staccato cough," PICK THIS!

## CHLAMYDIA PSITTACI

Bird exposure + atypical pneumonia could represent a Chlamydia psittaci pneumonia on the exam. Treat with doxycycline or macrolides.

## MYCOPLASMA PNEUMONIA

**M**ycoplasma pneumonia can cause an atypical pneumonia or bullous **m**yringitis. Some patients have also been known to get neurological symptoms following an infections. Diagnose with serum Ig**M** titers, NOT cold agglutinins because they only help to support the diagnosis if positive. Treat with a macrolide.

**MNEMONIC**: MMMMM for Mycoplasma! Get Ig**M** titers for the diagnosis, use a **M**acrolide for treatment, look in the ears for a bullous **M**yringitis and watch out for delayed "**M**eurologic" symptoms.

## HEMOPHILUS INFLUENZAE (aka H. FLU)

Although Hemophilus influenzae, aka H. flu, CAN cause bacteremia, pneumonia, meningitis, otitis, **epiglottitis**, sinusitis and conjunctivitis, it usually doesn't because of immunizations. For unimmunized patients, immigrants and asplenic patients, it can still cause severe disease. If the Gram stain is described, "**pleomorphic** organism" may be used as a descriptive phrase. Treat with **ceftriaxone**.

## BORDETELLA PERTUSSIS (aka WHOOPING COUGH)

Bordetella pertussis (aka whooping cough) patients are described as having bursts, or "paroxysms," of coughing. They cough so much they can't breath, and then they inspire deeply causing a WHOOP! Patients may have a HIGH WBC of > 20,000. Diagnose with a nasopharyngeal swab. Treat with **ERYTHROMYCIN**. All contacts (**even if immunized**) need to be given erythromycin for prophylaxis since immunity of the vaccination wanes. Hence the need for Tdap in teens now (more coming up later).

**PEARLS:** Consider this diagnosis in any patient with a **chronic cough**. Antibiotic treatment shortens the **early** stage of pertussis in which the patient is infectious and has URI-type symptoms (known as the catarrhal stage). It dose NOT decrease the "whooping," or paroxysmal stage.

**MNEMONIC**: If it ends in –ELLA, it's probably a Gram negative organism! Brucella, Shigella and Salmonella fit better elsewhere in this chapter.

## PSEUDOMONAS

Pseudomonas can cause skin infections, pneumonia, otitis externa, UTIs, osteomyelitis, sepsis, etc. For a history of a puncture wound leading to osteomyelitis, or a history of a cystic fibrosis (or ventilator patient) having a pneumonia. Anti-pseudomonal drugs include ceftazidime, cefepime, ticarcillin, carbenicillin, piperacillin, gentamycin and tobramycin. If the patient has an otitis externa, give ear drops.

**MNEMONIC**: Imagine the TAZ-MONIAN devil (cefTAZadime, pseudoMONas) dressed like a cool GENTleman (GENTamicin) and riding around in James Bond's CAR (CARbenicillin). The CAR is filled with cool gadgets like a PIPE bomb (PIPEeracillin) and a TICking time bomb (TICarcillin). Of course as he drives through a city, a woman THROWS A BRA (TOBRAmycin) at him.

* Cefepime is missing in this mnemonic.

# *FUNGAL & ATYPICAL BACTERIA*

**PEARL**: Any patient with sepsis due to fungemia that is unstable or in the ICU should be treated with AMPHOTERICIN B. If there is renal disease, give liposomal amphotericin.

## CRYPTOCOCCUS

Cryptococcus infections are associated with a history of bird exposure (pigeon's bird droppings) or travel to the Northwestern U.S. Can cause pneumonia or meningitis, ESPECIALLY in the immunocompromised. Diagnose with India ink staining. For meningitis or disseminated cryptococcus, treat with AMPHOTERICIN first, and then can switch over to Fluconazole. For isolated pulmonary disease (pneumonia), can use fluconazole alone.

**MNEMONIC**: Imagine a CRYING PT (CRYPTococcus) wearing a RED NECK COLLAR (RED = immunocompromised, NECK COLLAR = pain from meningitis). He has a PIGEON on his shoulder that is wearing a RED TURBAN (India ink).

## BLASTOMYCOSIS

Blastomycosis can present as a pneumonia or flu-like illness. Blastomyces is described as a broad-based budding yeast and is usually found near **water** . Can also cause skin lesions. Treat with azoles (itraconazole is the first-line agent).

**IMAGE**: http://i6.photobucket.com/albums/y228/usmle/Microbiology/BLASTOMYCOSIS.jpg?t=1241800200

**MNEMONIC**: It used to be called Chicago's disease. The solution was to move the sewage away from any source of drinking water!

## COCCIDIOIDOMYCOSIS

Coccidioidomycosis can present as a pneumonia (possibly with pulmonary nodules) or as flu-like symptoms in someone living/traveling in out west (Arizona, **California**, Texas, etc.). Treat with amphotericin or azoles (fluconazole or ketoconazole).

**MNEMONIC**: Used to be called California or Desert Disease.

## HISTOPLASMOSIS

Histoplasmosis presents as a flu-like illness. Typically cleared by most patients without intervention. Those who do not clear it can get CALCIFICATIONS that look like tuberculous disease on X-ray. Patients can also get pulmonary fibrosis, HEPATOSPLENOMEGALY (HSM,) and mediastinal/HILAR CALCIFICATIONS. Look for a history of exposure to bird droppings, or travel to **OHIO** or Mississippi. Treatment is with amphotericin or an azole (fluconazole or ketoconazole).

**MNEMONICS:** Used to be known as Cave Diease or Ohio Valley Disease.

**MNEMONIC**: HIS TOE is wrapped with RED TAPE. There's also a tag from the LIMITED on it and it seems to be leaking CALCIUM FIBERS (fibrosis) onto his sock!

**MNEMONIC IMAGE**:

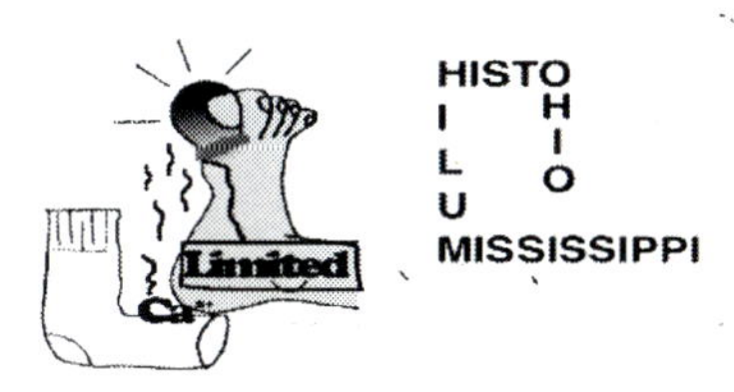


## (DOUBLE TAKE) ASPERGILLUS

In immunocompetent patients, aspergillus may present as Acute Bronchopulmonary **Aspergillosis** (ABPA). Patients seem to have an exacerbation of asthma that WORSENS despite "appropriate" treatment with steroids. Look for increased eosinophilia and lung infiltrates. Can lead to Chronic Eosinophilic Pneumonia as well (peripheral infiltrates on X-ray). In an immunocompromised patient, this can present as pulmonary disease with **nodular infiltrates**. Treat regular Aspergillus with **fluconazole**. Treat invasive Aspergillus with

**amphotericin**. Both may require long-term steroids. All Aspergillus patients should be in a negative pressure room (like MTB).

## MYCOBACTERIUM TUBERCULOSIS (aka MTB or TB)

Myco**bacterium** tuberculosis (aka MTB or TB) is a bacterium (not a fungus) that does not allow for Gram staining (that's why acid fast detection is used). It's presented here due to overlapping symptoms with the fungal infections. For newborns and young children, look for a prolonged illness with fever and cough. For older kids, look for fevers, chills, night sweats, weight loss, immigrant status, travel to an endemic area, hilar lymphadenopathy, an apical infiltrate, a pleural effusion and/or a **supraclavicular lymph node**. Diagnose by AFB smears of sputum/secretions, a positive PPD or a Quantiferon Gold. Regarding PPD READINGS, see below:

* **PEARL**: Ignore BCG vaccination status! Also, the PPD and treatment information is a little overwhelming. I have tried to simplify it, but if it's too much, MOVE ON. At the most, it would be worth 1 question. Other areas are MUCH more high yield.

* < 5 MM INDURATION: Negative for MTB
  - **PEARL**: If +induration and < 5 mm, consider an infection with ATYPICAL mycobacteria

* 5-10 MM INDURATION: Positive **IF** there are X-ray finding, or there is a history of "close contact" with someone who has recently got diagnosed with TB (newly +PPD), or the patient is immunocompromised. If none of the above conditions exist, a 5-10 mm induration in an otherwise healthy child is considered NEGATIVE for tuberculosis.

* > 10 MM: POSITIVE if there is ANY risk factor.

* > 15 MM is used as a POSITIVE only for the lowest risk patients with ZERO risk factors.

* **TREATMENT** AFTER A **+PPD SCREEN**:
  - NORMAL CHEST X-RAY: Single therapy with INH for 9 months
  - POSITIVE CHEST X-RAY: Positive means the presence of hilar lymphadenopathy, an infiltrate or a pleural effusion. Patient get 3 to 4 drug therapy.
    - TRIPLE THERAPY: Rifampin, Isoniazid and Pyrazinamide.
      - **MNEMONIC**: Must treat, or the patient will have to **R**est **I**n **P**eace!
    - QUADRUPLE THERAPY: Ethambutol is added to the regimen. The ethambutol is stopped after 2 months. RIFABUTIN

* **TREATMENT** FOR NEWBORNS WITH **+PPD MOMS**
  - If mom is asymptomatic **AND** has a negative chest X-ray, check a PPD in the baby every 3 months. If it turns positive, the baby will need treatment with INH for a year if the chest X-ray is negative, and triple to quadruple therapy if it is positive.
  - If mom has active disease **OR** a positive chest X-ray, baby is considered positive as well and gets full treatment (triple to quadruple drug regimen).

* **TREATMENT** FOR OLDER KIDS WITH **ACTIVE TB IN A HOUSEHOLD CONTACT**
  - PPD NEGATIVE & CXR NEGATIVE CHILD = INH prophylaxis for 12 weeks.
  - PPD POSITIVE & CXR NEGATIVE CHILD = INH prophylaxis for 9 months.
  - PPD POSITIVE & CXR POSITIVE CHILD = TRIPLE or QUADRUPLE therapy

- TREATMENT FOR TB MENINGITIS: Same meds + STEROIDS + STREPTOMYCIN

* **PEARLS:** The right supraclavicular nodes drain the mediastinum and the lungs, and are therefore more likely to show adenopathy in lung infections. The left side drains the thorax and the abdomen (lymphadenopathy is more likely to be lymphoma on the exam).

* **PEARL**: To check for how infectious a patient is, do NOT get a chest X-ray. Obtain sputum samples or gastric aspirates for AFB smear.

## *VIRUSES*

### COXSACKIE VIRUS & ENTEROVIRUS

Coxsackie virus and enterovirus are related, and BOTH can cause HAND, FOOT AND MOUTH disease. Look for a fever, sore throat and a maculopapular and vesicular rash on the palms and soles. Also look for small white lesions at the posterior oropharynx or ulcers on the tongue. Some patients ONLY have oral lesions. The skin lesions can be VERY tender. Self-limited. Can give steroids for severe pain.

* **PEARL**: If present, the oral lesions actually appear FIRST! Maybe it should be called MOUTH, HAND & FOOT DISEASE!

* **IMAGE**: http://bit.ly/M6dyGu

* **IMAGE**: http://i.quizlet.net/i/Ww0xK5g4ZZ3_J7xrwefBDQ_m.jpg

* **MNEMONIC**: MOUTH, HAND & FOOT DISEASE! Imagine the hands and feet being fine until the kid sticks them in his INFECTIOUS MOUTH!

* **MNEMONIC**: (image) http://images.inmagine.com/img/photoalto/paa300/paa300000040.jpg

* **COXSACKIE VIRUS**: Coxsackie virus also causes **HERPANGINA**. Look for high fever and oral lesions in the oropharynx.

- **PEARL**: It may be difficult to distinguish between the two Coxsackie syndromes. Keep in mind that herpangina has NO associated skin lesions, and may be associated with a HIGH fever and possibly a headache.

* **ENTEROVIRUS**: A different enterovirus serotype can cause high fevers, a rash and even "aseptic" or "viral" meningitis. Look for it in the SUMMERTIME.

### ADENOVIRUS

Adenoviruses can cause pharyngitis, URIs, conjunctivitis, otitis and **gastroenteritis (diarrhea, nausea, vomiting)**. No rash. Often occurs it he SUMMER.

**PEARL**: Pharyngitis can be exudative and look like Strep. If you're presented with a patient that has flu-like symptoms (or conjunctivitis, or a pharyngitis) and then gets GI symptoms, this could be your diagnosis! H1N1 can do that too, but you probably will not see that on the boards. If you do, look for very high fevers.

### ARBOVIRUS ENCEPHALITIS

The encephalitic clinical picture (confusion, lethargy) of arbovirus encephalitis will likely be tied to a mention of mosquitoes (California, St. Louis). Diagnose by getting virus-specific "**acute and convalescent**" antibody titers. That means you draw titers now and then again later to compare. PCR is NOT available.

**PEARL**: This is probably the ONLY condition on the exam where "**acute and convalescent**" titers is the correct answer.

**MNEMONIC**: Imagine sitting under a tree TREE (ARBO) that grows NOW & LATER candies which are falling on your HEAD and causing ENCEPHALITIS.

**MNEMONIC**: (image) http://www.pinatas.com/v/vspfiles/images/nowlater-product.jpg

## RESPIRATORY SYNCYTIAL VIRUS (RSV)

The chest X-ray of RESPIRATORY SYNCYTIAL VIRUS (RSV) BRONCHIOLITIS will show HYPERINFLATED LUNGS and DIFFUSE infiltrates. Diagnose with **direct immunofluorescence** testing on nasopharyngeal aspirates. Provide cool mist since albuterol may or MAY NOT help wheezing. Consider steroids for severe cases. Patients frequently need hospitalization due to severity of symptoms, comorbidities and possibly their social situation (family reliability).

* **SEVERE BRONCHIOLITIS**: Refers to pO2< 65 or pCO2 >40 on ABG, respiratory rate > 70 or pulse oximetry with saturations < 95%. This is associated with future asthma in about half of patients.

*__PEARL__: RSV is NOT the only bug to cause bronchiolitis! Paramyxovirus is second in line.

## EPSTEIN-BARR VIRUS (EBV)

Epstein-Barr virus (EBV) causes INFECTIOUS MONONUCLEOSIS. Look for fever, sore throat (often an **exudative** pharyngitis), very prominent and multifocal lymphadenopathy, malaise, URI symptoms and **atypical lymphocytes**. The fever can be high and long-lasting (1-2 weeks). Mono can even present with hepatomegaly and mild RUQ TTP. Diagnosis is made with EBV Ig**M** titers even if the monospot SCREEN is negative.

**PEARL**: Look at the entire clinical picture very carefully when dealing with a case of mononucleosis versus Strep. The monospot SCREEN can be POSITIVE for months after an asymptomatic mono infection. Also, both a Strep SCREEN and a Strep CULTURE be positive in CARRIERS.

**PEARL**: Differential diagnosis of atypical lymphocytes includes leukemia! Also, palatal petechiae MAY be present in both Strep **and** mono patients (more common with Strep).

## HUMAN HERPES VIRUS 6 (aka HHV-6)

Human herpes virus 6 (aka HHV-6) causes ROSEOLA. Look for a child 6 months – 2 years of age with sudden onset of HIGH fever. May also have neurological symptoms including SEIZURE or encephalopathy. The KEY is to look for a RASH A FEW DAYS AFTER THE FEVER SUBSIDES. The rash is usually on the neck, trunk and thighs.

**MNEMONIC**: Look for an infant or toddler that feels and looks like a ½ DOZEN (**6**) ROSY ROSES. Well, of course she does now that FEVER'S GONE and the patient finally presents to the doc!

**IMAGE**: http://upload.wikimedia.org/wikipedia/commons/7/75/Roseola_on_a_21-month-old_girl.jpg

## (DOUBLE TAKE) HERPES SIMPLEX VIRUS (HSV)

* **HERPES SIMPLEX VIRUS (HSV) AS AN STD:** The initial flare is often very painful. Pain may precede the presentation of lesions. Look for multiple, painful ulcers or vesicles on the labia or penis. Can have

lymphadenopathy. The vesicles are CLUSTERED on an ERYTHEMATOUS BASE. Lesions can also be ULCERATIVE. Diagnose by obtaining a viral culture or HSV PCR. The Tzanck smear is not specific for HSV. Treat with ORAL Acyclovir x 7 days (not topical). Treat babies with **IV** Acyclovir.

- **IMAGE**: http://www.coldsoreblisters.com/images/herpes1.jpg
- **PEARL**: HSV can be associated with a very painful infection called a HERPETIC WHITLOW (typically of a thumb or finger).
- **IMAGE**: http://img.medscape.com/pi/emed/ckb/pediatrics_general/960757-964866-1733tn.jpg

* **HERPES SIMPLEX VIRUS ENCEPHALITIS (HSV ENCEPHALITIS):** Look for fever, seizures and mentioning of the **temporal** lobe on CT brain. Treatment is STAT **IV acyclovir, followed by a lumbar puncture** to obtain fluid for PCR testing. EEG might show PLEDs (periodic lateralizing epileptiform discharges).

* **HERPES SIMPLEX VIRUS GINGIVOSTOMATITIS:** Oral and perioral/vermillion border lesions/vesicles. Gingiva is friable and malodorous. There is associated lymphadenopathy.

- **IMAGE**: http://img.medscape.com/pi/emed/ckb/pediatrics_general/960757-964866-1734tn.jpg

## (DOUBLE TAKE) VARICELLA ZOSTER VIRUS (CHICKEN POX)

The varicella zoster virus causes CHICKEN POX. A chicken pox lesion may be described as a "dew drop on a petal" during the vesicle phase. Lesions are said to come in "crops" at different times, and will therefore appear in different stages on the body (some vesicles, some crusted lesions). The rash goes to the **TRUNK and then to the FACE and EXTREMITIES**. It lasts for 7-10 days and leaves minimal scars.

**PEARL**: VZIG (VZV immunoglobulin) is given for prophylaxis to newborns if the mom developed symptoms within FIVE days prior to delivery. If symptoms started 6 days prior to delivery, NO PROPHYLAXIS NEEDED. Congenital varicella syndrome can result in low birth weight as well as CNS, eye and skin abnormalities.

**PEARL**: Any immunocompromised patient should avoid contact with patients who have a case of the Chicken Pox.

**NOTE:** It's doubtful you will need to know about the smallpox virus (Variola). In case you do, just know that the lesions all appear at the SAME TIME. So all lesions will look similar. Other facts include –> limited to face/extremities, last up to 3-4 wks and leave lots of scarring (why couldn't it be limited to the trunk and leave scars over there instead?)

**(DOUBLE TAKE) MNEMONIC**: Imagine a patient stuck in a NEGATIVE PRESSURE ISOLATION room. He's watching MTV and gets pissed because he's bored and all they ever show is reality shows. He grabs the phone and throws it at the "M.T.V." – Negative pressure isolation is required for **M**easles, **M**ycobacterium **T**uberculosis and **V**aricella. For VZV, droplet precautions are sufficient if only one dermatome is involved. As mentioned in the Aspergillus section, that too requires negative pressure isolation.

## HUMAN IMMUNODEFICIENCY VIRUS (HIV)

Consider Human Immunodeficiency Virus (HIV) in any child with cognitive impairment, FTT, opportunistic or frequent infections, thrush, hepatosplenomegaly, fevers, night sweats and weight loss. Can present in the first year with HIGH levels of immunoglobulins (which are dysfunctional), and later in life with abnormally LOW immunoglobulins. CD4/helper T cell count should be LOW. Primary mode of transmission for kids is vertical.

At birth, get immediate **DNA PCR** (**not** RNA PCR, which is for a viral load). Repeat DNA PCR at the 2-month, 4-month and 6-month well-child visit. If all are negative, patient probably neg. If they are all negative, the patient is probably fine. If any are positive, **repeat** to confirm. Maternal anti-HIV **antibodies** can persist for up to 18 months. Repeat anti-HIV antibodies at 12 or 18 months to see if they have cleared. Pregnant women with HIV should get **Zidovudine or Nevirapine** to possibly prevent the vertical transmission in the first place.

* CHEMOPROPHYLAXIS: For kids born to mothers who did not get **Zidovudine or Nevirapine** for prevention of vertical transmission, start Zidovudine (AZT) within 72 hours of birth (preferably within12 hours) and continue for 6 weeks. If 4 days have passed, it's too late and no prophylaxis is given.

* BACTRIM PROPHYLAXIS: Once a child is diagnosed with HIV, start PCP (aka PJP) prophylaxis regardless of what the CD4 count is.

* NEEDLE STICKS: If you get stuck with a needle form an HIV patient, you get 2 medications for prophylaxis.

* VACCINATIONS: MAY GIVE MMR, VZV and the FLU vaccines as long as there is only **mild** immunosuppression.

* DRUGS: It's doubtful the certification exam would ask you about a drug regimen, but the boards might ask about HIV medication side effects. They include pancreatitis, Steven Johnson Syndrome, liver toxicity, pneumonitis, CNS effects (neuropathy), neutropenia and renal problems (insufficiency and stones).

- **PEARL**: The "d" drugs (ddI & ddC) cause pancreatitis.
- **MNEMONIC**: Just imagine flipping the "d" upright to make a "p" for pancreatitis!

## (DOUBLE TAKE) RABIES VIRUS

Rabies is caused by a VIRAL infection in which the virus is transmitted through bites, scratches and contact with mucous membranes of infected animals, such as BATS, dogs, foxes, **RACCOONS** (most common in US), skunks, and WOODCHUCKS. If the history suggests a possible exposure (wild/aggressive animal), treat with standard wound care **PLUS** Human Rabies Immunoglobulin (HRIG) **PLUS** the 5 vaccine doses. If the animal is a pet, observation of the patient and animal is allowed without giving HRIG.

**PEARL**: Rabbits, rats and squirrels (rodents) are **NOT** associated with rabies.

**PEARL: For the boards, if the word "BAT" is mentioned (alive, escaped, whatever!), treat as an exposure!**

## MEASLES (aka RUBEOLA)

"COUGH, CONJUNCTIVITIS and CORYZA" are the classic symptoms of measles (aka rubeola). Coryza refers to rhinorrhea. Also look in the mouth for KOPLIK SPOTS and on the skin for a rash. The three C's come first, then the Koplik spots and the **LAST symptom to appear is the RASH. The rash starts at the head (around the hairline) and progresses down**. The rash resolves after about 5 days. The major cause of death is PNEUMONIA. The virus is transmitted via droplets. Patient requires negative pressure isolation because it's extremely dangerous. They are infectious starting FIVE days prior to the onset of symptoms and FIVE days after the infection resolves.

* **TREATMENT**:

- If < 3 days since exposure in an **unimmunized** child > 6 months old, give the **vaccine (MMR)**. Also give if vaccination status is unknown. If the MMR was given for prophylaxis and the child was 6 to 12 months of age, s/he will need it again at 1 year of age.
- If > 3 days since exposure in an **unimmunized** child, give measles immunoglobulin (**MIG**). Also give if vaccination status is unknown. Anyone given MIG will need a first-time vaccination, or possibly a **re**vaccination at 5 months.

* **PEARL**: Most measles exposures result in MIG being given when prophylaxis is indicated because the most obvious and scary symptom (the RASH) is the LAST thing to present. Late presentation to the doc means MIG is the only prophylaxis option for anyone exposed.

* **PEARL**: The MMR is a live vaccine, but measles is dangerous! So it SHOULD be given as post-exposure prophylaxis to HIV/immunocompromised.

* **NAME ALERT**: Note that this is not the GERMAN MEASLES caused by the RUBELLA VIRUS.

* **(DOUBLE TAKE) MNEMONIC**: Imagine a patient stuck in a NEGATIVE PRESSURE ISOLATION room. He's watching MTV and gets pissed because he's bored and all they ever show is reality shows. He grabs the phone and throws it at the "M.T.V." – Negative pressure isolation is required for **M**easles, **M**ycobacterium **T**uberculosis and **V**aricella. For VZV, droplet precautions are sufficient if only one dermatome is involved. As mentioned in the Aspergillus section, that too requires negative pressure isolation.

## (DOUBLE TAKE) RUBELLA VIRUS (aka GERMAN MEASLES)

In children, the Rubella virus causes "German measles" and presents with a mild fever and a maculopapular rash. The rash **starts on the face** and then spreads to the trunk and extremities within 24 hrs. The facial rash disappears as the body rash starts. The rash resolves within 3 days. For unimmunized children who get this infection, it's generally a benign viral illness that is self-limited. Diagnose with an IgM for acute cases.

* CONGENITAL RUBELLA: Most affected children were exposed during the first trimester. Eye issues are the most common finding. Can result in **cataracts**, **microphthalmia**, deafness, a patent ductus arteriosus (PDA), and Blueberry Muffin Syndrome.

- **PEARL**: Pregnant women should NOT get the MMR vaccine (it's live).
- **MNEMONIC**: Imagine looking at a PDA (personal data assistant) with an image of a BLUE BELL. The opening of the BELL looks like an EYE, and the hammer of the bell looks like a WHITE BALL. At the top of the bell, there is a tag hanging off of it that says "LIMITED."
  - **KEY:** PDA = patent ductus arteriosus, BLUE = blueberry muffin syndrome, BELL = ru-BELL-a, EYE SHAPED OPENING = microphthalmia and the WHITE BALL = a cataract.

* **NAME ALERT/MNEMONIC**: Measles, German measles, rubella and rubeola can cause confusion. You will not be given the terms rubeola or German measles on the pediatric boards. You will be given the names as they appear in the **MMR** vaccine (MEASLES, mumps and RUBELLA).

## MUMPS VIRUS

For mumps, look for a PAROTIDITIS associated with a **low-grade fever**. Difficult to make a mumps diagnosis if there is no fever to suggest infection, so also look for PANCREATITIS, ORCHITIS or even MENINGITIS (rare). This is a self-limited disease and requires no treatment.

**PEARL**: Mumps is the most common etiology of parotiditis. It's generally BILATERAL > unilateral. This is somewhat intuitive. MUMPS = Systemic illness = Bilateral parotiditis. Isolated parotiditis = Obstruction of only **one** of Stensen's ducts by a stone, or unilateral infection by Staph, etc.

## ***PARASITES/PROTOZOA***

### (DOUBLE TAKE) ERYTHEMA CHRONICUM MIGRANS

Erythema chronicum migrans is caused by BORRELIA BURGDORFERI, the spirochete that causes LYME DISEASE. Look for a large, flat lesion **(> 5 cm) that is annular and has a red border. It is located at the tick bite site in about 75% of patients. Classis description is a "bullseye" lesion.** The rash shows up 1-2 wks after the bite. Titers may still be negative during this period. Borrelia is transmitted via the Ixodes deer tick. IF the patient has an acute arthritis, disseminated erythema migrans, a palsy (BELL'S PALSY) or neuropathy, then treat with ORAL medication (**Doxycycline if >8 years old, or Penicillin or Amoxicillin if < 8 years old**). If the patient has **CARDITIS**, neuritis (encephalitis/meningitis) or **RECURRENT** arthritis, then treat with INTRAVENOUS medication **(PCN or Ceftriaxone)**. Arthritis is usually located at the large joints (especially the **knees**). Diagnosing using labs is often difficult. Obtain **Lyme antibody titers**. If positive, confirm with a Western blot. This is often a **clinical diagnosis**.

* **IMAGE**: (BULLSEYE LESION) http://web.princeton.edu/sites/ehs/bullseye.jpg

* **IMAGE**: (BELL'S PALSY) http://en.wikipedia.org/wiki/File:Bellspalsy.JPG

* **SIDE NOTES**

- BELL'S PALSY: Unilateral facial nerve paralysis (CN VII). Often idiopathic.
- The Jarisch-Herxheimer reaction results in fever, chills, hypotension, headache, myalgia, and exacerbation of skin lesions during antibiotic treatment of a bacterial disease (typically spirochetes). Due to large quantities of toxins released into the body. Classically associated with syphilis, but can also occur with Lyme disease. May only last a few hours

* **MNEMONICS:**

- From now on, think/say borreLIYME. "Don't ever throw a borLIYME to MY GRANy!" Or, "Don't ever borre-LIE to MY GRANy." Borrelia = borreLIYME. MY GRANy = Migrans.
- Imagine that BULL'S EYES are made of 2 bright neon-green LIMES! This should remind of you of the classic description.
- Imagine squeezing LYME into a CAN = **C**arditis, **A**rthritis, and **N**euritis.

### LEPTOSPIROSIS

Lepto**SPIRO**sis is caused by a **SPIRO**chete that doesn't stain well. Look for multisystem complaints. Some of them may include a **NONexudative conjunctivitis referred to as a CONJUNCTIVAL SUFFUSION**, a transient rash, abdominal pain and **headach** or evidence of a **non-bacterial meningitis**. May also present as the combination of liver and renal failure. It's transmitted through the urine of animals (dogs, cats and squirrels to name a few. Diagnostic testing depends on timing. During the first week, get blood cultures. After that, check the urine. Treat with penicillin.

**PEARL**: If you see the words **CONJUNCTIVAL SUFFUSION**, PICK THIS AND MOVE ON! A conjunctival suffusion basically looks like conjunctivitis (seen as red sclera), but without tearing or discharge.

**IMAGE**: http://www.leptospirosis.org/downloads/conjunctival-suffusion-02.png

**IMAGE**: http://www.leptospirosis.org/downloads/petechial-body-rash-01.png

**MNEMONIC**: Imagine that a DOG FIERY EYES LEPT into a POND and started doing a back stroke. When the owner LEPT in to play with the dog, his EYE hit the dog's GENITOURINARY APPARATUS (aka PENIS), and his ABDOMEN got hit with the dog's legs.

* **KEY:** DOG WITH FIERY EYES = animal reservoir & conjunctival suffusion, LEPT = LEPTospirosis, POND = from urine to water and then ingeste, EYE = conjunctival suffusion, GU APPARATUS = transmitted via urine and ABDOMEN = abdominal pain from liver involvement (might also remind you of renal failure).

## ENTAMOEBA HISTOLYTICA (aka AMEBIASIS)

For amebiasis, caused by Entamoeba histolytica, look for a history of travel to the Southwestern US or to a Native American reservation. Patients will have abdominal pain associated with diarrhea which can be watery, bloody or mucoid. Patients often have TENESMUS. Obtain and ultrasound to look for liver abscesses. Diagnosis with SEROLOGIC TESTING is preferred over stool smears. Stool examination for cysts is supportive, but it does not allow for definitive differentiation from several other parasites. Treat with Metronidazole for any signs of colitis and liver abscess. If metronidazole is not an option, choose iodoquinol.

**PEARL**: Tenesmus is the feeling of incomplete stool evacuation. It can be associated with pain, straining or cramping.

**IMAGE**: http://www.parasitecleanse.com/images/amoeba01.jpg

**MNEMONIC**: Imagine a METRO train filled with AMOEBAS playing TENNIS/TENESMUS

**MNEMONIC IMAGE**: http://bit.ly/odp0Nn

## (DOUBLE TAKE) TRICHOMONAS VAGINALIS

Trichomonas vaginalis is a protozoa. Look for yellow-green discharge, bubbly, **frothy,** motile flagella and pH > 4.5. Can be intensely pruritic. **Strawberry cervix**. Treat with one dose of metronidazole. Can cause nongonococcal urethritis.

**MNEMONIC**: Imagine 2 girls having this conversation. "Can you believe that dirty old man TRICKED MONA, our favorite BUBBLY stripper, into putting a STRAWBERRY into her va-jay-jay?!? Gross! I heard it still had a long green STEM on it!" Sorry for the gross mnemonic, but hopefully it helps you associate the stem with the motile TAIL/FLAGELLA of Trichomonas.

## BABESIOSIS

Babesiosis is unlikely to be tested. It's caused by a protozoa (Babesia). Usually asymptomatic or only mild fevers. In severe case, can look like malaria with high fevers, chills and a hemolytic anemia. Mild or severe cases can cause a sudden death. The key will be to look a non-autoimmune hemolytic anemia and to look for a smear with the classic "Maltese Cross" finding. Transmitted by Ixodes tick (same as Lyme disease and ehrlichiosis). Treat with quinine and clindamycin.

**IMAGE**: http://www.smom-za.org/images/babesia.gif

**IMAGE**: https://annals.org/data/Journals/AIM/19780/6FF1.jpeg

## CRYPTOSPORIDIUM

Cryptosporidium is a protozoa. Look for an outbreak of **watery** diarrhea lasting a LONG time (7 days to a month!). Occurs in immunocompetent and immunocompromised patients. Self-limited. Fecal oral transmission. Look for cysts in the stool, do direct fluorescence antibody (DFA) testing or do PCR for diagnosis. Treatment is primarily supportive.

## MALARIA

Malaria is rare in the U.S. Not commonly tested. Look for travel to an endemic area or migrant status. Symptoms will include cycles of feeling cold, then having harsh chills (rigors), and then high fever with diaphoresis (all within a 6 hour period). Patients can have hepatosplenomegaly and a hemolytic anemia. For the diagnosis may be given a blood smear. Each of the 4 species have characteristic appearances on the THIN & THICK BLOOD SMEARS/FILMS you would request. Plasmodium vivax and ovale can relapse so perhaps those would be the ones tested. Falciparum and malariae do not.

**MNEMONIC**: **V**ivax & **O**vale can relapse. "V.O.dka goes to the LIVER and hides… no wonder they can relapse!"

## TRYPANOSOMA CRUZI

Trypanosoma cruzi causes CHAGAS DISEASE. The mild illness initially goes undiagnosed in the child. The bug slowly damages the heart over many **years** until the patient eventually dies because of HEART FAILURE (CHF) or HEART BLOCKS. May present with some type of swelling and a heart block. Can also cause THROMBOEMBOLIC events. Transmitted by the REDUVIID bug in the Southern or Southwestern U.S. Also found in South America.

**MNEMONIC**: I TRYPPED over a CHAGGY carpet. When I fell, the left side of my CHEST landed on a RED BUG which started EATING MY HEART!

## TRYPANOSOMA BRUCEI

Trypanosoma brucei causes "African Sleeping Sickness." Doubtfully tested by the AMERICAN Board of Pediatrics (ABP) since it's not ever seen (< 1 case/year) in the U.S. Look for travel to Africa. Starts with a chancre and symptoms can include trouble sleeping (initially), edema, fever, poor appetite, tremors, irritability and headache. Kids can have seizures. Patients can eventually end up in a stuporous coma (hence the name). Transmitted by the tsetse fly.

**MNEMONIC**: I was having trouble SLEEPING so I went to the doctor. As I entered his office I **TRYP**PED on a **BRU**UUM (broom), fell, hit my head on his TSETSES and got knocked out into a STUPOROUS COMA! Finally… some sleep!

## *WORMS*

USE THE FOLLOWING ONLY FOR THE IMAGES! THEN MOVE ON!

**IMAGES**: http://www.fatfreekitchen.com/home-remedy/worms.html

**IMAGES**: http://www.parasitecleanse.com/GALLERY.HTM

## ENTEROBIUS (aka PINWORMS)

If the pediatric board exam mentions a child scratching his butt, the diagnosis is probably enterobius (pinworms). Diagnose with a TAPE TEST (apply 1 inch of the tape to anus in the MORNING, pull, apply to slide and examine for eggs). Treat with **MEBENDAZOLE.**

**PEARL**: One dose is all you need. If recurs, repeat x1 two weeks later. For all other worms, treatment usually requires multiple doses.

**MNEMONIC**: ENTER-obius = ENTERS-your-butt!

## (DOUBLE TAKE) ASCARIS LUMBRICOIDES

Look for a history of travel to a tropical climate or immigrant status if ascaris lumbricoides is suspected. Patients will have abdominal discomfort and pain. Can also have pancreatitis as well as INTESTINAL OBSTRUCTION. Can also have RESPIRATORY symptoms including a bronchitis that appears as SHIFTING INFILTRATES on serial chest X-rays and is associated with EOSINOPHILIA. Diagnose by examining stool for worms or barrel-shaped eggs. Treat with MEBENDAZOLE or PYRANTEL PAMOATE.

**NOTE:** Loffler Syndrome is the build up of pulmonary eosinophils in the lungs due to a parasitic infection. This is also called chronic pulmonary pneumonia. As mentioned, look for SHIFTING INFILTRATES on serial chest X-rays.

**IMAGES**: http://1.usa.gov/oQOt1s

**MNEMONIC**: **A SCAREEEE** (bunch of earthworm-like, **BENDY** creatures **CLOGGED UP** a **BARREL**.

- **Key: ascar**is, **e**osinophilia, me**benda**zole, intestinal obstruction & barrel-shaped eggs

**MNEMONIC**: Imagine **A SCAREEEE PYRATE** (**pyr**antel pamo**ate**) playing a **A NEW PIANO** (pneumonia) on his ship. As the waves get rough, the NEW PIANO starts to **SHIFT** all over the deck (shifting infiltrates).

## SCHISTOSOMIASIS (SCHISTOSOMA)

Schistosomiasis is caused by Schistosoma, a fluke. **FL**ukes are **FL**atworms. Variable presentation. For the pediatric certification exam, you would likely be presented with a child that has chronic, intermittent abdominal pain, possibly blood in the stool and possibly **hepato**splenomegaly (Schistosoma japonicum can result in both intestinal and liver schistosomiasis). Treat with praziquantel.

## TAENIA SOLIUM

Taenia solium is a tapeworm from PORK. Symptoms include GI discomfort. Can also result in **NEUROCYSTICERCOSIS**, which case the patient can present with SEIZURES. Treat with ALBENDAZOLE or praziquantel.

**IMAGE**: (CT) http://radpod.org/wp-content/uploads/2007/05/neurocysticercosis.jpg

**IMAGE**: (MRI) http://upload.wikimedia.org/wikipedia/commons/6/61/Neurocysticercosis.gif

## TAENIA SAGINATA

Taenia saginata is a tapeworm from BEEF. Often asymptomatic. Severe disease can result in abdominal pain, diarrhea or constipation, and a loss of appetite. Treat with praziquantel.

**MNEMONIC**: TAENIA SAGIN-UTA is from the "SAGGIN UTTA" of a cow!

**MNEMONIC**: (IMAGE) http://www.thenazareneway.com/vegetarian/udder.jpg

## (DOUBLE TAKE) TOXOCARA CANIS

Toxocara canis is a tapeworm that causes **VISCERAL** LARVA MIGRANS. Look for a child presenting with **multisystem** complaints. Usually affects the LUNGS and the GI TRACT, but can also affect the eyes. May present as abdominal pain in a child that has **hepatosplenomegaly** and wheezing on exam. Labs will show a HIGH LEUKOCYTOSIS with EOSINOPHILIA. Imaging will show lung infiltrates. The tapeworm is found in cats, dogs and dirt, so look for a kid that likes to eat dirt! It is often self-limited, but can be treated with albendazole (or mebendazole).

* **PEARL**: The words VISCERAL and MIGRANS in an answer choice should help you remember that this is a disease of the deep organs (viscera), and it migrates to multiple organs. Lung infiltrates may be noted to change, or "migrate," over time as noted on serial chest X-rays.

## HOOKWORM Ankylostoma

Hookworm causes **CUTANEOUS** LARVA MIGRANS. Look for a **SERPIGINOUS RASH** (snake-like) that **MIGRATES** and is associated with malaise, eosinophilia, GI symptoms and LUNG findings. Can get it through stepping on something. It then goes to the lungs, causes sputum production which is swallowed and then causes GI symptoms. RX Albendazole / Mebendazole

**IMAGE**: http://img.medscape.com/pi/emed/ckb/pediatrics_general/996090-998709-206.jpg

**MNEMONIC**: Imagine walking on a beach and stepping on a HOOK. You get a SNAKE-LIKE RASH that starts at your foot and ends at you CHEST (lungs). You then start coughing up some nastiness that you then have to SWALLOW!

## TRICHURIS

Trichuris is a whip worm that is associated with diarrhea and RECTAL PROLAPSE.

**MNEMONIC**: Imagine Indiana Jones saying, "My WHIP is going to TRICK YOU'RE A$$. Do you think I can do it?" He takes the WHIP, hits your bum and TRICKS YOU'RE A$$ into EJECTING his favorite hat! Unfortunately it's all WET.

* **KEY**: WHIP = Whipworm, TRICK YOU'RE A$$ = Trich-ur-is, EJECTING = rectal prolapse, and WET = Diarrhea

**MNEMONIC**: (IMAGE) http://www.coloring-pictures.net/drawings/IndianaJones/Indiana-uses-his-whip.gif

## FILARIASIS

Filariasis can be caused by a few different roundworms. Mostly a tropical disease. Look for ORCHITIS, epididymitis and inguinal adenopathy. Patients may have "elephantiasis" due to LYMPHATIC FILARIASIS.

**IMAGE**: http://bit.ly/qevrdc

## STRONGYLOIDES

Strongyloides causes abdominal discomfort and a pneumonitis. Treat with thiabendazole.

**MNEMONIC**: Imagine a STRONG LADY (strongyloides)able to do such STRONG THIGH BENDS (thiabendazole) that they crush her CHEST (pneumonitis).

## DIPHYLLOBOTHRIUM LATUM

DIPHYLLOBOTHRIUM LATUM = "Fill up a BROTH with that FISHy stuff" = raw fish. Causes B12 def.

# *SYNDROMES*

## GROUND GLASS PNEUMONIA

If a ground glass pneumonia is noted, consider Pneumocysistis pneumonia (aka PCP or PJP) if the patient is immunocompromised. PCP can be seen in patients with a history of HIV, Hyper IgM Syndrome and Severe Combined Immunodeficiency (SCID). In Hyper-IgM, look for a lymphoCYTOSIS and NEUTROpenia. In SCID, look for lymphoPENIA since there is a deficiency of T & B cells.

* **OTHER CONSIDERATIONS:** If the patient is a neonate, consider surfactant deficiency, respiratory distress syndrome (RDS) and Group B Streptococcus (GBS).

* **PEARL**: Look for underinflated lungs with air bronchograms on imaging for surfactant deficiency and GBS.

## ADOLESCENT + PNEUMONIA + LOW GRADE FEVER

If an adolescent presents with a pneumonia and a low-grade fever, choose **Chlamydia** (diagnose with immunofluorescent antibodies) or **Mycoplasma** pneumonia (diagnose with Ig**M**, and treat with a **M**acrolide).

## SPONTANEOUS BACTERIAL PERITONITIS (SBP)

Spontaneous bacterial peritonitis (SBP) is usually found in nephrotic children, so look for **low IgG.** Could also be cirrhosis. Most likely etiology is Streptococus pneumoniae (pneumococcus). Give Ceftriaxone **and Gentamicin** until there is a positive ID of the organism.

## SECONDARY PERITONITIS

Secondary peritonitis is due to trauma and perforation of the intestines. Most likely due to gram negative organisms and anaerobes. Ceftriaxone and metronidazole (or piperacillin-tazobactam) would be good choices. If the peritonitis is in a patient getting peritoneal dialysis, the cause is likely STAPHYLOCOCCUS EPIDERMIDIS (Staph epi). Use VANCOMYCIN to treat Staph epidermidis.

## TOXIC SHOCK SYNDROME (TSS)

For the pediatric boards, a question on toxic shock syndrome (TSS) would likely provide the triad of FEVER + HYPOTENSION + SKIN INFECTION. Can be cause by Staphylococcus OR Streptococcus infection. If asked to choose, which Streptococcus, choose GAS (Group A Strep). If due to a GAS infection, it may be a Strep *pyogenes* NECROTIZING FASCIITIS. If associated with Staph aureus, the patient may also have emesis, watery diarrhea or a foreign body (tampon). Mortality is higher with STREP.

**PEARLS:** You may not be given a classic scenario of an adolescent girl using a tampon for days. BUT, if the patient is an adolescent female with FEVER + HYPOTENSION, they might just be holding back a key piece of information.

## DENTAL ABSCESS

A dental abscess can present as swelling below the jaw or periorbital swelling

## NEONATAL FEVER

If a neonate (under the age of 30 days) presents with a fever, you MUST do a FULL septic workup. This includes blood cultures, urine culture and the dreaded LUMBAR PUNCTURE. A chest X-ray is only indicated if the child has respiratory symptoms.

## NEONATAL BACTEREMIA

Note that neonatal bacteremia this is NOT the same thing as the syndrome of neonatal sepsis.

* CHILD < 30 DAYS OLD: Differential includes Group B Streptococcus (GBS), E. coli, Streptococcus pneumoniae and Staphylococcus aureus. Strep is the most common overall.

* CHILD < 12 MONTHS OLD: Same as above + Salmonella.

* IMMUNOCOMPROMISED CHILD: Cancer and HIV patients tend to get very bad Gram negative infections. Think E. coli, Klebsiella and Pseudomonas.

## SINUSITIS

Sinusitis may present as a recent URI now turning into bad breath or a **persistent cough**. May also be associated with nasal congestion or rhinorrhea. An ethmoid sinusitis can lead to an orbital **cellulitis**. A frontal sinusitis can lead to a **brain abscess**.

* **PEARL`:** Children with sinusitis often do not have a headache. If given the history of UNILATERAL foul smelling nasal discharge, always consider **foreign body** as the diagnosis, even if the patient has cystic fibrosis!

* **ACUTE SINUSITIS** (< 4 WEEKS): Likely due to the usual bugs that cause otitis media, Haemophilus influenza, Moraxella catarrhalis and Streptococcus pneumoniae.

- **MNEMONIC**: Consider using "HMS" as the acronym to help you remember these bugs.

* **CHRONIC SINUSITIS** (> 4 WEEKS): Look for evidence of some other comorbidity, such as hay fever, an immunodeficiency, immotile cilia or a history of cystic fibrosis.

- < 90 DAYS: The usual HMS bugs. Treat with Amoxicillin.
- > 90 DAYS: Think **STAPHYLOCOCCUS AUREUS**. Treat with Cefazolin.

## PAROTIDITIS (aka PAROTITIS)

Parotiditis (aka parotitis) has multiple possible causes.

* MUMPS VIRUS: See VIRUSES above. Look for orchitis, pancreatitis or meningitis.

* STONE: Obstruction leads to INTERMITTENT parotid swelling

* STAPHYLOCOCCUS AUREUS: Look for **PUS** at Stensen's duct or a child that has HIGH fevers or is toxic-appearing.

* RECURRENT IDIOPATHIC PAROTIDITIS: Occurs for several days every 3-4 months. Idiopathic.

* CHRONIC PAROTIDITIS: Look for evidence of odd infections, including mycobacterium tuberculosis (TB), Bartonella henselae (cat scratch disease) and even HIV.

* SJOGREN'S SYNDROME: Dry eyes, dry mouth and parotiditis.

## MASTOIDITIS

Look for pain, swelling and redness in the posterior auricular area as an indication of mastoiditis. Caused by the "HMS" bugs of Haemophilus influenza, Moraxella catarrhalis and Streptococcus pneumoniae. Diagnose by doing a tympanocentesis, sending pus for culture and by getting a CT scan. Treat with IV antibiotics and with surgical debridement.

## OTITIS EXTERNA (aka SWIMMER'S EAR)

Otitis externa (aka swimmer's ear) results in pain with manipulation of the PINNA. Treat with **OFLOXACIN** drops (a quinolone) or Polymyxin. The quinolone drops will cover both **Pseudomonas and Staphylococcus**, which are the 2 most common pathogens. Acetic acid drops can be after swimming for prevention.

**PEARL**: Should also GIVE STEROIDS in order to decrease inflammation and edema in order to ALLOW the antibiotic drops to get in and do their job!

## ACUTE & RECURRENT OTITIS MEDIA

Amoxicillin is the first line of treatment for otitis media. If it's not responding to HIGH dose amoxicillin, it's probably a beta-lactamase producing organism like H. flu (non-typable) or Moraxella. Do NOT give a macrolide since they don't do much for beta-lactamase producing organisms. Give AMOXICILLIN-CLAVULANATE or a CEPHALOSPORIN instead.

* **PEARLS:** For a regular OM, you can choose to withhold antibiotics for up to THREE days. Once the patient is on antibiotics, THREE days is the maximum time you can wait to change antibiotics when there is no improvement. Meningitis is the most common intracranial complication of OM. MCC intracranial complication of OM

* **CHRONIC MIDDLE EAR EFFUSION**: If the patient is healthy and hears well, watchful waiting.

* **RECURRENT OTITIS MEDIA**: Four episodes within a 12 month span is ok if the child is healthy and hears well. Five or more "may" warrant myringotomy tubes. If the child is predisposed to effusions and infections, s/he may still get otorrhea with URIs. A possible complication of tube placement is a bloody granuloma.

## CHOLESTEATOMA

Cholesteatomas result in PAINLESS otorrhea due to DESTRUCTIVE lesions at the base of the skull. Can be associated with, or lead to, hearing loss. It can erode at important structures. Otorrhea will include debris containing SKIN and EPITHELIAL cells. Landmarks on otoscopic examination may be absent due to destruction. Treatment requires excision of the lesion.

## CHRONIC OTORRHEA & RECURRING OTORRHEA

The most common causes of chronic otorrhea and recurring otorrhea are **PSEUDOMONAS and STAPH.** Give OFLOXACIN **DROPS**.

**PEARL**: May give a similar presentation to cholesteatoma, but there should be NO skin or epithelial debris. Also, consider the presence of a concurrent cholesteatoma if it's not improving with appropriate antibiotics.

**MNEMONIC**: Most common bugs are not the HMS bugs. They are the same as the **otitis externa** bugs. Maybe because of the connection to the outside world?

## MENINGITIS, BACTERIAL & VIRAL

If you suspect a meningitis (unexplained fever, irritability, lethargy) in a patient with NO FOCAL DEFICITS, go straight for the lumbar puncture without waiting for the CT. If there is focality, give immediate antibiotics followed by a CT and then an LP.

* **BACTERIAL MENINGITIS**: Look for an elevated protein and a low glucose (< 40, or less than 2/3 the serum glucose). Look for a predominance of neutrophils in the CSF. The number of leukocytes should be > 100, but hopefully the board of pediatrics will be kind enough to provide one in the THOUSANDS.

- AGE < 30 DAYS: Ampicillin and gentamicin. The ampicillin covers Listeria.
- AGE 1-3 MONTHS: Consider changing over to vancomycin and ceftriaxone.
- AGE > 3 MONTHS: Definitely give vancomycin and ceftriaxone.
- **(DOUBLE TAKE) PEARL:** Ceftriaxone is still **great** for Streptococcus. When treating for meningitis, also give Vancomycin. The vancomycin is NOT being given for Staph aureus coverage. It's being given for possible Strep that is RESISTANT to ceftriaxone!

* **VIRAL MENINGITIS**: Will show a low WBC count (< 500) and should have a lymphocyte predominance. The CSF glucose level will be normal. The CSF protein level should be normal or slightly elevated. If there is concern for HSV encephalitis, START ACYCLOVIR.

- **PEARLS:** MENINGITIS + SEIZURE = IMMEDIATE ACYCLOVIR. Assume it's due to HSV, even if the CSF is 100% suggestive of a bacterial meningitis. Give anti-bacterial antibiotics as well. Send the CSF for HSV DNA PCR to diagnose.
- **PEARL**: WBCs do NOT belong in the CSF. If there are ANY, there's something going on. If it's due to a traumatic tap, then assume that 1 WBC is present for every 500 to 1000 RBCs. So if a patient has a traumatic lumbar puncture which shows 10,000 RBCs and 130 WBCs, you should assume there are over **100 extra WBCs**!

## *TORCH INFECTIONS*

**NOTE**: TORCH stands for **T**oxoplasma, **O**ther (Varicella, Syphilis), **R**ubella, **C**ytomegalovirus (CMV) and **H**erpes Simplex Virus (HSV). Perinatal TORCH infections can lead to severe fetal anomalies and even death.

### TOXOPLASMA GONDII

Toxoplasma gondii is a protozoa. Look for CNS and eye problems. The classic triad is **diffuse** intracerebral calcifications, hydrocephalus and uni- or bilateral **chorioretinitis**. Other findings may include adenopathy, thrombocytopenia, hepatosplenomegaly, jaundice and Blueberry Muffin Syndrome. Toxoplasma is transmitted via cat litter. Diagnose by getting Ig**M** titers or immunofluorescence. Treat with sulfadiazine or pyrimethamine.

**PEARL**: Just because a neonate looks good at birth, does not mean you can cross this answer off in an older child. This can still present with seizures, cognitive issues and even deafness.

**PEARL**: Toxoplasmosis can present exactly like CMV. A key differentiating factor is the IMAGING. The calcifications of toxoplasmosis are **DIFFUSE** and are referred to as "**RING-ENHANCING LESIONS.**" Those of CMV are **PERIVENTRICULAR** and **do not enhance**.

## (DOUBLE TAKE) VARICELLA ZOSTER VIRUS (CHICKEN POX)

The varicella zoster virus causes CHICKEN POX. A chicken pox lesion may be described as a "dew drop on a petal" during the vesicle phase. Lesions are said to come in "crops" at different times, and will therefore appear in different stages on the body (some vesicles, some crusted lesions). The rash goes to the **TRUNK and then to the FACE and EXTREMITIES**. It lasts for 7-10 days and leaves minimal scars.

**PEARL**: VZIG (VZV immunoglobulin) is given for prophylaxis to newborns if the mom developed symptoms within FIVE days prior to delivery. If symptoms started 6 days prior to delivery, NO PROPHYLAXIS NEEDED. Congenital varicella syndrome can result in low birth weight as well as CNS, eye and skin abnormalities.

**PEARL**: Any immunocompromised patient should avoid contact with patients who have a case of the Chicken Pox.

**NOTE:** It's doubtful you will need to know about the smallpox virus (Variola). In case you do, just know that the lesions all appear at the SAME TIME. So all lesions will look similar. Other facts include –> limited to face/extremities, last up to 3-4 wks and leave lots of scarring (why couldn't it be limited to the trunk and leave scars over there instead?)

**(DOUBLE TAKE) MNEMONIC**: Imagine a patient stuck in a NEGATIVE PRESSURE ISOLATION room. He's watching MTV and gets pissed because he's bored and all they ever show is reality shows. He grabs the phone and throws it at the "M.T.V." – Negative pressure isolation is required for **M**easles, **M**ycobacterium **T**uberculosis and **V**aricella. For VZV, droplet precautions are sufficient if only one dermatome is involved. As mentioned in the Aspergillus section, that too requires negative pressure isolation.

## (DOUBLE TAKE) SYPHILIS

Syphilis is caused by TREPONEMA PALLIDUM. Non-treponemal tests (+RPR or VDRL) can be FALSE positives so you need to **do a confirmatory treponemal test (FTA). If a mom was RPR+ but the FTA was not done, obtain the FTA in the baby**. FTA doesn't correlate with disease activity, the non-treponemal tests do, and they may eventually disappear. So once disease presence is confirmed with FTA, look at disease activity with non-treponemal titers to help guide management. If mom was treated and the baby's titers are lower than hers, it's safe to assume those are just the mom's IgGs that crossed placenta. There is NO NEED TO TREAT, just follow the titers. **If mom was treated < 1 month ago, TREAT.** If mom was given Erythromycin, TREAT because it doesn't cross the placenta.

* CONGENITAL SYPHILIS: Baby born with maculopapular rash, HSM, **PEELING SKIN, PERFORATED NASAL SEPTUM** or **HUTCHINSON TEETH**. These teeth are peg-shaped (cone-like) teeth but also have a **central notch** that is extremely specific for congenital syphilis. Treat with PENICILLIN (PCN).

* CONDYLOMA LATA: Refers to SECONDARY SYPHILIS, in which white-gray coalescing papules are seen.

* **PEARL:** If the FTA is positive but VDRL is negative, also consider LYME DISEASE (BORRELIA BURGDORFERI).

* **NAME ALERT/MNEMONIC**: Condyloma LATA (aka "condyloma FLATa," are much more FLAT than Condyloma ACUMINATA (found with HPV infections).

* **NAME ALERT/MNEMONIC**: Peg teeth are also found in patients with Incontinentia Pigmenti (aka "incontinentia PEGmentia")

* **IMAGE**: (PEG-SHAPED TEETH) - http://alturl.com/cysjv

* **IMAGE**: (HUTCHINSON TEETH) - http://alturl.com/jc6nf

## (DOUBLE TAKE) RUBELLA VIRUS (aka GERMAN MEASLES)

In children, the Rubella virus causes "German measles" and presents with a mild fever and a maculopapular rash. The rash **starts on the face** and then spreads to the trunk and extremities within 24 hrs. The facial rash disappears as the body rash starts. The rash resolves within 3 days. For unimmunized children who get this infection, it's generally a benign viral illness that is self-limited. Diagnose with an IgM for acute cases.

* CONGENITAL RUBELLA: Most affected children were exposed during the first trimester. Eye issues are the most common finding. Can result in **cataracts**, **microphthalmia**, deafness, a patent ductus arteriosus (PDA), and Blueberry Muffin Syndrome.

- **PEARL**: Pregnant women should NOT get the MMR vaccine (it's live).
- **MNEMONIC**: Imagine looking at a PDA (personal data assistant) with an image of a BLUE BELL. The opening of the BELL looks like an EYE, and the hammer of the bell looks like a WHITE BALL. At the top of the bell, there is a tag hanging off of it that says "LIMITED."
    - **KEY:** PDA = patent ductus arteriosus, BLUE = blueberry muffin syndrome, BELL = ru-BELL-a, EYE SHAPED OPENING = microphthalmia and the WHITE BALL = a cataract.

* **NAME ALERT/MNEMONIC**: Measles, German measles, rubella and rubeola can cause confusion. You will not be given the terms rubeola or German measles on the pediatric boards. You will be given the names as they appear in the **MMR** vaccine (MEASLES, mumps and RUBELLA).

## CYTOMEGALOVIRUS (CMV)

Cytomegalovirus (CMV) is generally benign in healthy individuals with symptoms similar to mononucleosis. Diagnosis in these healthy individuals can be made with Ig**M** antibody in the acute phase. Intrauterine exposure, however, can cause severe problems. The classic triad is uni- or bilateral **chorioretinitis**, **microcephaly** and **periventricular** intracerebral calcifications. Other possible findings include **hearing loss**, cognitive defects, seizures, thrombocytopenia (petechiae) and evidence of extramedullary hematopoesis (hepatosplenomegaly and Blueberry Muffin Syndrome). Diagnosis in neonates can be made by urine culture for CMV or PCR. Treat with **G**anciclovir > Acyclovir to possibly prevent future hearing loss.

**PEARLS:** The hearing loss caused in kids with CMV is a sensorineural hearing loss. HIV patients may present with a CMV pneumonia. The calcifications in CMV are **periventricular**, and those of toxoplasmosis are **diffuse**.

**MNEMONICS:** Regarding Ganciclovir versus Acyclovir, try to associate **C**MV treatment with the **G** drug (G & C look very similar). Or, "GCV for CMV?"

**MNEMONIC**: CMV = **CCC**! That should help you remember the classic triad of **C**horioretinitis, **C**alcifications and **C**ensorineural hearing loss.

## (DOUBLE TAKE) BLUEBERRY MUFFIN SYNDROME

Blueberry muffin syndrome represents extramedullary hematopoiesis. Can be seen in congenital **viral** infections like **Rubella**, Coxsackie, Cytomegalovirus (CMV), Herpes Simplex Virus (HSV) and Parvovirus. Can also be associated with congenital Toxoplasmosis (a protozoa).

**IMAGES**: http://alturl.com/pfdqj

## *ACUTE WATERY DIARRHEA*

**PEARLS:** Acute watery diarrhea generally does not warrant a big workup or any testing unless you have concerns for specific organisms, like C. difficile, Cholera or Giardia. NEVER choose a BRAT (bananas, rice, applesauce and toast). Always give an age-appropriate diet rather than prescribing bowel rest. This helps the GI mucosa stay as healthy as possible. When there is mild-moderate dehydration, a rehydration trial with Oral Rehydration Solution (ORS contains 2% glucose + 90 meq NaCl) should be tried first. Stool studies that are positive for WBCs, neutrophils, mucous or blood are more likely due to a BACTERIAL pathogen. ALWAYS treat Shigella (use Ceftriaxone or Bactrim).

**PEARL**: For the pediatric certification exam, you should **almost** never pick loperamide! If it's an organism that secretes toxins, stopping the diarrhea can be deadly. The one exception is traveler's diarrhea.

### ROTAVIRUS

Day 6 p 3 Viral

Look for rotavirus in younger children up to 3 or 4 years of age. Usually a WATERY diarrhea with no flecks of blood. CAN be associated with a fever and vomiting. Diagnose with antigen testing of the stool and treat supportively.

**PEARL**: Rotavirus is **NOT** associated with food and water intake. Do not pick this for a child that has recently been on a camping trip or a picnic.

### ADENOVIRUS

Look for fever, vomiting and diarrhea a few days after a picnic to indicate adenovirus. Adenovirus IS associated with food. Usually short-lived and resolves within a week. Treatment is supportive.

### NORWALK VIRUS

Look for Norwalk virus in an older kid with diarrhea who recently ate raw SEAFOOD.

**MNEMONIC**: Can you really imagine a child younger than 5 or 6 using chopsticks to eat sushi? I can't. Imagine taking a stroll along a "NORWALK" next to the ocean (instead of boardwalk). This should help you remember that NORWALK is associated with SEAfood.

### ESCHERICHIA COLI (E. coli)

Escherichia coli (E. coli) is a Gram negative bacteria that can cause multiple types of diarrhea.

BLOODY * **ENTEROHEMORRHAGIC E. COLI (EHEC)**: As the word "hemorrhagic" suggests, look for **BLOODY** diarrhea caused by a Shiga-like toxin. Also look for renal dysfunction and signs of hemolysis because this can cause Hemolytic Uremic Syndrome (HUS). Do **NOT** give antibiotics because that makes this worse!

- **(DOUBLE TAKE) MNEMONIC**: The symptoms/findings for Hemolytic Uremic Syndrome (HUS) can be recalled by using the mnemonic FAT RN. Fever, Anemia, Thrombocytopenia, Renal Failure and Neurologic Symptoms.

* **ENTEROTOXIGENIC E. COLI (ETEC):** This is your classic "traveler's diarrhea." Look for a non-bloody diarrhea associated with crampy abdominal pain. This can be associated with fever and vomiting. Look for it

in someone who recently went on a vacation. It's self-limited. Use a **macrolide** (azithromycin is preferred) if the diarrhea is persisting for greater than a week. Trimethoprim-sulfamethoxazole or doxycycline have too much resistance. Do NOT use quinolones for kids. **Loperamide is ok**! The CDC does not recommend antibiotic prophylaxis.

* ENTEROPATHOGENIC E. COLI (EPEC): Key phrase = POOR SANITATION. It's a non-bloody, acute or chronic diarrhea. Probably low yield for the pediatric boards.

* ENTEROINVASIVE E. COLI: VERY similar in presentation to like Shigella. Look for a BLOODY diarrhea associated with mucous and TENESMUS.

## SHIGELLA DYSENTERIAE

Shigella dysenteriae is a Gram negative organism that can cause SEVERE symptoms including fever, **bloody diarrhea, TENESMUS.** Some of the more serious complications include **rectal prolapse** and **seizures.** These patients will **LOOK SICK** and WBC may show a high neutrophil count and a **BANDEMIA**. Diagnose with stool culture. Treat with **trimethoprim-sulfamethoxazole** or **ceftriaxone**. A macrolides could be an alternative (azithromycin).

**PEARL**: ALWAYS TREAT! Tenesmus is the feeling of incomplete stool evacuation. It can be associated with pain, straining or cramping. It can also be seen in **Entamoeba histolytica.**

**MNEMONIC/PEARL**: Sing the following, "Shigella makes you SHAKE A LEG y'all!" Now imagine that Shigella is making you SHAKE A LEG as you are having a SEIZURE. You shake your leg so hard that your RECTUM prolapses, and along with it a TENNIS BALL comes out!

* **KEY:** SHAKE A LEG = Seizure, and the TENNIS BALL represents TENESMUS.

## SALMONELLA

* **SALMONELLA NON-TYPHI**: Look for a febrile patient with **green, malodorous diarrhea.** May be associated with blood in the stool. This is definitely associated with food, so look for a history of a recent **picnic**. Specific foods include ice cream, poultry and **eggs**. It can also be transmitted a pet IGUANA. This type of Salmonella is self-limited and does NOT require antibiotic treatment. In fact, treating can cause a **carrier** state and **prolong** shedding.

- **PEARL**: If it's diagnosed in an immunocompromised patient, treat with Ceftriaxone.

* **SALMONELLA TYPHI** (S. TYPHI): This causes TYPHOID **FEVER**. This causes fever, diarrhea, abdominal pain and "ROSE SPOTS." The PERSISTENT FEVER after the diarrhea has resolved is quite unique and is from persistence in liver, gallbladder and spleen. It's transmitted through the fecal-oral route and **REQUIRES treatment with Ceftriaxone or Cefotaxime**. If you need a study break, consider reading about Typhoid Mary at http://history1900s.about.com/od/1900s/a/typhoidmary.htm.

- **PEARL**: ROSE SPOTS occur on the chest and abdomen. They are small, flat, rose-colored macules.
- **IMAGE**: http://microblog.me.uk/wp-content/uploads/RoseSpots.jpg

## CAMPYLOBACTER JEJUNI

**Campylobacter jejuni has a similar presentation to Shigella**. Look for CRAMPY abdominal pain, fever and an inflammatory diarrhea that can be bloody. Diagnose by stool culture. Treat with Erythromycin.

**MNEMONIC**: Think of being at a CAMP near a FLAME (inflammatory diarrhea) when you begin having CRAMPY CAMPY symptoms.

## STAPHYLOCOCCUS AUREUS & BACILLUS CEREUS

The diarrhea associated with staphylococcus aureus or bacillus cereus starts QUICKLY! Look for symptom onset as quickly as 1 hour, and as long as 8 hours. Bacillus cereus is associated with ingestion of RICE. Diagnosis is primarily clinical and treatment is supportive since it resolves within 24-48 hours.

**MNEMONIC**: **8**acillus cereus & sta**8**h aureus. Onset by **8** hours.

**NEMONIC:** Ever heard of BC Headache Powder? It's basically WHITE POWDERED aspirin and caffeine. Associating BC Headache Powder with WHITE POWDER could help remind you that **B**acillus **C**ereus is associated with WHITE RICE ingestion.

## YERSINIA ENTEROCOLITICA

Ingestion of **raw pork** or **unpasteurized milk** resulting in fever and diarrhea should set of alarms about Yersinia enterocolitica. This can also cause a PSEUDOAPPENDICITIS, ERYTHEMA NODOSUM and a REACTIVE ARTHRITIS. It's a self-limited illness so there's no need to treat.

* **MNEMONIC**: Imagine an angry neighbor with a deep southern accent saying, "YER SON is a menace. He tossed RED CHITLINS on my yard that had an APPENDIX hangin' off of it!"

- **KEY:** YER SON = your son = Yersinia. Chitterlings are made of pork intestines. More commonly eaten in the south. If it's said with a strong southern accent, it's pronounced, "CHIT-LINS." APPENIX = appendicitis.

* **MNEMONIC**: (IMAGE) http://www.bailocom.it/images/suino_f.jpg

## CLOSTRIDIUM PERFRINGENS

Clostridium perfringens causes a **non-bloody** diarrhea fairly quickly after ingestion of RAW TURKEY or RAW BEEF. Onset is at about 10 hours or later.

* **MNEMONIC**: Imagine a macho man's man bragging about his stomach. He says, "You **C**, I **PREFER** to eat mystery **MEAT** out of a carved out BOWLING BALL that I've turned into a bowl."

- **KEY:** C. PREFER = C. perfringens. MEAT = raw meat, like beef or turkey. BOWLING BALL = 10 = approximate time of onset.

## CLOSTRIDIUM DIFFICILE (C. DIFFICILE or C. DIFF)

For Clostridium difficile (C. difficile or C. diff), look for diarrhea that's associated with mucus and probably BLOOD. They would almost have to give you a **history of recent antibiotic use** or "a recent illness." The illness may have required an antibiotic, such as **CLINDAMYCIN**, a beta lactam or a cephalosporin. Diagnose by sending multiple stool samples looking for the C. diff **TOXIN**. Treat with metronidazole.

## PEARLY DIARRHEA REVIEW

* BLOODY DIARRHEA REVIEW: Salmonella, Shigella, Campylobacter, EHEC and possibly C. difficile.

* TREATMENT FOR ACUTE DIARRHEAS: If you had to take a guess, Ceftriaxone would be a good guess for those needing treatment. Macrolides are good second options for the pediatric exam, and trimethoprim-sulfamethoxazole could be a third guess. For Clostridium, use METRONIDAZOLE.

- SALMONELLA TYPHI: Rose spots. Ceftriaxone or Cefotaxime.
- SHIGELLA: Profuse bloody diarrhea. Seizures. Rectal prolapse. Tenesmus. Trimethoprim, Ceftriaxone or a macrolide.
- CAMPYLOBACTER: Crampy pain. Erythromycin.
- C. DIFF: Metronidazole!

## *CHRONIC DIARRHEA*

### (DOUBLE TAKE) GIARDIA

Giardia presents with intermittent, watery diarrhea that has been going on for weeks. May be accompanied by abdominal distension, and may eventually cause malabsorption. Diagnose with a STRING TEST + ELISA, and **then** treat with Metronidazole.

**PEARLS:** History may include a camping trip or a child in daycare.

### CHRONIC NONSPECIFIC DIARRHEA (aka TODDLERS DIARRHEA)

Look for **undigested food!** Chronic nonspecific diarrhea (aka toddlers diarrhea) may present as normal stools in the morning followed by loose stool as the day progresses. The child will otherwise be healthy. It may be due to excessive carbohydrate intake. Provide rassurance and ask the family to decrease carbohydrate intake (especially juices) and INCREASE fat and fiber in the diet. They should also try to avoid very hot or cold foods.

### (DOUBLE TAKE) PROTEIN INDUCED ENTEROCOLITIS

Protein Induced Enterocolitis is **NOT** IgE mediated. Can cause problems from birth, but often presents as FAILURE TO THRIVE (FTT) after months of already being on **cow milk formula**. This **CAN** occur in breast fed babies, but only if enough whole protein from mom's diet enters the breast milk. Usually presents after switching from breastfeeding to **formula**. 40% also sensitive to soy.

* **MNEMONIC**: It's so much easier if you just call it "FORMULA INDUCED Enterocolitis." Side note: An old term for this is "Allergic Proctocolitis," but hopefully you will not see that on the pediatric exam.

* SYMPTOMS include diarrhea +/- emesis and possibly BLOODY stool or flecks while on formula, but **not on clears or non-whole protein containing formulas**.

* DIAGNOSIS: Best test to establish diagnosis is a flexible sigmoidoscopy with biopsy. NOT allergy testing.

* TREATMENT: If the mom is breastfeeding, try asking mom to stop drinking milk products. If that doesn't work, **change to HYDROLYSATE formula** (formula containing HYDROLYZED cow milk proteins) or to an AMINO ACID DERIVED FORMULA. Protein induced enterocolitis typically **resolves** once the GI tract matures.

## (DOUBLE TAKE) LACTOSE INTOLERANCE (aka LACTASE DEFICIENCY)

It is **not** common for kids < 5 years old to have lactose intolerance (aka lactase deficiency). So for the pediatric boards, if the child is less than 5 years of age, suspect a different diagnosis!

* SYMPTOMS: Diarrhea +/- abdominal pain.

* **HYDROGEN BREATH TEST**: Can be used to help diagnose a lactase deficiency (as well as bacterial overgrowth). Patient takes a carbohydrate load, if unable to digest because of a lack of lactase, the bacteria will digest the carbs and release hydrogen (measured in the breath).

* TREATMENT: SOY milk. It does not contain lactose. It contains sucrose.

* **PEARLS:** Stool is NOT malodorous and does NOT have food particles. These patients do NOT vomit and do NOT have an associated rash.

* **MNEMONICS:** Most of you probably know of at least one friend, relative or patient that had lactose intolerance. Did they ever have a rash or vomiting after eating desert? Never! I'm guessing they had LACTAID in their fridge (my residency roommate certainly did). "**LACT**ose comes from the **LACT**ating breasts of women & cows. NOT soy beans."

## INTESTINAL LYMPHANGIECTASIA

Intestinal lymphangiectasia is a **protein losing enteropathy**. Look for steatorrhea associated with a low serum protein, low serum gammaglobulin level, edema and possibly lymphopenia.

## FAT & CARBOHYDRATE MALABSORPTION

* FAT MALABSORPTION: If you are concerned about this (possibly due to a clinical vignette describing a fat soluble vitamin deficiency), obtain fecal fat testing.

* CARBOHYDRATE MALABSORPTION: Look for positive reducing substances on stool testing, or a positive hydrogen breath test.

## *ACUTE LYMPHADENOPATHY (< 3 WEEKS) IN THE HEAD AND NECK AREA*

**PEARLS**: For all lymphadenopathy evaluations in the head and neck area, always check HOW LONG the lymphadenopathy has been present! For a unilateral lymphadenopathy, consider choosing a BACTERIAL infection as your answer. ALL of the upcoming lymphadenopathy sections are VERY high yield. For the most part, highlighting has been turned off.

## STAPHYLOCOCCUS AUREUS & STREPTOCOCCUS PYOGENES (aka GAS or STREP PYOGENES)

* STAPHYLOCOCCUS AUREUS & STREPTOCOCCUS PYOGENES (aka GAS or STREP PYOGENES): Look for **cervical or submandibular lymphadenopathy in a neonate or infant**. May also present as a breast abscess.

- **PEARLS:** This is one of the few places where Staph is the correct answer for a neonate. Also, while GBS is a common answer for neonates, GROUP A STREP (GAS) is probably NEVER the answer for a neonate.

* GROUP A STREP (GAS or STREP PYOGENES): Look for an **older child** with **impetigo**, **pharyngitis** or even a single inflamed **lymph node**.

## PREAURICULAR LYMPHADENOPATHY

Consider adenovirus if you see a patient with conjunctivitis and preauricular lymphadenopathy. Mononucleosis may present with acute or chronic preauricular lymphadenopathy.

## EMPIRIC TREATMENT

Empiric treatment for acute lymphadenopathy in the head and neck area should aim to cover beta-lactamase producers. AMOXICILLIN-CLAVULANATE! Other appropriate options include cefazolin, clindamycin and erythromycin.

# *CHRONIC CERVICAL LYMPHADENOPATHY (> 3 WEEKS)*

## (DOUBLE TAKE) BARTONELLA HENSELAE

Bartonella henselae causes CAT SCRATCH DISEASE. Look for **DRAINING and TENDER** lymphadenopathy and possibly a history of cat (or dog) exposure. Resolves on it's own so there's no need to treat unless the patient is immunocompromised or has severe symptoms including hepatosplenomegaly. If so, treat with non-penicillin drugs, like macrolides, cephalosporins, doxycycline or trimethoprim-sulfamethoxazole.

**MNEMONIC**: Imagine BART simpson's cat being a TENDER kitty that's fine when it's LEFT ALONE, but it SCRATCHES you when you try to pet it. Solution is to LEAVE IT ALONE!

* **KEY:** TENDER = tender lymphadenopathy, LEAVE IT ALONE = no treatment needed.

## FRANCISELLA TULARENSIS

Francisella tularensis causes GLANDULAR or ULCEROGLANDULAR DISEASE (aka TULAREMIA) due to **pleomorphic** Gram negative organisms. Look for fever, hepatosplenomegaly, lymphadenopathy, **ULCERS** and a history of exposure to deer or rabbits. It's a tick borne illness and is treated with gentamicin, or with STREPTOMYCIN + tetracycline.

**MNEMONIC**: Instead of Tuluremia, say t-ULCER-emia.

## MYCOBACTERIUM TUBERCULOSIS (MTB or TB)

If a patient has been diagnosed with typical mycobacterium tuberculosis (MTB or TB) and is found to have lymphadenopathy, treat with medications alone. Do NOT excise the lesion UNLESS it turns into a draining abscess. Lymphadenopathy tends to be painless.

## ATYPICAL MYCOBACTERIA

Atypical mycobacteria include Mycobacterium avium-intracellulare (MAC) and Mycobacterium scrofulaceum. **Look for a freely movable node without many other symptoms**. First place a PPD. The result will show some induration, but it will NOT be large enough to diagnose TB. Once the induration is noted, **surgical excision is curative** and no antibiotics are required. Lymphadenopathy tends to be painless.

## BRUCELLOSIS

Brucellosis is caused by Brucella, a Gram negative organism. Look for fever, chronic cervical lymphadenopathy, myalgias and hepatosplenomegaly. There may also be mention of a farm or exposure ot unpasteurized milk products. Treat with trimethoprim, or if the child is old enough, tetracycline.

**OTHER:** Also consider Epstein-Barr Virus (EBV), Cytomegalovirus (CMV) and Toxoplasma gondii as causes of chronic lymphadenopathy.

**MNEMONIC**: Imagine a baby calf named BRUUUUUUUCE, that lives on a farm and loves to say MUUUUUUU and likes to suck on the **T**ee**T**s of his mother and drink UNPASTEURIZED MILK.

**MNEMONIC**: Imagine BROOOCE almighty sucking on a COW'S T-Ts and drinking UNPASTEURIZED MILK.

* **KEY**: BUUUCE/BROOOCE = Brucella. TeeTs and T-Ts = Trimethoprim and Tetracycline.

# *LYMPHADENOPATHY IN OTHER AREAS*

## (DOUBLE TAKE) LYMPHOGRANULOMA VENEREUM SEROVAR

Lymphogranuloma venereum serovar is an STD caused by Chlamydia trachomatis. Rare in the U.S. More common in tropical areas. Starts as small nontender papules or shallow ulcers which resolve. Eventually, a **TENDER UNILATERAL INGUINAL** lymph node appears. Pain is relieved when it ruptures. The node can **continue to drain** for months. Treat with DOXYCYCLINE or erythromycin.

## YERSINIA PESTIS

Yersinia pestis causes the PLAGUE. This is still found in the Southwestern U.S. Look for a fever associated with painful and swollen lymphadenopathy. Lymphadenopathy in the inguinal and axillary regions is especially common. Aspirate the lymph nodes to get a diagnosis and then treat with **streptomycin.**

**PEARL**: Look for a history of being in or near the woods, for example, a teenager who went hunting. Also, "bubo" means swollen lymph node. Hence the name, Bubonic Plage.

**MNEMONIC**: "PESTI" Plague.

# *NONTENDER LYMPHADENOPATHY*

**NOTE**: Please do not assume that all of the lymphadenopathies discussed outside of this section are *tender* lymphadenopathies.

## SPOROTRICHOSIS (aka ROSE PICKERS DISEASE)

For sporotrichosis (aka rose pickers disease), look for a single, PAINLESS, movable nodular lesion that enlarged, became fluctuant and then ulcerated. This is a slowly progressive disease and is caused by Sporothrix schenckii. After that additional painless, movable nodules appear along the lymphatic system.

**PEARL**: This used to be called Rose-picker's disease. Look for a history that includes a garden.

## MYCOBACTERIUM TUBERCULOSIS

The lymphadenopathy of mycobacterium tuberculosis tends to be painless.

## ATYPICAL MYCOBACTERIA

The lymphadenopathy of atypical mycobacteria tends to be painless.

## (DOUBLE TAKE) HODGKINS LYMPHOMA

Hodgkins lymphoma accounts for almost HALF of lymphoma cases and has a GOOD prognosis with GOOD cure rates. Signs and symptoms progress in a stepwise fashion. Look for a **TEENAGE** patient with **NONTENDER** CERVICAL or SUPRACLAVICULAR lymphadenopathy. It always starts this way! That will then be accompanied by fevers, fatigue, night sweats and weight loss. Later in the course, there is also generalized lymphadenopathy. On labs, patient may nave a normal or high LEUKOcyte count (WBC), but must have either **high or low LYMPHOcytes**. DIAGNOSE with an excisional node biopsy and noting **REED STERNBERG CELLS**, which are usually of B-cell origin.

**PEARL**: Your differential may include TB or Atypical Mycobacteria. For a NEGATIVE PPD, choose Hodgkins. For a slightly positive PPD (5-10 mm) choose Atypical. For a PPD that is > 10 mm, choose MTB.

**IMAGE**: http://hematopathology.stanford.edu/images/HP-Home-Image.jpg = RS Cells

**MNEMONIC**: **REED** Sternberg cells kind of look like OWLS' EYES, so imagine an OWL wearing glasses and **REED**ing a book on a farm. The farmer comes outside and takes a shot at the OWL. He misses the OWL but hits the book. The OWL flies away SAFELY (high cure rate). Whew… that owl really "HODGED" a bullet.

## NONTENDER SUPRACLAVICULAR LYMPHADENOPATHY

Choose MYCOBACTERIUM TUBERCULOSIS (MTB) or HODGKIN'S LYMPHOMA if presented with a nontender supraclavicular lymphadenopathy.

# ***MISCELLANEOUS ID RELATED TOPICS***

## SPLENECTOMY PATIENTS

Splenectomy patients are susceptible to the encapsulated "HNS" bugs, Haemophilus influenzae, Neisseria meningitidis and Streptococcus pneumoniae. Patients also do not mount much of a response to the Pneumococcal vaccines. So if you are given an immunized patient with sickle cell disease, or a splenectomy, consider Strep as a very real possibility for a bacteremia or sepsis.

**(DOUBLE TAKE) MNEMONIC**: A complete list of encapsulated organisms can be recalled by remembering that "Some Nasty Killers Have Some Capsule Protection." Streptococcus pneumoniase, Neisseria meningitidis, Klebsiella pneumoniae, Haemophilus influenzae, Salmonella typhi, Cryptococcus neoformans and Pseudomonas aeruginosa. Bruton's agammaglobulinemia and sickle cell patients are especially susceptible to encapsulated organisms.

## DFA & ELISA TESTING

Enzyme-linked immunosorbent assay (ELISA or EIA) and Direct Fluorescent Antibody (DFA) testing are more available and faster than Polymerase Chain Reaction (PCR) testing. If offered as a choice, pick DFA or ELISA over PCR for faster diagnosis of most viruses.

## GROWTH MEDIA & STAINING

* BORDETELLA PERTUSSIS: Use Regan-Lowe or Bordet-Gengou growth media.
* CORYNEBACTERIUM DIPHTHERIAE: Use Loeffler or Tinsdale agar
* NEISSERIA GONORRHEA: Use Thayer-Martin
* CRYPTOCOCCUS: Use India ink staining.

## LATEX AGGLUTINATION

Latex agglutination is probably a low yield topic, but if you have a high suspicion for GBS, Haemophilus, Neisseria or Streptococcus pneumoniae, this test could be somewhat useful as a second line test.

## DROPLET PRECAUTIONS

Droplet precautions are for patients with Neisseria meningitidis or the Influenza virus (flu).

**MNEMONIC**: DROPPIN' IN = **DROP**lets precautions for **I**nfluenza and **N**eisseria

## STACCATO, BARKY & PAROXYSMAL COUGHS

* **STACCATO**: If this term is used to describe an infection-related cough, pick CHLAMYDIA PNEUMONIAE.
* **BARKY**: If this term is used to describe the cough, pick CROUP. Which is most commonly from parainfluenza and other viruses.
* **PAROXYSMAL**: If this term is used to describe the cough, pick PERTUSSIS.

## (DOUBLE TAKE) APHTHOUS ULCERS

Aphthous ulcers are painful lesions found within the oral mucosa (buccal mucosa, lips and tongue) with a grayish-white base and a rim of erythema. Can occur in isolation or in association with Behcet's or Schwachman Diamond syndrome.

**IMAGE**: http://upload.wikimedia.org/wikipedia/commons/4/45/Aphthous_ulcer.jpg

**IMAGE**: http://upload.wikimedia.org/wikipedia/commons/2/2d/Aphtha2.jpg

**PEARLY NOTES**: Since the pediatric vaccination recommendations tend to change frequently, it's difficult for the American Board of Pediatrics (ABP) to include board questions on the most current guidelines. Your focus in studying this section should be the information that is not going to be controversial. Also, be sure to keep a strong focus on when vaccinations should NOT be given.

## SOME PERTINENT CDC LINKS

These links are here for your reference in case you get stuck and have questions. You do NOT need to go through them to pass the exam. If going through them will overwhelm you, don't. Some links in particular are more daunting than others.

* COMPLETE CDC RECOMMENDATION ON VACCINES (AVOID): http://1.usa.gov/qNbOJh
* CONTRAINDICATIONS TO VACCINES (GOOD): http://bit.ly/pR0E0r
* MISPERCEIVED CONTRAINDICATIONS (AVOID): http://1.usa.gov/r7HtCs
* IMMUNIZATION/VACCINATION SCHEDULE (0-6 YEARS OF AGE): http://1.usa.gov/qAoXMU
* IMMUNIZATION/VACCINATION SCHEDULE (7-18 YEARS OF AGE): http://1.usa.gov/pJyTcc
* IMMUNIZATION/VACCINATION CATCH-UP SCHEDULE (0-18 YEARS OF AGE): http://1.usa.gov/oKBTfg

## STEROIDS AND IMMUNIZATIONS

In general, the types of steroid regimens commonly prescribed are not a contraindication to immunizations. If a course of steroids is given for less than 14 days, then REGARDLESS OF THE DOSE given, ANY vaccination is allowed (including all live vaccines). If > 20 mg/day, or > 2 mg/kg/day of prednisone is prescribed for > 14 days, then vaccinations should be held for one month after the course is over.

**MNEMONIC:** If > **2**0 mg/day or > **2** mg/kg/day is given for > **2** weeks, then HOLD the vaccination for at least 1 month after the steroids have been stopped. For anything else, go ahead and give.

## PREMATURITY AND VACCINATIONS

Except for Hepatitis B and Rotavirus, vaccinations in premature babies are very similar to full-term babies. For Hepatitis B, if born to a Hepatitis B negative mom, the baby's first Hepatitis B vaccination can be delayed until 1 month of age or until hospital discharge. For the Rotavirus vaccine, since it is shed in the stool and is a risk to others it cannot be given while a premie is in the hospital. Regarding safely administering the rest of the immunizations, a premature baby may receive ALL scheduled vaccinations at 2 months of **chronological** age whether the baby is in the hospital or out of the hospital.

## LIVE VACCINES

The live vaccines have traditionally been Measles, Rubella, "Sabin" (aka polio - if **oral** OPV), Pox viruses (Varicella/VZV), Adenovirus (for military only), Mumps and Yellow Fever. Oral polio is not used in the U.S. so you're unlikely to be tested on it. The newer live vaccines are the inhaled Influenza (virus) vaccine (Flumist) and the Rotavirus as well. So the list includes **MMR, OPV, VZV, Adenovirus, Yellow Fever, inhaled Influenza and Rotavirus**.

**PEARL**: For the purposes of the exam, do not give any of these vaccines to a **pregnant teen** or someone who is **severely immunocompromised**. Consider these to be the truly immunocompromised patients with HIV and a low CD4 count, as well many of those patients with disorders reviewed in the Allergy & Immunology chapter.

**PEARL**: If an HIV patient is generally healthy, do NOT hold back any vaccines.

**MNEMONIC:** MR. SPAMY is ALIVE! This mnemonic includes the older live vaccines, including Measles, Rubella, Sabin (Polio if **oral** OPV), Pox viruses (Varicella/VZV), Adenovirus (for military only), Mumps and Yellow Fever. This does not include Flumist or Rotavirus!

**MNEMONIC:** MR. FARM-SPY is ALIVE includes all of the live vaccines. Creating a story around one or both of these mnemonics will help even more.

**MNEMONIC:** Here's a mnemonic that covers some (not all) of the DEAD vaccinations. "**RIP** Hepatitis **A** & **B**!" = Rest In Peace Hepatitis A & B = Rabies, Influenza, Polio (IPV), Hepatitis A and Hepatitis B.

## MEASLES, MUMPS, RUBELLA (MMR) & VARICELLA (VZV) PEARLS

* MMR & VARICELLA are both prohibited in patients with an immunocompromised state (SCID or AIDS) and in PREGNANCY. These are **NOT contraindicated in family members.**

* MMR & VARICELLA are both OK for children with HIV who are doing well.

* MMR & PPD: A PPD **can** be placed on the same day the MMR is give. However, if the MMR was already given LESS THAN 6 WEEKS ago, do not place the PPD yet.

* MMR contains some ingredients that can cause joint pain of the hands and knees for up to one month after administration.

## ROTAVIRUS VACCINE

The first rotavirus vaccine dose should NOT be given prior to 6 weeks of age OR AFTER 12 weeks of age. That means that if a child missed the 2-month visit and presents for the 4-month checkup, they can no longer be immunized for rotavirus.

**PEARLS:** Do NOT give the Rota vaccine to patients in the hospital. You CAN give it to children who have an immunocompromised FAMILY MEMBER in the home. Yes, it IS shed in the stool.

## INFLUENZA VACCINATION

The CDC has made the influenza vaccination recommendations very easy. Give the "flu vaccine" to everyone over 6 months of age, every year. High-risk patients should get priority. The INHALED INFLUENZA (FLU) VACCINE (brand name Flumist ®) can be given to children who are at least FIVE years of age, healthy and NOT PREGNANT, IMMUNOCOMPROMISED or an ASTHMATIC. The age restriction is due to the inability of a 4-year old child to coordinate the inhalation.

**PEARL**: If you have a question about injectable flu vaccination, the answer will likely be to VACCINATE. For the inhaled vaccination, you will need to look at the age and comorbidities.

**PEARL**: Remember, influenzA is the virus. H. flu is the Gram negative organism and is spelled influenzAE (HIB vaccine).

## HEPATITIS A VACCINE

Everyone gets he Hepatitis A vaccine. Starts at 1 year of age and 2 doses are given 6 months apart.

## HEPATITIS B VACCINE

Everyone gets the Hepatitis B vaccine. Start at birth and requires 3 doses.

**PEARL**: The last dose SHOULD NOT be given prior to 6 months of age. The IDEAL regimen for you to remember **is 0, 1 and then 6 months**. Here are a few reasons why:

- Must wait at least one month after dose #1 to give dose #2.
- Must wait at least **four** months after dose #2 to give dose #3.
- Dose #3 must NOT be given before the age of 24 weeks (6 months).
- SUMMARY: This means 0, 2 and 6 months is allowed. 0, 1 and 6 is allowed. 2, 4 and 6 is NOT allowed because four months have to pass between doses #2 and #3. 0, 2, 4 and 6 IS ALLOWED because 4 months passed between the 2-month shot and the 6 month shot. The 4-month shot was likely a combination vaccine and just went to waste (and is allowed by the CDC to do so).

**PEARLS:** If mom's Hepatitis B status is UNKNOWN, give the Hepatitis B **vaccine within 12 hours of birth**. Figure out mom's status, and if she is HBsAg POSITIVE, give the immunoglobulin (HBIG) no later than the $7^{th}$ day of life. If the status is already KNOWN to be positive, give BOTH the vaccine and the HBIG at birth.

## HUMAN PAPILLOMA VIRUS VACCINE (HPV)

The HPV vaccine requires 3 doses starting at 11 years of age on a 0, 1 and 6 month schedule (#3 has to be given at least 6 months after #1). It's recommended for **males** and females and can be administered as early as 9 years of age.

## MENINGOCOCCAL VACCINE (aka MENINGOCOCCUS VACCINE)

The meningococcal vaccine is recommended at 11 years of age for ALL kids, NOT just those starting at college and living in a dorm. Requires a booster. If started on time at 11 years of age, the booster is given at 16 years of age. Otherwise, at about 3 years later. For unimmunized freshman **living in a dorm**, give one shot and no booster.

**PEARL**: Minimum age is 2 years. Children < 11 year of age may receive the vaccine if they are at high risk (asplenic, **complement deficiency**, generally immunocompromised or traveling to an endemic area).

**PEARL**: If a CLOSE INTIMATE contact of your patient has meningitis, give your patient prophylaxis with **rifampin** or ceftriaxone within 24 hrs of diagnosis **regardless of the immunization status**. Close "intimate" contact includes nursery school, but **not** elementary school classmates or healthcare providers.

**PEARL**: Subtype B is Bad for Babies and has no vaccine. Most meningococcal meningitis cases in babies is due to this subtype.

## POSTEXPOSURE PROPHYLAXIS

The following discusses postexposure proplylaxis for N. meningitidis, H. influenzae B, pertussis, Hepatitis A, Hepatitis B and measles.

* RIFAMPIN PROPHYLAXIS: It's used for both **N. meningitidis** (aka meningococcus) and **H. flu** prophylaxis. Once prophylaxis is started, reddish-orange urine and tears may be noted.

- N. MENINGITIDIS: Only 1 case of exposure is required to start prophylaxis in contacts.
- H. INFLUENZAE B: TWO documented cases are required. If the immunization status is unknown, also give the Hib vaccination. If you are presented a question about the first child to be diagnosed the H. flu (the index case), then be sure to also immunize his/her household members.

* PERTUSSIS: Give a macrolide.

- **PEARL**: Give azithromycin for kids < 6 weeks of age due to concern for pyloric stenosis. For those > 6 weeks of age, give erythromycin. If the patient is unimmunized, also give the vaccine.

* HEPATITIS A: Given Hepatitis A immunoglobulin to the unimmunized **family members** only.

* VARICELLA (VZV): Give the immunoglobulin.

* MEASLES: For an exposure less than 72 hours ago, give the vaccine to unimmunized patients who are **at least 6 months old**. For exposures greater than 72 hours ago, give the immunoglobulin (MIG) and then DO NOT GIVE THE MMR for at least 5 months after that.

* HEPATITIS B NEEDLE STICK:

- IMMUNIZED: If the patient's immunity status is known to be up to date by titers, no prophylaxis is needed.
- UNCERTAIN IMMUNIZATION STATUS: CHECK for antibody. If negative, give the immunoglobulin (HBIG) **and also start** the full vaccination series. If the antibody is positive, do nothing.
- UNIMMUNIZED: Give the immunoglobulin (HBIG) **and also start** the full vaccination series.
- **PEARL/SHORTCUT:** CHECK FOR IMMUNITY. Then either do nothing, or give **both** the immunoglobulin (HBIG) and the Hepatitis B dose #1 of 3.

## TETANUS BOOSTER

Babies are immunized with **DTaP**. If a tetanus booster is being given before the age of 7, only the DT portion needs to be given. If a booster is being given **after the age of 7**, everyone now gets **Tdap** as the booster of choice.

* CLEAN WOUND: If the last shot was < 10 years ago, no need. If it was > 10 years ago, they are scheduled for a booster so go ahead and give it.

* DIRTY WOUND: If the last shot was > **5** years ago, **give a booster**. If the wound is dirty and the immunization status **unknown or incomplete**, give **IMMUNOGLOBULIN + BOOSTER.**

## VACCINE SCHEDULE REMINDERS

* 2, 4 & **SIX** MONTHS: **SIX vaccines are given.** Pediarix ® (DTaP, IPV, Hepatitis B) + Hib + Pneumococcal vaccine + Rotavirus vaccine.

- **PEARL**: The PCV-7 vaccine which is given to **all** kids. The **PPV-23 valent** pneumococcal vaccine is given to kids > 2 years of age if they have **chronic illness or asplenia**.

* 12 MONTHS: 6 vaccines are given (MMR #1, VZV #1, Hepatitis A #1, Hib #4 and the 2 boosters of PREVNAR #4 of 4, and DTaP #4 of 5).

* 4 YEARS OF AGE: Four vaccines are given around the age of 4 (MMR, Varicella, DTaP, **IPV**).

- **PEARL**: **MMR** # 2 is scheduled around 4-6 years of age, but it **CAN** be given just **1 month** after the 1st dose! **VZV** can be given after just **3 months**!

## CATCH-UP IMMUNIZATION SCHEDULE PEARLS

The CDC's catch up immunization schedule does NOT include every pediatric vaccine. For example, PCV7 and Hib are NOT recommended for kids over 5-years-old. Rotavirus vaccine is NOT given to anyone after the 12 weeks of age.

**PEARL/MNEMONICS:** You could be asked a question about how many **different** vaccinations/shots a child will need at a given age. You could also be asked how many shots a child will need in total for a given vaccine.

- **MNEMONIC:** SIX shots at SIX months of age and FOUR shots at FOUR years of age. For the 1 year shot, you could try to remember that SIX x 2 = 12. Therefore there are TWO major immunization milestones at which time SIX shots are given (SIX months of age and 6 x 2 = 12 months of age).
- **MNEMONIC:** 5-4-3-2-1 - Use this to help you remember the number of times various vaccinations are administered.
    - 5 = Total of 5 DTaP doses (last one at 5 years of age)
    - 4 = P = 4eumococcal & I4V = Total of 4 pneumococcal (last at 1 year of age) and IPV (last at 4 years of age) doses. A 4 also looks kind of like an H, which could remind you of Hib.
    - 3 = Total of 3 Hepatitis B doses (last at 5-6 months of age)
    - 2 = MMR, VZV, and Hepatitis A
    - 1 = Rhymes with NONE

## CROUP, TRACHEITIS & EPIGLOTTITIS

PLEASE REVIEW THE FOLLOWING PRIOR TO MOVING ON TO THE CROUP, TRACHEITIS & EPIGLOTTITIS SUMMARY TABLE:

**(DOUBLE TAKE) STACCATO, BARKY & PAROXYSMAL COUGHS**

* **STACCATO**: If this term is used to describe an infection-related cough, pick CHLAMYDIA PNEUMONIAE.

* **BARKY**: If this term is used to describe the cough, pick CROUP. Which is most commonly from parainfluenza and other viruses.

* **PAROXYSMAL**: If this term is used to describe the cough, pick PERTUSSIS.

## ***CROUP, TRACHEITIS & EPIGLOTTITIS SUMMARY TABLE***

| DISEASE | ETIOLOGY | SYMPTOMS | PEARLS | TREATMENT |
|---|---|---|---|---|
| **CROUP = VIRAL LARYNGOTRACHEO-BRONCHITIS** | **- PARAINFLUENZA** is the classic one.<br>- Can also be caused by RSV, Influenza A/B or other viruses. | **- BARKY**, coarse cough.<br>**- Stridor** (can be inspiratory and expiratory if severe)<br>**- Nontoxic**, but **do have low grade fever** & rhinorrhea (viral-syndrome) | - Winter<br>- Can rebound so hospitalize<br>- May prefer sitting up.<br>- May be slightly anxious, but NOT toxic appearing<br>- Usually in kids about 2 years of age<br>- Steeple sign on imaging: http://alturl.com/ti5p7 | **- COOL AIR/MIST**<br>- Use racemic epi if moderate-severe<br>- +/- STEROIDS<br>**- Bacterial tracheitis is a super-infection with Staphylococcus on top of the croup!** |
| SPASMODIC CROUP | - Unknown, but frequently associated with atopic disease. | - Same as above, EXCEPT **no viral syndrome symptoms** (no fever, no rhinorrhea). | - Recurrent problem.<br>- Often at night. | - Same as above. |
| BACTERIAL TRACHEITIS | - Staph superinfection of CROUP | **- COPIOUS MUCOID** secretions by a BACTERIA<br>- Look for a patient < 3 years of age with a croupy cough that QUICKLY deterioration.<br>- HIGH FEVERS, but **SUPINE**<br>- Often need intubation | **- NO DROOL**<br>- LOOKS BAD… **REALLY BAD**<br>- **PEARL**: If URI/barky for 1 day, think croup. If for a few days and **SICK**, think bacterial tracheitis.<br>- **MNEMONIC**: After all, what's going to be worse, a bacterial Staph infection or a "croupid" viral syndrome? (Pronounce croupid like stupid) | - Anti-Staphylococcus antibiotics<br>**- Nafcillin + cefotaxime (or cefuroxime)** |
| EPIGLOTTITIS | - HI**B (typable!)**.<br>- Can also be caused by Strep or Staph | **- FEVER, DROOLING, TONGUE PROTRUDING, RAPIDLY TOXIC, RED EPIGLOTTIS,** SITTING AND **LEANING FWD**<br>**- Hydrophobia**. +/- **biphasic stridor** (inspiratory AND expiratory) | - True emergency, rare now<br>- Thumbprint sign<br>- Be prepared to intub<br>- **IMAGE**: http://alturl.com/wgowy Look for a **lateral** neck X-ray!<br>- **MNEMONIC**: "A BIG LOT o' TIDY WHITIES that turned RED/BLOODY. HEY, MOP IT UP PHIL! The real docs are in a rush to go the OR!" = E-pig-lot-titis, RED epiglottis, Hae-mop-phil-us. | - Ceftriaxone or cefotaxime to cover Hib and Strep.<br>- INTUBATE IN THE OR |

# VACCINE CONTRAINDICATIONS

## ILLNESS

You may be asked if a vaccine is contraindicated in a child with a moderate to severe illness with or without fever. In such a child, all vaccines are relatively contraindicated and the physician is supposed to weigh risks and benefits.

## CHICKEN or EGG ALLERGY

**The Influenza (virus)** and **yellow fever** vaccinations are contraindicated if a patient has a CHICKEN or EGG allergy that has lead to an ANAPHYLACTIC REACTION:

**PEARL**: Giving the MMR is fine!

**MNEMONIC**: To remember which vaccinations are contraindicated in these poultry-ish allergies, try to member the phrase, "one **FLU** over the **YELLOW** cuckoo's nest of **EGGS**." It helps if you've read the book or watched the movie.

## DTaP CONTRAINDICATIONS

The DTaP contraindications are worth being familiar with for the pediatric boards.

* **ENCEPHALOPATHY or "PROLONGED" SEIZURE**: If a child has **encephalopathy** or a **"prolonged" seizure** within **1 week** of getting DT**aP**, the **PERTUSSIS (aP)** component is contraindicated in future vaccinations. Instead, **give DT**. Please remember that this is possibly a reaction to the PERTUSSIS component, and not he Tetanus component. Also, DO NOT order allergy testing because you may induce a seizure!

* **PROGRESSIVE NEUROLOGIC DISORDER**: If a child has a neurologic disorder that is considered progressive, or has a seizure disorder that is not yet under control, then DTaP is **relatively** contraindicated. If the neurologic disorder has been stabilized or is no longer progressive, you MAY give DTaP.

* **OTHER RELATIVE CONTRAINDICATIONS**: Having a **brief** seizure within 3 days, a high fever (>105') within 48 hours or having a shock-like state within 48 hours of previous administration.

- **PEARL/MNEMONIC**: If tested on one of these 3 relative contraindications You are more likely to be given a case in which the child had an issue 4-7 days from the date of previous vaccination. This would make the answer clear (give DTaP!). Therefore, keeping the FOURTH day in mind as the day when the child is in the clear might help with these 3 relative contraindications.

## GELATIN ALLERGY

If the patient has a gelatin allergy, the **MMR** is contraindicated.

**MNEMONIC**: Are you **GELL**in' with **M**agellin? Because if you're not, and you have a bad reaction to the GELLin', you better stay away from the **M**mr.

**MNEMONIC VIDEO**: http://www.youtube.com/watch?v=XmCyulruslY

## NEOMYCIN & POLYMIXIN ALLERGIES

The varicella (VZV), polio (IPV) and MMR vaccines all have **NEO**mycin and **POLY**mixin in them. So if a patient has an allergy to either of these components, the VZV, IPV and MMR are all contraindicated.

* **MNEMONIC**: Hopefully you've seen the Matrix and Along Came Polly. If so, imagine **NEO** meeting **POLY** under a **TREE** located at **ONE** Vaccine Avenue for a **FIRST** date/picnic.

- **KEY:** NEO = NEOmycin, POLY = POLYmixin. TREE = Peg for 1. This is to help you remember that all of the **NEW 1**-year old shots contain neomycin and polymixin. The fact that it's their FIRST date is also meant to remind you that it's the NEW 1-year old vaccines.

## STREPTOMYCIN ALLERGY

The polio vaccine (IPV) is contraindicated for anyone with a s**T**re**P**tom**Y**cin allergy.

**MNEMONIC**: s**T**re**P**tom**Y**cin = TPY kind of looks like IPV. If that doesn't work, how about the following chant said 5 times, "no STREPTOOOO no POLIOOOO!"

## ANAPHYLAXIS MANAGEMENT

If anaphylaxis does indeed occur after a vaccination, the patient **must get allergy skin testing** to the individual components that were in the vaccination.

## THIMEROSAL

If you see this as an answer choice, IGNORE IT. It's no longer used in the U.S.

## COMPLETE CDC GUIDE

In case you still have questions, 2 links are included below. One is the current CDC guide to contraindications and precautions to commonly used vaccines. The other is a guide to conditions commonly misperceived as contraindications to vaccination. My recommendation is that you memorize the above section and avoid getting too confused with the links below. They are included for you in case you have burning questions that just "have to be answered" before you can move on. The second link could be especially confusing.

* http://www.immunize.org/catg.d/p3072a.pdf

* http://www.cdc.gov/vaccines/recs/vac-admin/contraindications-misconceptions.htm

# INBORN ERRORS OF METABOLISM (IEM) & MISCELLANEOUS METABOLIC DISORDERS

**NOTE**: HIGH YIELD! This section is difficult. Focus on the absolutes for each category. If the chart is overwhelming, start on the next page and refer back.

| | Glucose | NH3 | Ketones | Acidosis? | Lactate | Notes & Names | Diagnosis/Treatme |
|---|---|---|---|---|---|---|---|
| **Organic ACIDemias** | LOW | HIGH (usually) | YES | + GAP **acid**osis from organic **acid** build up | HIGH<br><br>Except in Isovaleric acidemia. | - Presents by **DOL 2.**<br>- CAN have THROMBOCYTOPENIA and Granulocytopenia (→ low WBC)!<br>**ALL LABS ARE MESSED UP**<br>- Isovaleric academia (foot odor/seizure), Glutaric academia (odor), MMA, Propionic academia | **Dx:** Urine Organic Acid Levels<br>**Tx:** Hydration! Also char diet to a HIGH CARNITIN diet (help to get rid of the built up organic acids). |
| **Urea Cycle** | OK | HIGH NH3 & LOW BUN | NO | NO<br><br>Serum pH is normal or shows a mild **resp. Alkalosis** | -<br>Usually normal, but can be slightly elevated | - **Hypotonia**!<br>- Can have RESP. ALKALOSIS<br>- Hyperammonemia without ketosis or liver dysfunction<br>- Ornithine Transcarbamylase (OTC), Citrullinemia, Argininosuccinic Aciduria (ASA) | **Dx**: **Low Arginine** in all. citrulline and high urine orotic acid in OTC. No arginine in citrullinemia.<br>**Tx**: Restrict protein intak Benzoate & phenylbutyr (NH3 scavengers). |
| **Fatty Acid Metabolism**<br><br>**Mitochondrial disorder (energy problem)!** | LOW<br><br>Because the hypoglycemia starts the chain of events! | OK or HIGH | **NO**<br><br>Can't make any! | MILD<br><br>From anaerobic metabolism and protein breakdown | HIGH Lactic Acid & Uric Acid | - Induced by physiologic stress, illness & **fasting** (low glucose).<br>- Since it is a disorder of fatty acid metabolism, the fatty acids cannot be broken down. That means no ketone production. During fasting states, or when carbohydrate reserves have been used up, there is a **NONKETOTIC Hypoglycemia!**<br><br>- MCAD | **Dx**: Low fasting glucose **without** appropriate rise ketones. Obtain Carnitin or Acyl-carnitine levels.<br><br>**Tx**: IV GLUCOSE |
| **GSD I** (von Geirke)<br><br>**Mitochondrial disorder (energy problem)!** | LOW | -<br>Usually OK | YES | + GAP acidosis<br><br>+Lactic acidosis & +Ketoacidosis | HIGH Lactic Acid & Uric Acid<br><br>Depends on anaerobic metabolism & protein breakdown | - GSD 1 is a true glycogen storage disease. Patients are unable to breakdown glycogen or perform gluconeogeneis. They depend on fats, protein and anaerobic metabolism for energy. Glycogen builds up in organs causing organomegaly.<br><br>- GSD 2 = "pompLay" = a Lysosomal issue of glycogen accumulation (heart, liver). It's **not** a glycogen breakdown issue so there's no hypoglycemia, ketonuria or acidosis. | **Dx**: Low fasting glucose with + ketones.<br><br>**Tx**: Cornstarch to allow slow carbohydrate breakdown |
| **Amino Acidopathy** | **OK**<br><br>Except in MSUd (Low) | **OK**<br><br>Except in MSUd (High) | **NO**<br><br>Except in MSUd (High) | **NO**<br><br>Except in MSUd (High) | **OK**<br><br>Except in MSUd (due to Ketoacidosis from protein breakdown) | - For each specific disorder, there is a specific amino acid which cannot be broken down.<br>- PKU (smell), Alkoptonuria (black urine), Tyrosinemia, Homocysteinuria (clots, Marfanoid)<br><br>- MSUd is a disorder in which ketoacids formed during the catabolism of Valine, Isoleucine and Leucine cannot be broken down. Patients are hypoglycemic at presentation because that is when there is a need for protein catabolism. | **Dx**: Look for high levels a specific amino acids i the serum and urine<br><br>**Tx**: Limit intake of spec amino acids based on t disorder |
| **GaLactosemia** | LOW | OK | NO | **NO** | - | - Galactose-1-phosphate **uridyltransferase (aka GALT)** deficiency, so glactose (a product of lactose) can't be broken down, and galactose-1-phosphate builds up and deposits in the liver, kidney and brain. Patient's often **present with GNR sepsis (especially E. coli).** | **Dx**: Increased galactos phosphate level, decre GALT activity in RBC's<br>**Tx**: Lactose & gaLACTOSE-free diet |
| **HypER-GLYCINEmia** | OK | OK | NO | NO | - | Unique because most labs are normal! Look for seizure activity (burst suppression on EEG), lethargy or "hiccups." | **Dx**: High GLYCINE<br>**Tx**: Limited tx options. Restrict diet. Poor prog |

# ***INBORN ERRORS OF METABOLISM (IEM) PEARLS***

**NOTE**: Use this section in conjunction with the inborn errors of metabolism table **as a quick review** of some of the IEMs. A full discussion of each disease **will** be discussed separately.

## INBORN ERRORS OF METABOLISM (PEARLS)

Inborn errors of metabolism usually present with an acute onset of symptoms such as lethargy, vomiting, tachypnea, confusion or seizures. Look for the ABSENCE of a fever in most of these patients. Keep in mind, though, that galactosemia patients may actually present with Gram negative rod sepsis (especially E. coli).

## ORGANIC ACIDEMIAS (PEARLS)

In orrganic acidemias, an enzyme deficiency leads to difficulty in breaking down certain organic acids from particular amino acids and fatty acids. Look for **virtually every lab mentioned in the IEM summary table to be abnormal** (hypoglycemia, hyperammonemia, ketoacidosis and lactic acidosis). Patients can also have thrombocytopenia and granulocytopenia. Patients present EARLY at about DOL 2.

## UREA CYCLE DEFECTS (PEARLS)

The primary role of the urea cycle is to clear ammonia (NH3). Due to the defect, there is a massive build of ammonia, but the rest of the labs are fine! Look for ammonia levels in the **hundreds** and hypotonia, that's it! No acidosis, but patient can possibly have a mild RESPIRATORY alkalosis.

## FATTY ACID METABOLISM DISORDERS (PEARLS)

Patients present due to fatty acid metabolism disorders when there is a stressor, or meals are spread out and carbohydrate reserves become depleted. That's when the inability to metabolize fatty acids due to a disorder results in symptoms. Look for hypoglycemia (because that's the trigger), metabolic acidosis (lactic acid), hyperammonemia (from protein break down) but **NO KETOACIDOSIS** because the fatty acids cannot be broken down **into** ketones!

## STORAGE DISEASES (PEARLS)

The storage diseases usually present later in infancy, or early in childhood, with SLOWLY PROGRESSIVE issues. This is because it takes months to years for the intra-organ build up of metabolites up to occur.

* **GLYCOGEN STORAGE DISEASES I (GSD I)**: Presents during fasting, or when a child's meals have been spread out. In an older child it may present when s/he is sleeping more.

## MITOCHONDRIAL DISORDERS (PEARLS)

Elevated lactic acid and pyruvate are great markers for mitochondrial disorders. These disorders have to do with fat or carbohydrate metabolism, and the patients are often dependent on anaerobic metabolism (aka anaerobic respiration) at the time of presentation. Also have increased URIC ACID because depend on protein for energy. Can have hepatomegaly, if so then think mitochondrial issues = GSD (+ketones) or FA metabolism (no ketones).

## AMINO ACIDOPATHIES (PEARLS)

Amino acidopathies occur when isolated amino acids cannot be broken down. Since the problem is limited to only a particular amino acid, there is no significant elevation in ammonia levels. Also, there is no acidosis because there are no fuel issues (carbohydrate and fat metabolism is working fine).

* **MAPLE SYRUP URINE DISEASE (MSUD)**: This one is an exception, and does result in hypoglycemia, hyperammonemia and acidosis. Symptoms start the very first week and progress to severe neurological issues within 2-3 weeks if left untreated (encephalopathy, seizures). Baby's urine should smell like maple syrup.

## GALACTOSEMIA (PEARLS)

Galactosemia is caused by galactose-1-phosphate uridyltransferase deficiency, or GALT deficiency. Lactose breaks down into glucose and ga**Lactose**. Due to the GALT deficiency galactose-1-phosphate cannot be broken down. Look for an elevated galactose-1-phosphate level in the serum or tissues (builds up in the liver, kidney and brain). Can also look for decreased GALT activity in RBCs. Patient's often present with **Gram negative rod sepsis (especially E. coli)** before the diagnosis is even made. Treat with a LACTOSE and a gaLACTOSE-free diet.

## HYPERGLYCINEMIA (PEARLS)

Hyperglycinemia findings include an acute encephalopathy, possible hiccups, seizures with a mention of "burst suppression" on EEG, and all normal labs.

## NEWBORN SCREEN (NBS)

The standard newborn screen (NBS) includes testing for phenylketonuria (PKU), hypothyroidism, sickle cell anemia and galactosemia in EVERY U.S. state and in Puerto Rico.

## AMMONIA LEVEL

A normal ammonia level is < 50. For urea cycle disorders, look for a VERY high level.

## INHERITANCE PATTERN

Most of these diseases are autosomal recessive. Hunter and Ornithine Transcarbamalase are 2 exceptions.

**MNEMONICS:** "X marks the spot" for a Hunter. Ornithine transcarbamala**X** is how you should pronounce it from now on. Hurler's is autosomal recessive. (DOUBLE TAKE) Remember, "PAT HAS WACK GAS that makes me **HURL** in the BACK SEAT of an AUTOmobile" from the Genetics section?

# *ORGANIC ACIDEMIAS*

## ORGANIC ACIDEMIAS OVERVIEW

Organic acidemias are caused by enzyme deficiencies that lead to difficulty in breaking down certain organic acids from particular amino acids **and** fatty acids. **Virtually every lab mentioned above will be abnormal.** Look for a **METABOLIC ACID**osis **+ KETOSIS + Lactic ACIDosis +/- Hyperammonemia** (due to a of mild effect on urea cycle) +/- elevated bilirubin. Patients can also have **thrombocytopenia** and **granulocytopenia (WBC may be low)**. These patients **present EARLY at about DOL 2**. Diagnose by obtaining URINE organic acid levels. Treat immediately with HYDRATION and then prescribe a HIGH CARNITINE diet to help get rid of the build up organic acids. Some of the disorders include Isovaleric ACIDemia, Glutaric ACIDemia, Methylmalonic ACIDemia (MMA) and Proprionic ACIDemia.

* **PEARL**: Anytime you see evidence of an ACIDosis in the vignette, and the word ACID in an answer choice (e.g. Isovaleric ACIDemia), start looking for evidence to support this diagnosis. Of course, they could always leave out names and just expect you to choose a diagnostic test or treatment. Isovaleric Acidemia is the one more likely to be tested. Also, anytime you see an abnormally HIGH respiratory rate, assume there is a metabolic acidosis! That may be the only clue they give you.

* **MNEMONIC**: Imagine going grocery shopping for **ORGANIC** food at a Whole Foods store. As you walk by the TOILET PAPER (TP) section, you look down and notice a piece of tile that has a beautiful white KITE ON IT. You then slip because the floor is wet due to AMMONIA cleaner that the janitor forgot to wipe up. Your head hits the tile with the white KITE ON IT and you bleed all over the beautiful design. Your blood forms the shape of 2 DOLLS playing with KITES. A manager comes and wraps your head with TP/TOILET PAPER. He then tries to clean the white KITE with his secret formula of white **LACT**aid milk + white **GRANULATED SUGAR**.

- **KEY**: TP/TOILET PAPER = Thrombocytopenia. KITE ON IT = Ketosis/Ketones. AMMONIA = **Possible** Hyperammonemia. 2 DOLLS = Presentation by DOL 2. LACTaid milk = Lactic Acidosis. GRANULATED sugar = Granulocytopenia with possible leukopenia (WBC). SUGAR = Hypoglycemia.

## ISOVALERIC ACIDEMIA

Isovaleric acidemia findings include an organic acidemia presentation with a history of a SWEATY FEET smell, poor feeding and SEIZURE. Patients are especially prone to getting INFECTIONS so look for a fever. Treatment should include PROTEIN RESTRICTION.

* **PEARL**: This one DOES NOT HAVE LACTIC ACIDOSIS!

* **MNEMONIC**: It helps if you know someone named Eric for this one. Rename this as **iso-WALL-ERIC** in your mind. Imagine ERIC leaning on a WALL by himself at lunch time. He's DEPRESSED because he has SMELLY/SWEATY FEET. He likes a girl, but every time he gets near her, his SWEATY FEE make her have a SEIZURE. So he continues to stand there alone at lunch time with NO APPETITE while everybody else eats STEAK for lunch and chats about all the fun they're going to have at the CARNIVAL.

- **KEY**: DEPRESSED = Lethargy. NO APPETITE = Poor feeding. STEAK = Protein. CARNIVAL = Carnitine.

* **MNEMONIC**: (IMAGE - Eric Cartman from the Simpsons) http://alturl.com/o6q2b

## GLUTARIC ACIDEMIA

Glutaric acidemia can also present with the smell of SWEATY FEET.

**MNEMONIC**: If you can connect ERIC with SMELLY FEET, you're set. Glut-ERIC.

## METHYLMALONIC ACIDEMIA & PROPIONIC ACIDEMIA

In methylmalonic acidemia and propionic acidemia, children present with poor feeding, vomiting, dehydration, lethargy, possibly hypotonia and tachypnea. May look like pyloric stenosis. Labs will show the elevated ammonia, acidosis and elevated lactate. Diagnose with urine organic acids.

* MMA: Treat with B12!

- **MNEMONIC**: Connect MMA with B12 by remembering that you obtain an MMA level to diagnose B12 deficiency!

* PROPIONIC ACIDEMIA: Treat with Biotin.

# ***UREA CYCLE DEFECTS***

## UREA CYCLE SUMMARY

**IMAGE**: (UREA CYCLE SUMMARY) http://alturl.com/hgna8

Nitrogen waste presents as NH3 (ammonia) to the urea cycle.

* AMMONIA **and** ORNITHINE (a urea cycle end product) present to CARBAMOYL PHOSPHATE SYNTHETASE, which converts them into CITRULLINE using the catalyst ORNITHINE TRANSCARBAMYLASE (OTC).

- **PEARL**: OTC deficiency is the disorder most likely to be tested because it's the most common, and unlike the others, it's X-linked (ornithine transcarbamola**X**). If you're finding this section overwhelming, focus on this disorder and note that without OTC, there can be NO CITRULLINE!

* ARGININOSUCCINIC SYNTHETASE (ASS) converts Citrulline into ARGININOSUCCINIC ACID (aka ARGININOSUCCINATE in the image)

* ARGININOSUCCINATE LYASE converts Argininosuccinic Acid into ARGININE

* ARGINASE converts Arginine into ORNITHINE **AND** UREA

* ORNITHINE TRANSCARBAMYLASE then works again with CARBAMOYL PHOSPHATE SYNTHETASE to catalyse the reaction between AMMONIA and ORNITHINE to create CITRULLINE!

## UREA CYCLE DEFECTS INCLUDE...

Ornithine Transcarbamylase deficiency (OTC is the most common urea cycle defect and is **X-linked**), Citrullinemia, Argininosuccinic Acid**uria** and Carbamoyl Phosphate Synthetase deficiency. Look for vomiting, lethargy, **HYPOTONIA** and eventually comma. Labs should show a MASSIVE hyperammonemia (possibly greater than 500!) with a **RESPIRATORY ALKALOSIS** (look for an ABG). There is NO ketosis and usually NO lactic acidosis. If you are given a **LOW ARGININE** or a **LOW BUN** you will know it's a urea cycle defect. To DIAGNOSE, check **serum** citrulline, arginine, argininosuccinic acid (ASA) and **URINE Orotic acid**. TREAT with increased caloric intake but **protein restriction.** Also give ammonia scavengers (**Benzoate & Phenylbutyrate**). A defect in the cycle prior to arginine production may also respond to treatment with arginine supplementation.

* **PEARLS:** LOW amino acids levels (Citrulline, ASA, Arginine) represent a lack of production. HIGH amino acid levels represent a build up due to an enzyme defect that is further down in the pathway.

* **MNEMONIC**: Imagine a child riding a FLOPPY tri**CYCLE** on the wet kitchen floor that has a ton of yellow AMMONIA cleaner on it that looks like UREAN.

- **KEY:** FLOPPY = Hypotonia. CYCLE = Urea cycle. AMMONIA = Hyperammonemia. UREAN = Urine to remind you of the Urea Cycle.

* **MNEMONIC**: All the **-INE's** are part of the urea cycle (citrulline, arginine, ornathine, glutamine). Plus carbamoyl phosphate.

## ORNITHINE TRANSCARBAMYLASE DEFICIENCY

In ornithine transcarbamylase deficiency, the serum is completely void of citrulline and arginine because OTC is not catalyzing the reaction needed to make citrulline! This is the most common defect and it's X-linked. Much more symptomatic in boys than the girl carriers who only get symptoms when they are sick. Look for a **HIGH urine Orotic Acid** (all of the built up carbamoyl phosphate gets turned into this).

**MNEMONICS:** For **orni**thine transcarbamyla**X** deficiency**,** think **PORNO**thine transcarboxylase deficiency (or PORNO THEME transcarboxylase deficiency). Who watches EROTIC (Orotic) movies? Boys! That should remind you that it's an X-linked disorder. "Erotic" should remind you to look for an elevated OROTIC acid.

## CITRULLINEMIA

Look for a **very high citrulline level** to indicate citrullinemia because it is not being broken down into ArgininoSuccinic Acid (ASA) and then Arginine, due to a defect in Argininosuccinic Synthetase (ASS).

## ARGININOSUCCINIC ACIDURIA

In argininosuccinic aciduria, the defect is at Argininosuccinate Lyase. Look for a moderately elevated citrulline and an elevated Argininosuccinic Acid level of course.

## UREA CYCLE LAB SUMMARY (TABLE)

| | OTC | CITRULLINEMIA | AS Acid**uria** |
|---|---|---|---|
| CITRULLINE LEVEL | **LOW** | HIGH | ~HIGH |
| ARGININOSUCCINIC ACID LEVEL | LOW | **LOW** | HIGH |
| ARGININE LEVEL | LOW | LOW | **LOW** |
| URINE OROTIC ACID LEVEL | **HIGH** | NORMAL | NORMAL |

# ***MITOCHONDRIAL DISORDERS***

## MITOCHONDRIAL DISORDERS OVERVIEW

The mitochondrial disorders include **fatty acid metabolism disorders** and **glycogen storage diseases** (GSD). These are disorders of ENERGY metabolism, therefore anaerobic metabolism kicks in and will be **elevated LACTIC ACID** and **PYRUVATE** levels. **Uric acid** levels are also increased because of increased protein catabolism. Look for evidence of **hearing loss with an elevated lactate and hepatomegaly**. Patients can also have strokes. Confirmatory diagnosis requires gene analysis for a mitochondrial disorder.

**PEARL**: Associate the word "mitochondrial" with energy since it is the power house of the cell. Don't think about mitochondrial inheritance. These are autosomal recessive.

## FATTY ACID METABOLISM DISORDERS

The fatty acid metabolism disorders include deficiencies in MCAD (medium chain acyl-coa dehydrogenase), LCAD (long chain acyl-coa dehydrogenase) and VLCAD (very long chain acyl-coa dehydrogenase). The problem here is that the mitochondria is not able to breakdown fatty acids. Patients present when there is a stressor, or meals are spread out and carbohydrate reserves become depleted, resulting in an inability to metabolize fatty acids. Look for hypoglycemia (because that's the trigger), **metabolic acidosis (lactic acid +/- uric acid from protein break down)**, hyperammonemia (from excessive protein break down) but **NO KETOACIDOSIS** because the fatty acids cannot be broken down **into** ketones! Urine will be negative for reducing substances (since this is not a carbohydrate issue), negative for ketones, and the serum will have normal organic acids. There may be a transaminitis and possible **hepatomegaly** (but no splenomegaly). Diagnosis requires a carnitine and acylcarnitine profile. These only present when there is hypoglycemia, therefore they do not present in the first week of life.

**PEARL**: If given an answer choice with a disease that has "CARNITINE" in the name of it, that IS a fatty acid oxidation disorder.

**MNEMONIC**: Ever heard of a FATTY LIVER? Imagine the inability of fatty acid break down being so bad that all of the fat deposits in the liver causing it be a palpable FATTY ACID LIVER.

## GLYCOGEN STORAGE DISEASES OVERVIEW

**NOTE: GLYCOGEN STORAGE DISEASES (GSD)** include von Gierke's disease, Pompe's disease, Cori's disease and McArdle's disease. We'll focus on the two most commonly tested. Since these are storage diseases, they do not present in the first week of life.

**MNEMONIC**: If you can remember **VoPoCoMc**, then you now have a mnemonic for the four GSDs. Also, McArdle's = M = Mitochondrial = Muscle problems. Having said that, they seem to be getting away from the eponyms for these particular disorders.

## GSD I (aka VON GIERKE'S DISEASE)

**GSD I (aka VON GIERKE'S DISEASE) is the** most commonly tested of the glycogen storage diseases. There is a glucose-6-phosphat**ase** deficiency, which results in an **inability to breakdown glycogen**. Glycogen then deposits in the liver, kidneys and pancreas resulting in hepatomegaly. Some patients have a "doll-like" face. It only present when there is hypoglycemia (when meals are spread), therefore will not present in the first week of life. Labs show a fasting **hypoglycemia, ketoacidosis (from fat break down), lactic acidosis and hyperuricemia**. Treat inpatients with continuous feedings. Treat outpatients with a LOW carb diet and frequent **cornstarch** (since that can break down slowly and thus prevent hypoglycemia). A high carb diet would be wrong because it would just result in MORE GLYCOGEN being made and deposited!

- **PEARL**: Glycogen stores cannot be broken down, so glucagon is useless for hypoglycemia.

## GSD II (aka POMPES DISEASE)

GSD II (aka Pompes Disease is **NOT** a mitochondrial disorder, it is a **LYSOSOMAL disorder** in which cellular breakdown of glycogen cannot occur. Deposition occurs in the liver, muscles and heart, so look for **organomegaly**. Symptoms can begin at less an a month of age, or not until many years later as a teen. For babies, look for a floppy/**HYPOTONIC baby** with FTT and **macroglossia**. Death can occur due to respiratory failure. There is NO hypoglycemia in GSD II because other glycogen stores CAN be broken down. The problems in this disease arise due to the deposition. A key finding in this disease is CARDIOMEGALY from deposition in the heart. Look for left axis deviation (LAD) on an EKG.

* **MNEMONIC**: PUMP-LAY's disease! PUMP = HEART and LAY should remind you that this is a LYsosomal disease!

* **MNEMONIC**: Have you heard of Pompe's restaurant? Well, imagine being at POMPE's restaurant (or PUMP-LAY's restaurant) where there's so much food that you have to roll out your HUGE TONGUE to put all of the food on it. After that, you end up completely HYPOTONIC/FLOPPY because of a food coma!

## AMINOACIDOPATHIES

**NOTE**: Aminoacidopathies are disorders in which isolated amino acids cannot be broken down. Since the problem is limited to only a particular amino acid, there is **no significant elevation in ammonia levels** (the exception is MSUD). Also, there is **no acidosis** because this is not a fuel issue (carbohydrate and fat metabolism is working fine, but once again MSUD is the exception). The disorders include Phenylketonuria (PKU), Alkaptonuria, Maple Syrup Disease (MSUD), Homocysteinuria and Tyrosinemia.

### PHENYLKETONURIA (PKU)

Phenylketonuria (PKU) is a deficiency of phenylalanine hydroxyl**ase,** so **phenylalanine can't be broken down into tyrosine**. Patients are noted to have a **musty/mousy odor and LIGHT colored skin and hair**. If left untreated, patients will be asymptomatic for a few months. They can then start to present with developmental delay, the musty/mousy odor, skin changes (light skin/hair, possible eczema like rash) and eventually might even carry a diagnosis of "autism" because of severe mental retardation. The disease is also associated with septal defects. All states test for this on the newborn screen, which looks for an elevated phenylalanine level. Your first step in a neonate should be to **REPEAT testing** because some kids have a TRANSIENT elevation in their phenylalanine level. If it's abnormal again, the child still might be ok because it may just be that the enzyme hasn't matured yet. Consult a geneticist to sort it out. The goal should be to **start treatment by 14 days** of life, otherwise **severe mental retardation can occur**. For kids with PKU, **TYROSINE** levels will be low (or nonexistent), so tyrosine becomes an **essential amino acid**. Some children require Tyrosine supplementation in the diet. The general treatment, though, is a **LOW PHENYLALANINE DIET**. OVER-TREATING (meaning zero phenylalanine) can result in lethargy, rash and diarrhea (don't get it confused with pellagra!). Also, if a **pregnant** patient has poor dietary control, her child can be born with microcephaly, congenital heart defects (CHD) and cognitive defects.

**PEARL**: Look for a child that looks like they might have albinism. Also, keep in mind that the newborn screen is only valid AFTER protein intake.

**MNEMONIC**: Imagine walking into a smelly basement. You say, "PEE-YEW!" Just then you see a LIGHT-SKINNED MOUSE with WHITE HAIR running by. You scream, "EEW," and run away.

* **KEY:** PEE-YEW = PKU. LIGHT COLORS = Light skin and hair. EEW = PKU again. You could consider adding a TYRE (for tyrosine) to the story, and imagine that the mouse runs out of an old tire in the corner.

## ALKAPTONURIA (aka ALCAPTONURIA)

Alkaptonuria (aka alcaptonuria) is a problem with **phenylalanine AND tyrosine metabolism**. Can present with "black" urine in a diaper since urine turns dark after long exposure to air. There is excess **homogentisic** acid which can deposit in **joints** and on **heart valves.** Check for elevated homogentisic acid levels in urine. Treat by limiting intake of phenylalanine and tyrosine.

**MNEMONIC**: Instead of ALKAPTONURIA, rename it to "ALL-BLACK-TONED-URIA!"

## MAPLE SYRUP URINE DISEASE (MSUd, aka BRANCHED-CHAIN KETOACIDURIA)

Maple Syrup Urine Disease (MSUd, aka branched-chain ketoaciduria) presents **EARLY** (first week) with sweet smelling urine, tachypnea and lethargy. If left untreated, it can result in severe neurologic issues in 2-3 weeks, including encephalopathy and seizures. MSUD is the exception amongst this group and **DOES have hyperammonemia, ketoacidosis and hypoglycemia**. It can look like an organic acidemia, so be careful! Branched chain amino acids cannot be broken down, so high levels of **Valine, Isoleucine, Alloisoleucine and Leucine** levels are found in the serum. The defective enzyme (complex-branched chain alpha-ketoacid dehydrogenase – just an FYI, low yield) is also partly responsible for the breakdown of ketones, which is why there is **ketosis and ketonuria**. Treat with a special diet free of branched-chain amino acids.

* **MNEMONIC**: HYPOGLYCEMIA occurs because all of the sweetness is in the SWEET SMELLING MAPLE SYRUP PEE!

* **MNEMONIC**: Imagine a college named MSU (Maple Syrup University). At MSU, the kids get TACHYPNEIC and TIRED/LETHARGIC during the VERY FIRST WEEK of college because they're too busy partying. One guy even got hazed this week by some frat guys who gave him a chem class VIAL of URINE with MAPLE SYRUP IN IT. They told him the VIAL was full of beer!

- **KEY:** VERY FIRST WEEK = Early Presentation. VIAL = Valine, Isoleucine, Alloisoleucine and Leucine. Consider using "LEUCI loves her MAPLE SYRUP!" as well.

## (DOUBLE TAKE) HOMOCYSTEINURIA

Homocysteinuria is an aminoacidopathy with physical features that are Marfanoid, but this includes COGNITIVE DEFICITS and neurologic deficits. Also, like Marfan Syndrome, there can be displacement of the lens. However, the displacement is DOWNWARDS or POSTERIORLY. These patients are hypercoagulable so there may be a history of DVT. It can be diagnosed by noting high levels of homocysteine in the urine. Treat with pyridoxine or diet high in cysteine and low in methionine.

**MNEMONIC**: (Image of Homoerectus) http://bit.ly/roDnZq

**MNEMONIC:** Imagine a proud and TALL HOMOerectus next to a fire. He suddenly develops CLOTS in his BRAIN that cause him to FALL DOWN, face forward so that he injures his EYES. He now has NEUROLOGIC DEFICITS because of the CLOTS in the brain, and his ocular lens is DOWNWARDLY displaced.

**MNEMONIC:** This sounds intuitive, but how do you treat someone that is loosing an amino acid in their urine? Give them more of it! Treat with **cysteine.**

# CARBOHYDRATE METABOLISM DISORDERS

## DISORDERS OF CARBOHYDRATE METABOLISM

Carbohydrate metabolism disorders include Galactosemia, Hereditary Fructose Intolerance, Aldolase deficiency = Often result in hypoglycemia, hyperammonemia, acidurias, organic acidemias

## GALACTOSEMIA (aka GALACTOSE-1-PHOSPHATE URIDYLTRANSFERASE DEFICIENCY or GALT DEFICIENCY)

Galactosemia is caused by galactose-1-phosphate uridyltransferase deficiency, or GALT deficiency. It can look like an early version of lactase deficiency and can present with vomiting and diarrhea. The more serious symptoms include lethargy, hepatomegaly, jaundice, hypoglycemia and seizures. Also, patient's often **PRESENT** with **Gram negative rod sepsis or meningitis (especially E. coli)** before the diagnosis is even made. As a reminder, lactose breaks down into glucose and ga**Lactose**. Due to the enzyme deficiency mentioned, **galactose-1-phosphate** cannot be broken down. Lactose is in breast milk and formulas, so this presents EARLY (first week). Test for an elevated **galactose-1-phosphate level** in the serum or tissues (builds up in the liver, kidney and brain). Can also test for decreased **GALT activity** in RBCs. You can also test for **non-glucose reducing substances** in the urine. Treat with a LACTOSE & gaLACTOSE-free diet, such as **SOY MILK** (the primary carbohydrate in soy milk is sucrose or corn syrup). Untreated galactosemia will result in **CATARACTS**, severe cognitive defects/mental retardation and liver disease.

**MNEMONIC**: Since patients must have a lactose free diet, they cannot have any LACTOSE containing milk that comes from LACTATING breasts (cow or human). Soy is fine. So that means SOY is good for patients with **galactoSOYmia**.

## HEREDITARY FRUCTOSE INTOLERANCE

Children with hereditary fructose intolerance can present with vomiting, seizures, HYPOglycemia, hepatomegaly or jaundice after eating high fructose meals. They are unable to breakdown fructose and often develop a tendency to AVOID SWEETS. Look for a child who has a SEIZURE right after eating. Treat with a low fructose diet.

# LYSOSOMAL STORAGE DISEASES

**PEARL/MNEMONIC**: The made up word "**HuNiTaG**" can help you remember the names of the most common lysosomal storage diseases tested by the American Board of Pediatrics: Hurler, Hunter, Niemann-Pick, Tay-Sachs and Gaucher. So if you're sure it's a lysosomal storage disease on the exam but forget which disease is what, maybe this will help you make a better guess.

## MUCOPOLYSACCHARIDOSES (MPS)

Hurler Syndrome (aka Hunters Syndrome) and Hunter Syndrome (aka Hunters Syndrome) are the mucopolysaccharidoses (MPS) to DEFINITELY KNOW. The MPS disorders are due to a missing or dysfunctional lysosomal enzyme needed to break down a long carbohydrate (**glycos**aminoglycans). This results in the accumulation of dermatin, keratin and heparin sulfate in the brain, bones, connective tissue and other organs. Since this is an accumulation/deposition disease, it takes time to present. Look for **PROGRESSIVE** physical deformations and **cognitive** problems. Also look for **coarse features, thick eyebrows**, **organomegaly,** progressive joint contractures, **growth deceleration** and even **progressive** deafness. It's diagnosed by showing excessive **urine mucopolysaccharides**. Note that they are all autosomal recessive except for **Hunter's** syndrome. Many of the features are the same for these diseases. See below for some of the unique features of each:

* **PEARL**: Hurler and Hunter syndromes are the ones most likely to be tested.

* **HURLER SYNDROME** (aka MPS I or HURLERS SYNDROME): Children have **growth deceleration, coarse facial features, CORNEAL CLOUDING**, hepatosplenomegaly, pes cavus (high plantar arches) and hirsutism.

- **IMAGE**: http://alturl.com/rv2q4 (coarse features, thick eyebrows)
- **IMAGE**: http://alturl.com/9aw5y
- **MNEMONIC**: Please refer to the "PAT HAS WACK GAS that made me HURL" mnemonic to remind yourself that this is an autosomal recessive disorder. The next part could be, "I HURLED so hard that I WENT BLIND (from corneal clouding)!"

* **HUNTER SYNDROME** (aka MPS II or HUNTERS SYNDROME): Look for a child with **coarse facial features, skeletal anomalies** and **growth deceleration** that has evidence of an **X-LINKED** disorder (affected maternal uncles). There is NO corneal clouding.

- **IMAGE**: http://blog.timesunion.com/mdtobe/files/2010/12/mps.jpg
- **PEARL**: Since this is the only one that's X-linked, it's a very "testable" area for the pediatric boards. Use the mnemonics to nail down the inheritance patterns to make sure you don't get the names confused.
- **MNEMONIC**: "X" marks the spot for a HUNTER. Now imagine a **HUNTER with GOOD VISION** who is **SHORT** and has an abnormal **SPINE** which helps him go through the jungle unseen. He's extremely skilled and always makes the kill by aiming for an "X" on his prey's forehead.

* SANFILIPPO SYNDROME (MPS III): Progressive reduction in cognitive abilities.

* MORQUIO SYNDROME (MPS IV): Skeletal involvement, corneal clouding but normal intelligence.

* I-CELL DISEASE: This is a mucolipidosis (don't memorize these long words!) but it has features very similar to HURLER syndrome. Patients can have coarse facial features, corneal clouding, skeletal problems, hip dislocation, club feet, joint contractures, hepatomegaly or splenomegaly.

- **PEARL**: The I is for inclusion-cell disease. If they mention "inclusion," pick this. Otherwise, you're probably better of picking Hurler.

## ***SPHINGOLIPIDOSES***

### TAY-SACHS DISEASE (aka TAY-SACH DISEASE)

**Tay-Sachs disease (aka Tay-Sach disease) is a Hexosaminidase A enzyme** (a lysosomal enzyme) deficiency resulting in **progressive neurologic deficits**. Patients can develop normally until about **9 months** but then begin to show signs of lethargy and hypotonia. They are also found to have macrocephaly, an **exaggerated startle reflex** and a **cherry red spot** on retina. Children usually die by the age of four. You are

**REQUIRED TO SCREEN all kids born to Ashkenazi Jews for this, and may do so by amniocentesis or chorionic villus sampling.** It's autosomal recessive, which means both parents are carriers.

**PEARL**: **There is no organomegaly.**

**MNEMONIC**: Do you know anyone named Tay? Tate? Taz? Imagine TATE becomes famous and develops a HUGE HEAD (macrocephaly) from his success. He's now a smooth and LOOSE talking player. Everything is fine until a jealous PREGNANT ASHKENAZI JEWISH RABI shoves a NEEDLE in his belly (amniocentesis) and a CHERRY in his EYE. He is HUGELY STARTLED (startle reflex) by how his fame has done him in.

### GAUCHER DISEASE (aka GAUCHERS DISEASE)

Patients with Gaucher disease (aka Gauchers disease) have **hepatomegaly**, short stature, thrombocytopenia, easy bruisability, short stature, **osteosclerosis and lytic lesions with bone pain**.

**MNEMONIC**: Rename it OUCHers disease since patient complain of bone pain.

**MNEMONIC**: Rename it GROUCH-ers disease. Imagine some GROUCHY old lady complaining of bone pain and easy bruising. She's so dramatic about it all that she starts wrapping her arms and legs in TOILET PAPER (TP = Thrombocytopenia) to show you how much pain she has!

### FABRY DISEASE (aka FABRYS DISEASE)

Fabry disease (aka Fabrys disease) findings include opacities of the eye, vascular disease of the kidney, heart or brain, and ORANGE-colored skin lesions.

**PEARL**: There is no organomegaly. Though not included in the HuNiTaG mnemonic, it's a very common lysosomal disorder so it's likely high-yield.

**MNEMONIC**: Imagine Mr. Kanye West wearing his famous WHITE SUNGLASSES (opacities) and an ORANGE-colored suit (skin) made of ROUGH (rash) **FABRY**c.

### NIEMANN-PICK DISEASE

Niemann-Pick disease is a **sphingomyelinase deficiency** that results in accumulation of **sphingomyelin** in macrophages within the liver and lungs. Patients have **neurologic problems**, **hepatosplenomegaly** and a **cherry red spot** on the macula.

**PEARL**: If you are given a patient with a cherry red spot on the macula and **hepatomegaly**, this is your answer! Fabry and Tay-Sachs do not have organomegaly!

## *MISCELLANEOUS DISORDERS & PEARLS*

### HYPOGLYCEMIA DIFFERENTIAL

When the cause of hypoglycemia is not clear, consider getting other labs such as urine ketones, reducing substances, urine organic acids and serum lactic acid.

* **(DOUBLE TAKE) INFANT OF A DIABETIC MOTHER (IDM)**: Infants of diabetic mothers (IDM) will have CARDIAC, CARDIAC, CARDIAC, and NEURAL TUBE DEFECTS! Patients are especially prone to having septal defects. May also have coarctation of the aorta, hypertrophic obstructive cardiomyopathy (HOCM),

truncus arteriosus, dextrocardia and more. Can also have rib or vertebral column abnormalities, hydrocephalus, macrosomia/LGA size (with birth trauma), small Left colon, duodenal atresia (double bubble… or "gaseous distention in both the stomach and duodenum"), apnea, hypOcalcemia, hypOmagnesemia, hypOphosphatemia, hyperbilirubinemia, polycythemia, vascular thromboses and RDS (look for tachypnea).

- **PEARL**: The hypoglycemia may not occur until ONE or TWO days after birth.
- **PEARL**: For hypocalcemia, look for jitteriness, a prolonged QT, Chvostek's sign (tapping the facial nerve elicits a twitch) or Trousseau's sign (carpopedal spasm noted when the wrist is clasped).

* HYPERINSULINISM: Look for a BIG BABY in whom the hypoglycemia gets better with glucagon, but only transiently. All growth parameters are likely to be high (around the 95th percentile). Treat with DIAZOXIDE, which decreases insulin production and increases cortisol release. Calcium channel blockers do not work in children with hypoglycemia due to hyperinsulinism.

* ADRENAL INSUFFICIENCY: Hypoglycemia + Ketones +/- Electrolyte abnormalities

* HEREDITARY FRUCTOSE INTOLERANCE: Hypoglycemia + Ketones + Emesis after a meal containing fructose.

* ORGANIC ACIDEMIAS: ALL LABS ARE MESSED UP! + EARLY symptoms (by DOL 2)

* FATTY ACID METABOLISM DEFECT: Hypoglycemia WITHOUT Ketosis!

* GSD I (von Gierke): Hypoglycemia + Ketones + Lactic Acidosis + Elevated Uric Acid + Normal Ammonia.

* GALACTOSEMIA: Hypoglycemia + Hepatomegaly + Non-glucose reducing substances in the urine + FTT + Early presentation due to lactose in mom's milk.

## (DOUBLE TAKE) INFANT OF A DIABETIC MOTHER (IDM)

Infants of diabetic mothers (IDM) will have CARDIAC, CARDIAC, CARDIAC, and NEURAL TUBE DEFECTS! Patients are especially prone to having septal defects. May also have coarctation of the aorta, hypertrophic obstructive cardiomyopathy (HOCM), truncus arteriosus, dextrocardia and more. Can also have rib or vertebral column abnormalities, hydrocephalus, macrosomia/LGA size (with birth trauma), small **L**eft colon, duodenal atresia (double bubble… or "gaseous distention in both the stomach and duodenum"), apnea, hypOcalcemia, hypOmagnesemia, hypOphosphatemia, hyperbilirubinemia, polycythemia, vascular thromboses and RDS (look for tachypnea).

**PEARL**: The hypoglycemia may not occur until ONE or TWO days after birth.

**PEARL**: For hypocalcemia, look for jitteriness, a prolonged QT, Chvostek's sign (tapping the facial nerve elicits a twitch) or Trousseau's sign (carpopedal spasm noted when the wrist is clasped).

## PURINE & PYRIMIDINE DISORDERS

Purine and pyrimidine disorders may cause recurrent infections, developmental delay/cognitive defects and failure to thrive (FTT).

* LESCH-NYHAN SYNDROME: Look for **SELF MUTILATION** + **choreiform** movements + mental retardation + elevate uric acid levels (can also get gout). This is X-linked.

- **MNEMONIC**: LeX luthor in the Superman series was little crazy. Call it "le**X**-nyhan" to remember that it's X-linked and to consider it in male patients.

- **MNEMONIC**: Rename it "LET'S EAT MY HAND (lesch-nyhan) syndrome because it won't stop DANCIN'!" Dancin' = Choreiform movements.

* ADENOSINE DEAMINASE (ADA) DEFICIENCY: Look for recurrent infections and developmental delay/cognitive defects. It's common in patients that have other disorders like SCID and certain rheumatologic conditions.

## (DOUBLE TAKE) WILSONS DISEASE

Wilsons Disease (aka Wilson's Disease) is an autosomal recessive disorder resulting in excess copper accumulation, especially within the liver and brain. Accumulation in the liver can lead to **hepatomegaly**, spider nevi, esophageal varices and a Coombs-negative hemolytic anemia. Accumulation in the brain can lead to **neurologic changes** including tremors, poor school performance, ataxia, abnormal eye movements and spasms. On eye exam, a Kayser-Fleischer ring may be visible. Copper levels in the serum are **low**, but high in the **tissues**. Diagnose by LIVER BIOPSY. Treat with PENICILLAMINE, a copper chelator.

**PEARLS**: **Diagnose** by LIVER BIOPSY. Abnormal eye movements and a Fleisher ring may be present, but there is no visual disturbance. Kayser-Fleischer rings are seen in 90% of symptomatic patients, and almost 100% of patients with neurologic manifestations. **Screen** family members with CERULOPLASMIN levels. Ceruloplasmin is made in the liver and is the primary copper carrying protein. If the level is **LOW**, that suggests Wilsons Disease because excess copper is not being incorporated into ceruloplasmin, and is therefore still in the TISSUES. Therefore, supportive labs may include a low serum ceruloplasmin, high tissue copper levels, high urine copper levels and low serum copper levels.

**IMAGE**: http://www.gourmandizer.com/wilsons/kf2.html

**IMAGE**: http://upload.wikimedia.org/wikipedia/commons/0/00/Kayser-Fleischer_ring.jpg

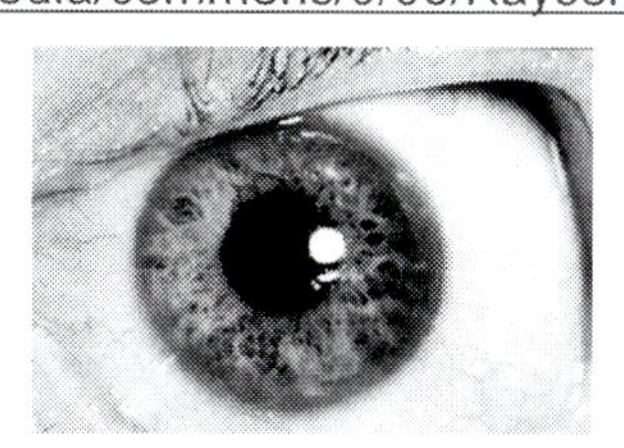

**MNEMONICS**: Treat with a "**COPPER PENNY-**cillamine." Also, ever heard of Wilsons? The leather company that makes baseball gloves? See the image below to note the strong resemblance in color to a COPPER PENNY. This should help you remember that Wilsons Disease has to do with copper, and that it is treated with "PENNY-cillamine."

## MENKES KINKY HAIR SYNDROME

In Menkes kinky hair syndrome, look for low copper levels and low ceruloplasmin levels (like Wilsons Disease). Also has high TISSUE copper. Patient's get kinky and twisted hair. It's X-linked.

**MNEMONIC**: X-linked. Who watches XXX movies? Kinky boys!

## SMITH-LEMLI-OPITZ SYNDROME

In Smith-Lemli-Opitz syndrome, there is defective cholesterol synthesis. Associated with cognitive deficits/mental retardation, microcephaly and various organ malformations (may have ambiguous genitalia). Also has **2-3 toe syndactyly** (fused) in which they look like a "Y". Since it is a cholesterol metabolism issue, there will be an elevated 7-dehydrocholesterol.

**IMAGE**: http://newborns.stanford.edu/images/syndactylyDAClark.jpg

**MNEMONIC**: smYth lemlY opYtz since they look like a "Y"

## CHERRY RED SPOT DIFFERENTIAL

The differential for a cherry red spot includes Tay-Sachs Disease, Niemann-Pick Disease and Farber's Disease (Farber's Disease is low yield).

* TAY-SACHS: Ashkenazi Jews. Macrocephaly.

* NIEMEN PICKS: Hepatomegaly and neurologic problems.

* FARBER'S DISEASE (aka DISSEMINATED LIPGRANULOMATOSIS): Probably low yield. Progressive disease. Fatal within the first few years. Look for a baby with joint pain and skin nodules presenting within the first week of life.

- **MNEMONICS FOR THE ABOVE DISORDERS:**
  - Imagine a CHERRY FARMER going to NIEMEN marcus in his overalls to try and sell a SACH of cherries to them! Wow, talk about rough economy.
    - **KEY:** Cherry, Farber, Niemen Picks, Tay-SACHS.
  - CHERRY FARMER'S disease = Imagine a CHERRY FARMER working so hard that he gets ARTHRITIS and SWOLLEN BUG BITES all over his skin.

* RETINAL ARTERY OCCLUSION: Presents as painless visual loss with a **cherry red** spot at macula.

- **(NAME ALERT) RETINAL VEIN OCCLUSION:** Painless visual loss with multiple hemorrhages noted on ophthalmologic on exam, referred to as having a "thunder and lightning" appearance. Sometimes also referred to as "blood and thunder" or "pizza pie."
  - **MNEMONIC**: Venous blood is dark. Thunder and lightning occur at night.
- **(NAME ALERT) RETINAL DETACHMENT:** Visual loss with what looks like cobwebs or a curtain coming across a patient's vision.
  - **MNEMONIC**: Imagine the retina being really far back in the eye in an area where COBWEBS webs are growing on old CURTAINS.

## GENERAL IEM PEARLS & RECAPS

* INBORN ERRORS OF METABOLISM (IEM): For these disorders, look for sudden onset of lethargy, vomiting, tachypnea, seizures and a lack of fever.

* ORGANIC ACIDEMIAS: Look for TONS of messed up labs. Hyperammonemia, Ketosis, possible Thrombocytopenia, Lactic Acidosis, possible Granulocytopenia and Hypoglycemia.

* UREA CYCLE DEFECTS: Look for hypotonia and hyperammonemia, but that's it! No acidosis, but possible respiratory alkalosis.

* MITOCHONDRIAL DISORDERS: Includes fatty acid defects and GSD I. Since the powerhouse (mitochondria) is defective, look for evidence of anaerobic metabolism (elevated lactate and pyruvate). Uric acid is also elevated from protein break down. These are fuel problems, so HEPATOMEGALY is common. "Only a MOM can LACTATE." Mito = Mom.

* FATTY ACID OXIDATION DEFECTS: Symptomatic once glucose/glycogen reserves are depleted. Fatty acid catabolism is defective so there is NO ketosis, but there is acidosis (from lactate). There's also hyperammonemia, possibly due to excessive protein breakdown for energy need.

* STORAGE DISEASES: **Slowly** progressive. Present in late infancy or early childhood.

* AMINO ACIDOPATHIES: No hyperammonemia because urea cycle is fine, and there's not excessive protein break down. No acidosis because there's no sugar, fat or energy/fuel issue.

- * MSUD is the exception! Acidosis + Hypoglycemia + Hyperammonemia

* HYPERGLYCINEMIA: Normal labs + Acute Encephalopathy + Hiccups + Burst suppression on EEG.

# ACID-BASE DISORDERS

## A GUIDE TO CALCULATIONS & SHORTCUTS FOR ACID BASE DISORDERS

**NOTE:** In order to do acid-base questions, you have to know some of the fundamentals. The upside is that if you learn the rules below, you'll get some very quick points on the exam because these questions tend to be short and simple. Also, there ARE some shortcuts that can help you answer certain questions within seconds!

* LOW PH = ACIDOSIS!
* HIGH PCO2 = RESPIRATORY ACIDOSIS!

## THE ULTIMATE ABG CALCULATOR BIBLE!

* To evaluate an ABG, start by looking at the pH, pCO2 and HCO3 to identify a primary disorder (if possible). Use 40 as your normal PCO2 and 24 as your normal HCO3.
* Next, look at the chemistry. A normal anion gap can be up to 12. Know that a patient's upper limit of normal for the anion gap decreases by 2.5 for every 1 g/dL drop in albumin. So if s/he has an albumin of 1.4, the gap should be no more than 7. That means that a patient with an albumin of 1.4 and an anion gap of 10 has a positive anion gap.
* If you diagnose a **RESPIRATORY DISORDER**, you must calculate compensation using the PCO2.
  - If the bicarb level is higher than what your calculation predicted, then there's an additional metabolic alkalosis.
  - If the bicarb level is lower than what your calculation predicted, then there's an additional metabolic acidosis.
* **RESPIRATORY ACIDOSIS**:
  - **(DOUBLE TAKE) PEARL**: If given an ABG and you note a respiratory acidosis, ALWAYS calculate for compensation. You do this by looking at the increase in PCO2 from the baseline of 40, and checking to see if the compensatory increase in bicarbonate (compensatory metabolic alkalosis via bicarb retention) is appropriate. If the bicarbonate level is lower than expected, then you have an additional primary metabolic acidosis. If it's higher than expected, then there is an additional primary metabolic alkalosis. Keep in mind that the compensatory increase in bicarbonate could be 1 per PCO2 increase of 10, or 3.5 per PCO2 increase of 10. That depends on whether or not you are dealing with an acute or a chronic respiratory alkalosis.
* **RESPIRATORY ALKALOSIS**:
  - **PEARL**: If given an ABG and you note a respiratory alkalosis, ALWAYS calculate for compensation. You do this by looking at the decrease in PCO2 from the baseline of 40, and checking to see if the compensatory decrease in bicarbonate (compensatory metabolic acidosis via bicarb excretion) is appropriate. If the bicarbonate level is even lower, then you have an additional primary metabolic acidosis. If it's higher than expected, then there is an additional primary metabolic alkalosis. Keep in mind that the compensatory decrease in bicarbonate could be 2 per PCO2 increase of 10, or 5 per PCO2 increase of 10. That depends on whether or not you are dealing with an acute or a chronic respiratory alkalosis.

| Acute | R. Acidosis | ↑ | HCO3 | 1 | For every 10 | ↑ | of PCO2 | **MNEMONIC**: **A**cute then chronic, **A**cid then base, up (**A**scending) then down (Descending), 1/2/3.5/5 |
|---|---|---|---|---|---|---|---|---|
| | R. Alkalosis | ↓ | HCO3 | 2 | For every 10 | ↓ | of PCO2 | |
| Chronic | R. Acidosis | ↑ | HCO3 | 3.5 | For every 10 | ↑ | of PCO2 | |
| | R. Alkalosis | ↓ | HCO3 | 5 | For every 10 | ↓ | of PCO2 | |

* **METABOLIC ACIDOSIS**; If you diagnose a metabolic acidosis, ALWAYS do this stuff!
  - Calculate the anion gap.
  - Now calculate $\Delta$ GAP (the difference between the upper limit of a normal gap, and the gap your patient has). The delta gap = Calculated AG – 12. So if the anion gap is calculated at 16, the $\Delta$ GAP is 16 – 12 = 4.

- In a metabolic acidosis, the bicarb should be low. Is the measured bicarb greater than the upper limit of normal (24)? If so, there is a concurrent metabolic alkalosis! (SHORTCUT)
- Now calculate the Δ Bicarb (the difference between the upper limit of a normal bicarb level, and the gap your patient has). The delta bicarb = 24 – Measured Bicarb. So if the measured bicarb level is 20, the Δ Bicarb is 24 – 20 = 4.
- The drop in bicarb (Δ Bicarb) should equal the Δ GAP if there is only a gap metabolic acidosis. If the Δ Bicarb = Δ Gap, there are no other disorders. You're done.
- If the bicarb is **lower** than expected, then there's an additional NON GAP metabolic acidosis that is driving the bicarbonate level down.
- If the bicarb is **higher** than expected, then there's an additional metabolic alkalosis driving the bicarbonate level up.
- For any metabolic acidosis, you should also check for appropriate respiratory compensation by using Winter's formula.
    - Expected value of PCO2 = (1.5 x Bicarb) + 8 +/-2.
    - If the values match, respiratory compensation is adequate.
    - If the measured PCO2 is higher than the predicted PCO2, there is also a primary respiratory acidosis.
    - If the measured PCO2 is lower than the predicted PCO2, there is also a primary respiratory alkalosis.

*** ABG SHORTCUTS**

- If the bicarb is not normal, there is SOME TYPE of acid base disturbance.
- If the bicarb level is low, it could be due to a gap or a non-gap acidosis.
- If the anion gap (AG) is > 20, that means there is at least a primary metabolic GAP acidosis
- If only one answer choice has that option, you're done!
- If the chloride is low, there is at least a metabolic alkalosis.
- So if you see a low chloride, there is at least a metabolic alkalosis. In a metabolic alkalosis, the bicarbonate level should be HIGH. If it's not high, then there is also a metabolic acidosis.
- If only one answer choice has metabolic alkalosis and metabolic acidosis, you're done!
- A PCO2 > 55 means there is at least a primary respiratory acidosis present (because the body is unable to compensate for a metabolic alkalosis beyond 55).
- A Bicarb > 40 means there is at least a primary metabolic alkalosis (because the body is unable to compensate beyond 40 for a respiratory acidosis).
- A Bicarb < 12 means there is at least a primary metabolic acidosis (because the body is unable to compensate beyond 12 for a respiratory alkalosis).

## ***ACID-BASE DISORDERS & PEARLS***

### ACIDOSIS

The only evidence of a primary metabolic acidosis may be **tachypnea** (occurring as a respiratory compensatory mechanism).

### ANION GAP

An anion gap represents the unmeasured ions (phosphorus, organic acids, proteins, lactic acid, sulfate, etc.) outside of the Na+, Cl- and CO2- that are in the routine chemistry panel. A normal gap is 12 or less. Up to 16 might be ok in a newborn.

### ANION GAP METABOLIC ACIDOSIS

An anion gap metabolic acidosis can be noed in conditions including Methanol ingestion, Uremia, Diabetic ketoacidosis, Paraldehyde, **IRON**/INH ingestion, Lactic acidosis, Ethanol/Ethylene Glycol and Salicylates. Also Theophylline. There should be a NORMAL chloride and LOW bicarb. Can also be found in acid ingestion, renal failure (because kidneys cannot excrete acids) and various inborn errors of metabolism (due to ketosis and lactic acidosis). Can also occur with Fanconi Syndrome and Medullary Cystic Disease (**extremely** low bicarb with high BUN and Cr).

**MNEMONIC**: **Fill the ANION GAP with MUDPILES.** The D = DKA, **not** DIARRHEA!

**PEARL**: The urea cycle defects have NO acidosis. They can, however, have a primary RESPIRATORY alkalosis.

## NON-ANION GAP METABOLIC ACIDOSIS

Non-anion gap metabolic acidosis conditions include Ureterostomy, Small bowel fistula, Extra chloride, **DIARRHEA** (most common cause), **Carbonic anhydrase inhibitors (acetazolamide), Adrenal insufficiency**, **Renal Tubular Acidosis**, and Parenteral nutrition/Pancreatic fistula/PosthypOcapnea. Look for **HYPERCHLOREMIA** and a LOW BICARB. Here are a few mechanisms to keep in mind:

* Extra chloride: Too much saline causes a hyperchloremic non-gap metabolic acidosis
* Diarrhea: Acidosis occurs from bicarbonate loss.
* RTA: Mechanism varies.
* Carbonic Anhydrase Inhibitor: Promotes renal bicarbonate loss.

**PEARL**: -CAPNEA refers to how much CO2 is in the blood and is NOT referring to how fast someone is breathing. So, hyp**O**capnea means there is **low CO2** in the blood, and this occurs due to **HypERventilation**. Therefore you get a RESPIRATORY ALKALOSIS and a compensatory nongap metabolic ACIDOSIS. Hyp**ER**capnea means there is **high CO2** in the blood, and this occurs due to **hypOventilation**. Therefore you get a RESPIRATORY ACIDOSIS and a compensatory metabolic ALKALOSIS. Make sure you look carefully at the terminology in the question stem or vignette.

**MNEMONIC**: "Acid-azolamide"

**MNEMONIC**: NON-GAP CRAP! Several stool related issues cause a non-gap acidosis (diarrhea, small bowel fistula, ostomies).

**MNEMONIC**: USED CARP (Ureterostomy, Small bowel fistula, Extra chloride, **DIARRHEA, Acetazolamide/Adrenal insufficiency**, **Renal Tubular Acidosis**, and Parenteral nutrition/Pancreatic fistula/Poshypercapnea.

**MNEMONIC**: If you do not know your normal values for routine chemistries, a non-gap acidosis can be tricky. If you just calculate the gap and see that it's normal, you could get in trouble. You should look closer at the chloride and bicarbonate levels. So, what I'm saying is that the chemistry looks good on the outside (i.e. gap looks good), but it's actually falling apart on the inside (low bicarb and high chloride), LIKE A **USED CARp**!

**MNEMONIC**: Building on the one above mnemonic, a USED CARp can only travel on smooth roads WITHOUT GAPS/potholes!

## RENAL TUBULAR ACIDOSIS (RTA)

All types of renal tubular acidosis (RTA) **a NORMAL ANION GAP** metabolic acidosis.

**PEARLS:** You don't need to know about RTA IV for the pediatric certification exam (it's not really seen in children).

## RENAL TUBULAR ACIDOSIS TYPE I (RTA I, aka CLASSIC DISTAL RTA)

The collecting duct is main site of hydrogen ion secretion. In renal tubular acidosis type I (RTA I, aka classic distal RTA) the **serum acidosis occurs due to hydrogen ions (H+) not being excreted** as NHCl. Therefore the **urine is more ALKALINE** than it should be because the **body is holding on to too much acid**. Look for a HIGH/BASIC urine **pH of > 5.5** and LOW urine NHCl. Since chloride is being kept in the body, there will be a HIGH serum chloride level (**hyperchloremic nongap metabolic acidosis**). Also look for **HYPOKALEMIA**.

**PEARL**: If you see a NON-ANION GAP + a urine pH of > 5.5, PICK THIS! If you're only given the elevated urine pH, ASSUME there's a serum acidosis! RTA I can be founding Sjogrens, Rheumatoid Arthritis and Lupus (SLE). It can also lead to stones and therefore urinary tract obstruction! Keep all of these in mind as a possible answer choice. Lastly, keep this very **high on your differential for a child that that is having FTT**.

**MNEMONIC**: In case you get these mixed up, think of it this way... RTA I was the 1st one to be discovered so it works pretty much like the scientists who discovered it thought it would. The renal tubules are dysfunctional and not excreting acid. Therefore, the serum is ACIDOTIC while the urine is ALKALINE. Classic!

**MNEMONIC**: RTA I = 1+ = $H^{1+}$ = an $H^{1+}$ secretion defect.

## RENAL TUBULAR ACIDOSIS TYPE II (RTA II, aka PROXIMAL RTA)

Renal tubular acidosis type II (RTA II, aka proximal RTA) is caused by a **partial** BICARBONATE reabsorption defect. That means **some** of the bicarb that is supposed to be kept is actually **lost in the urine**. Bicarb is a base, so the loss of this base results in a **serum acidosis**. Here's the trick part, the urine is **not alkaline as you would expect**. Since NHCl (an acid) secretion is normal, and **some** of the base is still being reabsorbed, the urine is pH ends up being LOW. So look for a **urine pH of < 5.5**. In other words, enough bicarb is lost to create a serum acidosis, but enough NHCL/ACID waste is being normally excreted to create an acidic urine with pH < 5.5. Like RTA I, this one also has a **HYPOKALEMIA**. RTA II can be seen in multiple myeloma and paroxysmal nocturnal hemaglobinuria (these last 2 facts are likely low yield).

**MNEMONIC**: RTA II = 2 = Bi = BIcarb problem (partial)

**PEARL**: Fanconi syndrome can look like an RTA, but Fanconi syndrome is a GAP acidosis. Also, acetazolamide is a carbonic anhydrase Inhibitor which inhibits bicarb reabsorption. It also creates a metabolic acidosis ("ACID-azolamide"), but it **alkalinizes** the urine.

## RENAL TUBULAR ACIDOSIS TYPE IV (RTA IV)

Renal tubular acidosis type IV (RTA IV) is LOW YIELD. But, if you must know, look for a NONGAP acidosis and **HYPERKALEMIA**. Both RTA I & II have **HYPO**kalemia. NHCl secretion is intact, so urine pH is < 5.5. This is low yield because it's almost never seen in children. Can be see in hypoaldosteronism and diabetes mellitus.

**MNEMONIC**: If you take th IV in RTA IV and turn the V on it's side, you kind of get I<, which looks like a K and should remind you of ↑<+, hyperkalemia for RTA IV.

**MNEMONIC PEARL:** RTAs and diarrhea both cause a non-gap metabolic acidosis. If given the urine chloride value, it can help. Urine chloride is LOW in RTAs (< 15) and HIGH in diarrhea (> 15). If you understand the RTA section, you will know this is not correct, but the mnemonic to follow could help you remember this subtle point. **RTA** is at the level of the **kidney**. As one would imagine, the kidney can't get rid of H**CL** (or NH**Cl**), so

**urine Cl is LOW (<15)**. In diarrhea, the kidneys are just fine, so they can get rid of **HCL** so urine Cl is HIGH. In all likelihood, you will only come across RTA I on the exam, so you should be fine.

## METABOLIC ALKALOSIS

The best way to categorize a metabolic alkalosis is by whether or not it is a chloride responsive metabolic alkalosis, or chloride unresponsive metabolic alkalosis.

**PEARL**: If you see a HYPOKALEMIA in a patient with a metabolic alkalosis, FIX THE POTASSIUM FIRST! Otherwise it probably will not correct..

* **CHLORIDE RESPONSIVE**: Look for a urine chloride of < 15. If you see that, then assume that the body is STARVING for chloride and holding on to as much as possible. You need to **give normal saline**!

- VOMITING: Due to loss of HCL
- DEHYDRATION: Causes a contraction alkalosis through multiple mechanisms in the kidney in an effort to compensate. It's especially bad when you are already vomiting out HCL!
- DIURETICS: Occurs with loop diuretics and thiazides, but only if there is insufficient dietary chloride (like in a patient on NaCl/salt restriction). If a metabolic alkalosis is given in a patient on a diuretic, the patient will also likely have a concurrent hypokalemia.
    - **PEARLS**: **Loop diuretics** can block reabsorption of Na+, K+ and Cl- and cause **hypo**natremia, **hypo**kalemia and **hypo**chloremia. So they usually cause a hypochloremic hypokalemic metabolic alkalosis. BUT, furosemide can also cause hypernatremic dehydration (we won't go into why), hypercalciuria without hypocalcemia (which can result in stones/nephrocalcinosis) and hypomagnesemia.
    - **THIAZIDES (like hydrochlorothiazide or chlorthalidone):** Cause sodium and chloride losses. Can also cause hypokalemia. Bicarb is **not** lost.
- **PEARL**: Laxatives, Diuretics & Excessive Vomiting all cause **hypOkalemic hypOchloremic metabolic ALKALOSIS (KNOW THIS!)**

* **CHLORIDE UNRESPONSIVE**: Look for a urine chloride of > 15. If you see that, then assume that the body has plenty of chloride and has no interest in getting more (therefore, saline will not help).

- CUSHING'S SYNDROME: More to come
- HYPERALDOSTERONISM: More to come
- BARTTER, GITLEMAN & LIDDLE SYNDROMES: If a metabolic alkalosis is presented with a high urine chloride and hypokalemia, keep these syndromes in mind. If any of these *are* in your answer choices, keep the pearls/rules below in mind.
    - BARTTER SYNDROME (aka BARTTER'S SYNDROME): LOW blood pressure.
    - GITLEMAN SYNDROME (aka GITLEMAN'S SYNDROME): NORMAL blood pressure.
    - LIDDLE SYNDROME (aka LIDDLE'S SYNDROME): HIGH blood pressure because of an overactive renin-angiotensin-aldosterone system.
    - **MNEMONIC**: The blood pressure follows the alphabet for these! BGL, B – G – L = LOW – NORMAL – HIGH.

## RESPIRATORY ACIDOSIS

A respiratory acidosis can be caused by anything that causes bradypnea (a decrease in respiratory rate) or a defect in ventilation perfusion exchange of carbon dioxide. The list includes, but is not limited to, several neurologic and neuromuscular disorders, late asthma (obstruction) and pulmonary emboli.

**(DOUBLE TAKE) PEARL:** If given an ABG and you note a respiratory acidosis, ALWAYS calculate for compensation. You do this by looking at the increase in PCO2 from the baseline of 40, and checking to see if the compensatory increase in bicarbonate (compensatory metabolic alkalosis via bicarb retention) is appropriate. If the bicarbonate level is lower than expected, then you have an additional primary metabolic acidosis. If it's higher than expected, then there is an additional primary metabolic alkalosis. Keep in mind that the compensatory increase in bicarbonate could be 1 per PCO2 increase of 10, or 3.5 per PCO2 increase of 10. That depends on whether or not you are dealing with an acute or a chronic respiratory alkalosis.

## RESPIRATORY ALKALOSIS

A respiratory alkalosis can be caused by basically anything that causes tachypnea. Sometimes this is due to hypoxia. The list includes, but is not limited to, early asthma, pneumonia, one of the aspirin phases, high altitude, fever and anxiety/hyperventilation.

**PEARL**: If given an ABG and you note a respiratory alkalosis, ALWAYS calculate for compensation. You do this by looking at the decrease in PCO2 from the baseline of 40, and checking to see if the compensatory decrease in bicarbonate (compensatory metabolic acidosis via bicarb excretion) is appropriate. If the bicarbonate level is even lower, then you have an additional primary metabolic acidosis. If it's higher than expected, then there is an additional primary metabolic alkalosis. Keep in mind that the compensatory decrease in bicarbonate could be 2 per CO2 decrease of 10, or 5 per CO2 decrease of 10. That depends on whether or not you are dealing with an acute or a chronic respiratory alkalosis.

# FLUIDS & ELECTROLYTES

## *MAINTENANCE IV FLUIDS (MIVF) & DEHYDRATION*

### MAINTENANCE IV FLUIDS (MIVF)

There are multiple formulas that help to estimate the maintenance IV fluid (MIVF) requirement in a child. Become familiar with one and expect to use it on the pediatric boards because you are guaranteed to need this on the exam. Plus, the are quick and easy points!

* ONLINE HELP: http://netscut.templaro.com/fen/holliday_segar.html

* HOURLY IV FLUIDS: **4** ml/kg for the first 10 kg + **2** ml/kg for the second 10 kg + **1** ml/kg for each additional kg

- **MNEMONIC**: 4/2/1. 4 divided by 2 = 2. 2 divided by 2 = 1. Hence, 4/2/1.
- **SHORTCUT:** For patients > 40 kg, the hourly IV fluid rate = **40 + WEIGHT IN KG**. For a 40 kg patient, the hourly rate is 80 ml/hour. For a 50 kg patient, the hourly rate is 90 ml/hour using the formula above. Using the shortcut, 40 + wt in kg = 90! For an obese 100 kg adolescent with diabetes, the hourly rate is 140 ml/hour using the formula above. Using the shortcut, 40 + wt in kg = 140!

?- **DAILY FLUID REQUIREMENT**: 100 ml per day for each of the first 10 kg + 50 ml/day for each of the second 10 kg + 20 ml/day for each kg beyond that (e.g., a 15 kg patient needs 1250 ml/day).

- **PEARL**: FEEL FREE TO USE THE HOURLY IV FLUID RATE MULTIPLIED BY 24!
- **MNEMONIC**: **Use 100/50/25 instead**. It's easier to remember! 100/2 = 50. 50/2 = 25. Hence, 100/50/25! IT'S ALSO SIMILAR TO THE HOURLY RATE MNEMONIC AND IS THE SAME MNEMONIC, WITH THE SAME NUMBERS AS THAT OF CALORIC INTAKE (100/50/25)!

* RENAL FAILURE PATIENTS: (Likely low yield) Keep strict ins/outs and replace urine cc for cc with D5 0.2 NS, and give an additional 1/3rd of the calculated fluid requirement by weight (the latter is for insensible losses). Use a low sodium solution (D5 0.2 NS) because the total body sodium is normal and this is truly for maintenance fluid/water only.

### DEHYDRATION

If **mild-moderate dehydration**, a trial of oral replacement should be tried first if the patient can tolerate small amounts of fluid (preferably with something like Oral Rehydration Solution which has 2% glucose + 90 meq NaCl per liter). If admitting into the hospital, give MIVF + FLUID DEFICIT. Give ½ of the deficit over 8 hours and rest over the next 16 hours. If the patient is 10% or 15% dehydrated, give boluses of fluid first and then subtract the total amount from the boluses from the 8 hour fluids.

* FLUID DEFICIT IN LITERS = % DEHYDRATION x WT IN KG.

* 5% DEHYDRATION = Decreased tears

* 10% DEHYDRATION = Sunken eyes & fontanelle. Poor turgor (assume 10% if hypERnatremia), dry mucus membranes, elevated heart rate.

* 15% DEHYDRATION = Poor capillary refill + signs of **shock**

## GASTROENTERITIS

If the dehydration from gastroenteritis is mild - moderate (dry MM, tachycardic), ORS and a regular diet at home is a great choice as long as the child passes a fluid trial/challenge.

## HEAT STROKE

Heat stroke results in a very HIGH temperature (> 105) and hot/dry skin. There is end organ damage from cytokine and endotoxin release. Give ICE packs to the groin and axilla.

**PEARL**: HEAT EXHAUSTION does not present with a very high temperature and the patient is sweaty. Kind of like you are after a really long jog (exhausted and sweaty, but not febrile).

# *ELECTROLYTES*

## (DOUBLE TAKE) HYPERCALCEMIA

Hypercalcemia is defined as a calcium level > 12. Hypercalcemia shortens the ST segment, therefore it also shortens the QT interval. Polyuria occurs due to an osmotic diuresis. Patient's can also get nausea, vomiting and change in mentation and possible abdominal pain. Initial treatment for hypercalcemia should be IV HYDRATION. Possible etiologies include:

* FAMILIAL **HYPO**CALCI**URIC HYPER**CALC**EMIA**: The disorder is fairly benign. Nothing to do.
* WILLIAMS SYNDROME (aka William's Syndrome)
* VITAMIN D & VITAMIN A ingestion
* THIAZIDE DIURETICS
* SKELETAL DISORDERS: Including dysplasias, skeletal **immobilization and skeletal/body casting**. For hypercalcemia related to immobilization, treat with **IVF and loop diuretics**.
    - **PEARL: For hypercalemia due to immobilization, treat with IV fluids and loop diuretics**
* HYPERPARATHYROIDISM
* MALIGNANCY
* **MNEMONIC:** "Bones, stones, abdominal groans and psychiatric moans." This refers to the classic mnemonic used to describe the complications of hyperparathyroidism.
    - BONES: Osteoporosis, osteomalacia, pathologic fractures, osteitis fibrosis cystica (aka Von Recklinghausen's disease)
    - STONES: Nephrolithiasis, nephrocalcinosis and nephrogenic diabetes insipidus.
    - ABDOMINAL GROANS: Nausea, vomiting and constipation.
    - PSYCHIATRIC MOANS (or OVERTONES): Coma, delirium, depression, fatigue and psychosis.

## (DOUBLE TAKE) HYPOCALCEMIA

For **hypo**calcemia, look for a calcium level < 8.5 or, or an ionized calcium level < 4.5. **Hypo**calcemia is much less common than hypercalcemia, but has much greater morbidity associated with it and is thus tested more often on the pediatric boards. Can range from minimal symptoms due to mild hypocalcemia from hyperventilation to HORRIBLE outcomes. Symptoms of hypocalcemia include: **Paresthesias, Tetany**, Carpopedal Spasm/Tetany (also known as **Trousseau's sign**), **Chvostek's sign** (abnormal reaction/tetany of the facial nerve when tapped), Seizures that do not respond to benzodiazepines, Laryngospasm (resulting in tachypnea. Can look/sound like Croup), **prolonged QT**.

* **MNEMONIC**: Calcium gluconate is **given** to stabilize the most important **muscle** in the body, the HEART! If you can remember that a low calcium results in numerous "muscle symptoms/signs," that should help you on the exam.

* **MNEMONIC**: Most of us have had an anxious patient in the emergency room who complains of transient numbness and tingling in their hands and/or feet... it's due to transient hypocalcemia from hyperventilation. If you can remember this association, you can remember that all of these tetany-like symptoms are related to hypOcaclemia!

* **MNEMONIC**: The **CH**vostek Sign (aka **CH**vostek's Sign) has to do with tapping the **Cheek** and looking for a twitch. **CH**eeky **CH**vostek sign! If all else failed with an anxious and hyperventilating patient, wouldn't you want to slap their cheek?

* **MNEMONIC**: Carpopedal spasm due to hypocalemia = Trousseau's sign = "TROUSER'S" Sign = Think of it affecting the extremities!

* **EARLY HYPOCALCEMIA**: Refers to within the first 3 days of life. Your differential should include asphyxia, **IDM** (Infant of a Diabetic Mother), **maternal** hyperparathyroidism and IUGR.

* **LATE HYPOCALCEMIA**: Refers to the after 7 days, or during the 2nd week of life or later.

- DIGEORGE/22Q11 DELETION
- VITAMIN D DEFICIENCY
- HYPOPARATHYROIDISM: Causes low calcium and high phosphorus
- PSEUDOHYPOPARATHYROIDISM: Autosomal dominant disorder in which receptors ar resistant to PTH. Serum PTH will be high. Calcium is low, phosphorus is high. Also look for clinodactyly (short/stubby 4th fingers/toes), developmental delay and moon facies.
- HYPERPHOSPHATEMIA
- RENAL FAILURE: Can't hydroxylate Vitamin D to 1,25, so less calcium is absorbed.
- NEPHROTIC SYNDROME: The low albumin causes the measured calcium to look low. Get an IONIZED calcium to confirm. Or, calculate the corrected calcium by using the shortcut of adding 0.8 to the calcium level for every drop in albumin of 1 g/dL.
- HYPOMAGNESEMIA: Be sure to CORRECT the magnesium!
- ALKALOSIS: Shifts ionized calcium to the protein-bound form, resulting in less of the active form.
    - **HYPERVENTILATION:** Causes a transient respiratory ALKALOSIS, hypOcalcemia and paresthesias!
- RHABDOMYOLYSIS: Initially there can be hypocalcemia. During the later recovery phase, there can actually be hypercalcemia. Calcium deposits into the muscles (let's assume that's the reason for the hypocalcemia).
- ETHYLENE GLYCOL: Look for mention of calcium oxalate crystals in the urine.
    - **MNEMONIC**: Just assume that the calcium lost in the urine as calcium oxalate crystal causes the hypocalcemia.

- (DOUBLE TAKE) CALCIUM O**X**ALATE CRYSTALS: There are two types of calcium oxalate crystals. One is square and looks like it has an X on it. That one very common. Calcium oxalate stones are the most common ones in humans. and the other is elongated and looks like a rod. The long one is the one usually associated with **ethylene glycol poisoning**.
    - **IMAGE**: http://meded.ucsd.edu/isp/1994/im-quiz/images/caloxal.jpg
        - **MNEMONIC**: calcium o**X**alate crystals look like a big X on a crystal.
    - **IMAGE**: http://alturl.com/mg3xv (This one is related to ethylene glycol and kind of looks like a long can of beer, or needles).

## HYPOKALEMIA

Hypokalemia can result in **constipation**, lethargy and weakness. May be caused by losses of potassium in stool or emesis. If there is **hypomagnesemia**, REPLACE THE MAG!

* **PEARL**: If you see a patient with hypokalemia **and** hypomagnesemia, have a high suspicion for renal dysfunction, such as acute tubular necrosis due to **cisplatin**/platinum drugs.

* LOW YIELD/GUNNER FACTS: Can result in T-wave flattening, U waves, ST segment depression and PVCs.

## HYPERKALEMIA

Hyperkalemia is **much** more serious than hypokalemia and can potentially cause SUDDEN DEATH due to arrhythmias. Look for **peaked T-waves** on an EKG. Treat with calcium (calcium chloride or calcium gluconate) to immediately limit the chances of lethal arrhythmias. Also give either a loop diuretic (furosemide, bumetadine) and/or sodium polystyrene (oral or rectally). As a temporizing measure only, you can give insulin to shove the potassium into cells, but DON'T forget to also give glucose to prevent hypoglycemia. If all else fails, or the patient has extreme symptoms (weakness, arrhythmias), dialyze. Potential causes are listed below. Management is more important.

* **RHABDOMYOLYSIS**: Look for elevated CK level, POSITIVE BLOOD on the urine dipstick with NO or few RBCs on urine microscopy. It's actually MYOGLOBIN that gives the false positive. As cells lyse, intracellular potassium is released and **hyper**kalemia ensues. Calcium deposits into muscles, and assume that's why you can also get **hypo**calcemia. Treat with aggressive hydration and possibly sodium bicarbonate to alkalanize the urine.

* **RENAL FAILURE**: Dialyze

* **POTASSIUM SPARING DIURETICS**: Includes spironolactone and triamterene. Never give with an ace inhibitor. Low yield topic.

* **ALDOSTERONE DEFICIENCY**: This can be a primary disorder, or due to **AD**DISONS **D**ISEASE. Aldosterone helps retain sodium and excrete potassium. When absent, the classic picture of HYPERKALEMIA + HYPONATREMIA ocurrs. Any time you see that combination on a chemistry panel, have a HIGH suscpicion that there is an aldosterone problem.

* **METABOLIC ACIDOSIS**: Assume it's because the extra hydrogen ions are going into the cells and potassium is coming out to maintain neutrality.

* **BURNS**: From cell lysis.

* **MASSIVE BLOOD TRANSFUSIONS**: From cell lysis.

* **RTA IV:** Low yield

## HYPONATREMIA

Hyponatremia is MUCH more common than **hyper**natremia. Look for a sodium level less than 130 mEq/L. If the level is < 120, watch out for **seizures**! Seizures can occur as free water starts moving into CNS cells at about 125 meq/L. For babies, remember not to give free water until at least 6 months of age. That's when kidneys can handle the job of getting rid of excess water. Regarding diagnosis of **hypo**natremia, always calculate the fractional **excretion** of sodium (F<u>e</u>Na) if given the labs to do so. Also look for serum and urine osmolality values. Regarding correction of sodium, remember that rapid correction (faster than 12 meq/day) can result in central pontine myelinolysis.

* <u>**PEARL**</u>: **FeNa+ = [($U_{Na}$/$U_{Cr}$) / ($P_{Na}$/$P_{Cr}$)** = How much Na+ is **EXCRETED** as compared to reabsorbed. So **a HIGH FeNa = HIGH excretion of Na+ into the urine!**

* <u>**PEARL**</u>: It's possible that you will only be given a urine sodium level. For the exam, **use a urinary sodium of 20 as your cutoff** for the LOWER LIMIT of normal. So if a urine sodium level is < 20, the kidneys are either holding on to sodium tightly, or they are not holding on to water tight enough. If you get confused, it's probably in your best interest to assume that a low urine sodium reflects low/decreased sodium excretion (i.e., a low FeNA).

* <u>**PEARL**</u>: **Formula to correct = (desired Na – measured Na) x Weight in kg x 0.6.** Then **ADD 3 meq/kg** as the daily **maintenance** amount of sodium needed. This will give you the total amount of sodium needed for the next 24 hours to reach the **desired sodium** level you input into the equation, which should never be greater than 12 above the measured sodium.

- **SHORTCUT:** If the desired change in sodium is 12 meq over 24 hours, then 12 x 0.6 = 7.2, and a simpler formula for the correction of sodium = (7 x Weight in kg) + (3 meq x Weight in kg).

* DEHYDRATION:

- The dehydrated patient's blood is concentrated, so expect a **HIGH serum osmolality**.
- The Kidneys are working to retain sodium/water, so expect a **LOW FeNa**. This is a KEY POINT. Do not assume that a concentrated urine means an elevated urine sodium.
- The urine will be concentrated, so expect a **HIGH urine osmolality**.

* DIURETICS: May vary depending on the diuretic, but in general, if a diuretic is at fault for hyponatremia, that means it's forcing sodium out. Expect a **HIGH FeNa** and probably a **LOW urine osmolality** since free water is being forced out as well.

* ACUTE TUBULAR NECROSIS & RENAL FAILURE: Expect a **HIGH FeNa** because the kidneys are not able to do their job of holding on to sodium.

* GASTROENTERITIS & DIARRHEA: This usually does NOT cause a hyponatremia **until** the patient is only given (or is only tolerating) **hypotonic liquids** (like free water). The patient is volume depleted due to GI losses, so the kidneys hold on to as much sodium as they can. Therefore, the **FeNa and the urine Na+ are LOW** (very low, often < 10).

* PSYCHOGENIC POLYDIPSIA: Refers to patients who just tend to drink a lot of water on a fairly chronic basis. Serum becomes dilute, and the kidneys are excreting as much water (using sodium to do it) as

possible. So look for the serum osmolality **and** urine osmolality to be low. Since the kidneys are pushing water out with sodium, the **FeNa should be HIGH**.

* ACUTE WATER INTOXICATION: This refers to drink lots of water over a **very short time**. Look for a healthy, afebrile child having change in mentation or a seizure after being around a body of water (tub/pool). Think of it as a VERY acute issue, so total body sodium is normal. This is a **hyper**volemic hyponatremia, though the child will not be edematous and will **look** euvolemic. Urine sodium will again be high since there's no drive to retain it.

* NEPHROTIC SYNDROME: Patients are edematous due to 3rd spacing from hypoalbuminemia. The total body water (TBW) is normal, BUT the intravascular department is depleted. Patients can get hyponatremia. Assume it's the same as being dehydrated. **FeNa and urine sodium are LOW** (very low, often < 10).

* SYNDROME OF INAPPROPRIATE ADH SECRETION (SIADH): Excess ADH results in excess water retention and an INCREASED total body water (and sometimes BP). Urine output may be decreased to < 1 cc/kg/day. There is serum **hypo**smolality (<300 osms) and **hypo**natremia because of retained water. Urine is concentrated/**hyper**osmolar (>300 osms). **Urine sodium is > 25 (normal to high)**, so the **FeNa is high.** Possible etiologies of SIADH include brain tumor/bleed, pneumonia or other lung disease, cancer, infection, recent surgery and Guillain Barre. It's a diagnosis of exclusion, so other etiologies of hyponatremia should be ruled out first. Treat with fluid RESTRICTION. Use Demeclocycline (diuretic) or Fludrocortisone if fluid restriction is not working and you note that the sodium level is persistently < 120.

- **MNEMONIC**: Assume the FeNa is high in SIADH because there is no drive to hold on to sodium given the hypervolemic state. Part of the kidney must hold on to water due to the ADH, but there is no problem with sodium excretion, so that part of the kidney is trying to EXCRETING sodium in an effort to excrete water.

* PSEUDOHYPONATREMIA: Total body sodium is normal, but elevated triglycerides or proteins cause an increased oncotic pressure which pulls excess water into the intravascular space. Also, triglycerides and proteins are big, and sodium is measured as the amount of sodium per overall volume (not per volume of water). This could be seen in a nephritic with hyperlipidemia.

* DIABETIC KETOACIDOSIS (DkA): Hyperglycemia results in increased osmolality, and thus increased oncotic pressure. This leads to water being drawn in and causing a hyponatremia. No treatment is needed since the sodium level will correct as the glucose corrects with insulin. To check if the drop in sodium is appropriate, add 1.6 meq/L to the sodium level for every 100 mg/dL of glucose above 100 to see if the final sodium value is normal. If it's still low, then there's also something else going on. So for a glucose of 300, the sodium should've dropped by only 3.2.

* CEREBRAL SALT WASTING: This is **low yield** and complicated. Look for increased urine output, increased salt excretion (and thus an elevated FeNa and urine sodium) and hyponatremia. The key finding (which is confusing) is that the serum osmolality is **higher** than the urine osmolality.

## HYPERNATREMIA

Hypernatremia is much less common then hyponatremia. Correct at no more than 12 meq/L/day to avoid an intracranial hemorrhage due to fluid shifts resulting in the tearing of bridging blood vessels. Can also get pulmonary edema from fluid shifts. The most common causes of hypernatremia include diabetes insipidus

(DI), excessive sweating and increased intake. Hypernatremia can also occur due to diarrhea related dehydration when more water is lost than sodium. If a patient is noted to have **hyper**natremic dehydration, assume s/he has at least 10% dehydration.

## DIABETES INSIPIDUS (DI)

If a baby has diabetes insipidus (DI), the baby will instinctively avoid solutes/formula so he or she ends up with **FTT**. If the urine SG is > 1.008, this diagnosis is highly unlikely. GET IMAGING of the brain once the diagnosis is confirmed and obtain all of the pituitary hormone levels (FLATPiG).

* CENTRAL DIABETES INSIPIDUS: Insufficient ADH leads to constant water loss. Look for is **hyper**natremia **when the patient does not have access to free water**. Polyuria occurs, and the dilute urine has a **very low urine osmolality**. There is NO glycosuria in Diabetes Insipidus. Serum osmolality will be HIGH once the patient is not drinking enough free water. Treat with DDAVP/ADH by mouth or intranasally.

- **PEARL**: If patient is an infant, s/he is dependent on an adult to provide enough free water to **prevent HYPERNATREMIA**. The child cannot walk to the kitchen when s/he feels that strong sense of thirst, and thus can get hypernatremic during some of the longer breaks between feeds. A 10-year old child, though, has the ability to respond to thirst and obtain free water in the middle of the night. Polyuria (and **nocturia**) will exist in BOTH examples, but only the baby will get hypernatremic. The key is whether or not a patient has access to free water.

* NEPHROGENIC DIABETES INSIPIDUS: Usually **X-Linked** recessive, so look for a male patient. He will have evidence of DI (massive urine output) but will have NO response to DDAVP/ADH supplementation. Water is lost and the labs will show hypernatremia. Treat with **hydrochlorothiazide** and salt restriction

**MNEMONIC**: nephrogeni**X** diabetes insipidus = **X**-linked recessive. Imagine that the kidneys are replaced by two **X**'s which can no longer do their job of holding on to water.

# NEPHROLOGY

## *THE URINALYSIS*

### MICROSCOPIC HEMATURIA

Microscopic hematuria is defined as greater than **5 RBCs** per HPF. Unless there are over symptoms, DO NOT get too aggressive initially. Repeat the urinalysis in 2 weeks and if there is still a problem, then start a workup. Look for systemic symptoms and a **family history** of any renal disease. The first step for a question on the pediatric boards is to get a **URINE CALCIUM & URINE CREATININE**. If that option is not available, you can choose to obtain imaging (renal ultrasound), serum BUN, serum creatinine, coagulation studies, platelet count, ANA, anti-DS DNA, compliment levels or sedimentation rate (ESR) depending on the clinical picture. If the values of the urine calcium and urine creatinine are given, calculate a urine calcium to creatinine ratio on the random sample of urine. If it's > 0.25, do a **hypercalciuria** workup (see below). If the value is **< 0.25** on the random sample, get a renal **ultrasound** to look for structural anomalies.

* **PEARL**: On imaging, you would especially be looking for evidence of ureteropelvic junction (UPJ) obstruction (dilated calyces). This would have to be confirmed with a Mag 3/furosemide scan showing **delayed excretion** to make the diagnosis.

* **PEARLS:** CYSTOSCOPY IS RARELY NEEDED for hematuria in pediatrics. So don't pick it. Also, during any imaging if there is a structural issue noted in one kidney, check the other one too. Lastly, if you are given a vignette in which a track athelet has a positive urine dipstick for blood, there are only a few RBCs per HPF on microscopy, the patient has MYOGLOBINURIA.

* HYPERCALCI**URIA**: This is probably your answer for persistent hematuria in an otherwise healthy kid. Obtain a urine calcium to creatinine ratio. If the value is **> 0.25**, the patient has **hyper**calci**uria**. Can be a stand alone problem, or due to loop diuretics (furosemide). If the ratio is > 0.25, calculate the ratio **again** from a 24-hour urine collection to see if it's > 4.0. If so, **then** get a renal ultrasound to look for a stone because this CAN be associated with nephrolithiasis (symptoms may include dysuria and abdominal pain).

- **NAME ALERT/PEARL**: Hypercali**uria** is not the same thing as **familial** hypercalc**emic** **HYPO**calci**uria**. This is **not** associated with hematuria and is benign. The name says it all for this other disease. Look for a family history of it, hypercalc**EMIA** and HYPOcali**uria** (the urine calcium to creatinine ratio would be **less than** 0.25).

### PROTEINURIA

Proteinuria is often benign and due to **orthostasis** (meaning children get it when they have been upright for a while). May also get **transient proteinuria** from general medical illnesses and dehydration. If it's said to be 1+, just repeat in 2 weeks. If 2+ or greater, you HAVE to do a workup. First, order an AM void. If there is **NO PROTEIN** noted on the AM void, it's probably due to orthostasis (which CAN indeed cause 2+ proteinuria). If there is no proteinuria, you should order a creatinine next just to make sure the renal function is ok. **If there is proteinuria** on the AM void, obtain a **random** protein to creatinine ratio. If the ratio is **> 0.2, there is true renal disease** and the next step is a renal **biopsy**.

**PEARLS:** 24-HOUR URINE PROTEIN COLLECTION is no longer recommended. Just do a spot protein:creatinine ratio. Also, don't get your numbers confused. Urine **protein**:creatinine cut off is 0.2. Urine **calcium**:creatinine cut off is 0.25.

## WBC CASTS

WBC casts are associated with INTERSTITIAL NEPHRITIS and PYELONEPHRITIS.

## RBC CASTS

RBC casts are associated with GLOMERULONEPHRITIS

## URINARY CRYSTAL IDENTIFICATION

Urinary crystals may be identified by shape and color. The type of crystal noted can help you identify a disorder.

*** URIC ACID CRYSTALS: Seen in breast-fed babies and is harmless!** It can also be associated with rapid cell turnover and hyperuricemia (as seen in lymphomas).

- **PEARL: Look for a breast fed baby presenting with a pink spot in the diaper! Uric acid turns to a salmon or pink colored powder when it dries. It scares parents to heck, but only requires reassurance.**
- **IMAGE**: http://alturl.com/zjk4n
- **IMAGE**: http://newborns.stanford.edu/images/urates2.jpg

* CYSTINE CRYSTALS: CYSTEINURIA cause nephrolithiasis. Stones are made of cystine and cystine crystals can be seen on urine microscopy.

- **IMAGE**: http://meded.ucsd.edu/isp/1994/im-quiz/images/cystine.jpg
- **MNEMONIC**: "6-teinuria." The cystine crystals are hexagons.

* STRUVITE CRYSTALS: Crystals are associated with urinary tract infections from urease producing organisms that cause **staghorn calculi**.

- **IMAGE**: http://meded.ucsd.edu/isp/1994/im-quiz/images/struvit.jpg
- **MNEMONIC**: The crystals look like coffin lids. They also look like "U" in strUvite crystals.

* (DOUBLE TAKE) CALCIUM O**X**ALATE CRYSTALS: There are two types of calcium oxalate crystals. One is square and looks like it has an X on it. That one very common. Calcium oxalate stones are the most common ones in humans. and the other is elongated and looks like a rod. The long one is the one usually associated with **ethylene glycol poisoning**.

- **IMAGE**: http://meded.ucsd.edu/isp/1994/im-quiz/images/caloxal.jpg
    - **MNEMONIC**: calcium o**X**alate crystals look like a big X on a crystal.
- **IMAGE**: http://alturl.com/mg3xv (This one is related to ethylene glycol and kind of looks like a long can of beer).

# UROLOGY, OBSTRUCTIONS & MASSES

## URETEROPELVIC JUNCTION OBSTRUCTION (UPJ OBSTRUCTION)

Ureteropelvic junction obstruction (UPJ obstruction) is the most common cause of urinary obstruction in children. Often suggested on ultrasound by dilated renal calyces, pelvis or frank hydronephrosis. Children may be born with signs of **oligohydramnios**. After a **postnatal ultrasound** to confirm the findings, a **voiding cystourethrogram (VCUG)** should be done to rule out other anomalies and to look for reflux from the bladder into the renal pelvis due to posterior urethral valves (PUV). Once that's been ruled out, a **mag-3 furosemide renal scan** (aka renogram) is done to look for slow flow of urine from the renal pelvis into the ureter. That is the definitive test to make the diagnosis.

**PEARL**: The mag-3 furosemide obviously uses a diuretic, and should only be ordered in patients with UNILATERAL hydronephrosis. If the patient has BILATERAL hydronephrosis, do NOT order it because you will make the problem worse.

## VESICOURETERAL REFLUX (VUR)

Vesicoureteral reflux (VUR) is another cause of hydronephrosis. It can be diagnosed by a **voiding cystourethrogram (VCUG)** by showing reflux of contrast from the bladder into the **ureter** +/- renal calyx. Children are at risk for renal failure, hypertension and urinary tract infections. The goal of treatment is to prevent hypertension and renal failure. Once diagnosed, patients should be given antibiotic prophylaxis to prevent UTIs (use amoxicillin or trimethoprim). **Grade 1-2 VUR usually resolves spontaneously within 5 years**. Grade 3-5 VUR requires surgical correction if there is no improvement after a year.

* LOW YIELD FACT: Deflux is a minimally invasive procedure that is done at the time of diagnosis. If after a year there is no improvement, THEN the true surgical correction is done.

## POSTERIOR URETHRAL VALVES (PUV)

When posterior urethral valves (PUV) are present, they are located in the **urethra** and face the wrong way. This causes **bilateral** hydronephrosis. Renal ultrasound shows decreased renal parenchyma. Needs a STAT urologic consultation for surgical correction. It's a bad disease and most children will still go on to having ESRD within FIVE YEARS.

**PEARL**: This is ONLY IN MALES. Also, remember to think about terminology. This is a urETHRal problem and has nothing to do with the urETERs. Look for a newborn MALE with a midline abdominal mass. If you're luck, they just tell you that the bladder is distended and that the child dribbles a weak stream of urine.

* **MNEMONIC**: Girls have a tiny urethra. Boys have a penis. The longer urethra with multiple PUVs would obviously be the one that is symptomatic.

**PEARL**: This is associated with **PRUNE BELLY SYNDROME.**

* (DOUBLE TAKE) PRUNE BELLY SYNDROME: Prune belly syndrome results in weak abdominal musculature (due to deficient abdominal muscles), cryptorchidism (**missing** testicles), undescended testicles and **urinary tract anomalies.** The urinary tract anomalies include vesico**ureter**al reflux (VUR), posterior urethral valves (PUV) and renal dysplasia. The renal anomalies often result in **hydronephrosis,**

**oligo**hydramnios and pulmonary hypoplasia. Children have a poor prognosis and most unfortunately die within a few months.

- **PEARL**: Look for a description of a **newborn** with weak abdominal muscles, "undescended testes" (may actually be missing) and a midline abdominal mass (distended bladder or hydronephrosis).

## NARROW FEMALE URETHRA

If a "narrow female urethra" is presented on the exam, there is no need to workup or intervene.

## ABDOMINAL MASS AT BIRTH

If a newborn is noted to have an abdominal mass at birth, it's probably due to a **renal mass**. Get a renal ultrasound and look for evidence of **hydronephrosis** or and **medullary cystic kidney disease** (MCKD).
**PEARL**: If they say **MIDLINE** lower abdominal mass, they mean the BLADDER!

## MEDULLARY CYSTIC KIDNEY DISEASE (MCKD)

Medullary cystic kidney disease (MCKD) is usually a unilateral disease and the affected kidney is essentially considered to be nonfunctioning kidney. The **contralateral** kidney has more than a 25% chance of having vesico**ureter**al reflux (VUR), so once the MCKD diagnosis is made, start antibiotic prophylaxis FIRST and then move on to getting the **VCUG** to look for VUR. The workup is similar to hydronephrosis and you do have to eventually also do the mag-3 furosemide scan because there's about a 50/50 chance that the child has at least some type of issue in the "good" kidney. They could have **ureter**opelvic junction obstruction (UPJ obstruction), vesico**ureter**al reflux (VUR) or posterior **urethral** valves (PUV).

## URETEROCELE

Ureterocele is the most common cause of urinary retention in girls. May be associated with abdominal pain, dysuria and/or hematuria (could sound similar to a UTI!). Look for a filling defect on the IV PYELOGRAM (IVP). The filling defect is where the mass is located.
**IMAGE**: http://bit.ly/Wu6kva
**IMAGE**: http://bit.ly/10RyS6L
**IMAGE**: http://www.meddean.luc.edu/lumen/MedEd/urology/vcurtcl1.jpg

# *INFECTIONS*

## URINARY TRACT INFECTION (UTI or PYELONEPHRITIS)

If a urinary tract infection (UTI or pyelonephritis) is diagnosed and the urinalysis shows POSITIVE NITRITES, this is virtually diagnostic of a GRAM NEGATIVE organism (E. coli, Klebsiella pneumoniae & Proteus). If nitrites are negative, it could be Gram positive but (Enterococcus or Staph saprophyticus). For most UTIs, outpatient therapy. If the patient is not septic and can tolerate orals, give **trimethoprim or a 3rd generation cephalosporin (cefixime, cefdinir)**. May also use **amoxicillin-cavulonate (but NOT amoxicillin alone**!). Early treatment is key to prevent renal damage. If the patient has nausea and vomiting from pyelonephritis, they cannot tolerate outpatient therapy and should be hospitalized. Inpatient treatment should be with **IV ceftriaxone**, IV cefotaxime or IV ampicillin and gentamycin.

* FEBRILE UTI: Any child with a **febrile** UTI should get a renal ultrasound to look for any glaring structural problems (cysts, hydronephrosis, etc.). Next, the child would get a VCUG (**at least a week after UTI resolution and when a repeat urine culture is negative**).

- **PEARL**: Febrile UTI in a child is associated with a VERY HIGH rate of vesicoureteral reflux (VUR).

* RECURRENT URINARY TRACT INFECTIONS (RECURRENT UTI: Do the workup above, and if there is no abnormal finding, try prophylactic antibiotics (amoxicillin or trimethoprim or ok).

## (DOUBLE TAKE) POST STREPTOCOCCAL GLOMERULONEPHRITIS (PSGN, aka POST INFECTIOUS GLOMERULONEPHRITIS)

Post streptococcal glomerulonephritis (PSGN, aka post infectious glomerulonephritis) is a Group A Streptococcus (GAS) syndrome. Look for **HEMATURIA + PROTEINURIA +/- swelling +/- HTN**. There may be a history of a skin infection 1 month before, or a throat infection 1-2 weeks earlier. Labs will show **a low C3 and NORMAL C4 +/- renal impairment**. Biopsy will show LUMPY BUMPY IgG deposits (biopsy is not required to make the diagnosis). Treat with IVF and a loop diuretic if there's HTN. Generally has a GOOD prognosis, but may consider steroids or cyclophosphamide for cases that are not improving. C3 should be followed until it's back to normal.

**PEARL: Antibiotics given for a Group A Strep infection can prevent rheumatic fever, but they CANNOT prevent PSGN. Also, PSGN can occur from pharyngitis OR skin infections, but rheumatic fever only occurs after a PHARYNGITIS.**

**PEARL**: If the C3 level does not come back down to normal after 6 weeks, **consider a different diagnosis**, such as **MPGN (membranoproliferative glomerulonephritis has a low C3 and C4)** or lupus nephritis (+ANA, low C3 **and C4**). If the renal impairment is severe, obtain a biopsy to look for rapidly progressive glomerulonephritis (RPGN).

## HEMOLYTIC UREMIC SYNDROME (HUS)

Hemolytic uremic syndrome (HUS) can be associated with DIARRHEA caused by enterohemorrhagic E. coli (EHEC). Symptoms include fever, **hemolytic anemia**, thrombocytopenia, renal failure and neurologic changes are common. Look for schistocytes, burr cells or helmet cells, **uremia + thrombocytopenia + purpura or ecchymoses (called TTP-HUS)**, hematuria, oliguria, elevated creatinine, **altered mentation,** seizures, lethargy, coma or hypertension (HTN). The hemolysis is **Coombs negative and the complement levels are normal**. Hemolysis is caused by verotoxin (Synsorb can be given to bind it – zero yield). The bug have been contracted from bad meat or milk ingestion.

**(DOUBLE TAKE) MNEMONIC**: The symptoms/findings for Hemolytic Uremic Syndrome (HUS) can be recalled by using the mnemonic FAT RN. Fever, Anemia, Thrombocytopenia, Renal Failure and Neurologic Symptoms.

# ***INTRINSIC RENAL DISEASE***

## RENAL FAILURE

Obviously, the creatinine goes up in renal failure since it's still being made but not excreted. If still making some urine, the urine sodium will be HIGH (> 25) due to renal salt wasting from dysfunctional tubules. Also, there is eventually decreased urine output with creates less of a drive for the kidneys to retain sodium given increased total body water. When there's decreased urine output, a patient that drinks water beyond the insensible losses will get hyponatremia.

**PEARLS:** Symptoms of renal failure may include FTT, a normocytic anemia due to decreased erythropoietin production, hypertension, vitamin D deficiency (and thus **hypo**calcemia and elevated phosphorus), secondary hyperparathyroidism due to hypocalcemia, neurologic changes from uremia and metabolic acidosis from bicarb losses and decreased acid excretion.

**PEARLS:** For patients with end-stage renal failure (ESRD), always keep strict ins/outs and replace the measured urine output cc for cc with D5 0.2 NS, and give an additional 1/3 of the calculated daily fluid requirement by weight (the latter is for insensible losses). Use a low sodium solution (D5 0.2 NS) because the total body sodium is normal and this is truly for maintenance fluid/water only.

## OLIGURIA

A child with acute tubular necrosis and oliguria (voiding < 300-500 cc/day) has a poor prognosis.

## RENOVASCULAR DISEASE

If given the history of a female teen with new onset hypertension, FIBROMUSCULAR DYSPLASIA (a type of renovascular disease) is likely the diagnosis. She likely has RENAL ARTERY STENOSIS (RAS). Never give an ACE inhibitor to a patient with RAS.

## GLOMERULONEPHRITIS

In order to choose an answer that has the word glomerulo**nephritis** in it, make sure there is at some **hematuria**. Other findings can include an elevated BUN, proteinuria, oliguria and HTN.

## IGA NEPHROPATHY

IGA nephropathy results in recurrent episodes of hematuria. May be worse during or after a URI or GI infection. During an episode, urinalysis will show RBC casts and some ridiculously high number of RBCs. In between episodes there can be microscopic hematuria. Labs will show **absolutely no compliment issues**. Diagnose by obtaining a **renal biopsy** and noting **IgA and IgG mesangial deposits** (and C3). The condition is the most common cause of primary glomerulonephritis, but the **prognosis varies** and this **can be self-limited.**,

**PEARLS:** PROTEINURIA is a BAD prognostic sign. Do NOT order an IgA level because it can be high **OR** normal. Therefore a normal IgA level doesn't help you. Since this can come on after a URI, it can get confused with post-streptococcal glomerulonephritis (PSGN). Remember that PSGN has a low C3 level, and IgA nephropathy has **normal complement levels**.

## (DOUBLE TAKE) POST STREPTOCOCCAL GLOMERULONEPHRITIS (PSGN, aka POST INFECTIOUS GLOMERULONEPHRITIS)

Post streptococcal glomerulonephritis (PSGN, aka post infectious glomerulonephritis) is a Group A Streptococcus (GAS) syndrome. Look for **HEMATURIA + PROTEINURIA +/- swelling +/- HTN**. There may be a history of a skin infection 1 month before, or a throat infection 1-2 weeks earlier. Labs will show **a low C3 and NORMAL C4 +/- renal impairment**. Biopsy will show LUMPY BUMPY IgG deposits (biopsy is not required to make the diagnosis). Treat with IVF and a loop diuretic if there's HTN. Generally has a GOOD prognosis, but may consider steroids or cyclophosphamide for cases that are not improving. C3 should be followed until it's back to normal.

**PEARL**: **Antibiotics given for a Group A Strep infection can prevent rheumatic fever, but they CANNOT prevent PSGN. Also, PSGN can occur from pharyngitis OR skin infections, but rheumatic fever only occurs after a PHARYNGITIS.**

**PEARL**: If the C3 level does not come back down to normal after 6 weeks, **consider a different diagnosis**, such as **MPGN (membranoproliferative glomerulonephritis has a low C3 and C4)** or lupus nephritis (+ANA, low C3 **and C4**). If the renal impairment is severe, obtain a biopsy to look for rapidly progressive glomerulonephritis (RPGN).

## MEMBRANOPROLIFERATIVE GLOMERULONEPHRITIS (MPGN)

Membranoproliferative glomerulonephritis (MPGN) findings include hematuria, proteinuria and high blood pressure. The key lab findings include a low **C3 AND a low C4**. Confirmatory diagnosis is via biopsy showing TRAM TRACK lesions.

**MNEMONIC**: MEMBERS OF PARLIAMENT (MPGN) riding a TRAM on TRACKS leading to a circus LOOP OF FIRE (LOOP = Lupus since SLE patients can get MPGN. Low yield.)

## RAPIDLY PROGRESSIVE GLOMERULONEPHRITIS (RPGN)

Rapidly progressive glomerulonephritis (RPGN) is low yield. Glomerulonephritis findings/symptoms + **crescents** on biopsy. Give Immunosuppressives and plasmapharesis.

## NEPHROTIC SYNDROME

Nephrotic syndrome can present with generalized edema, periorbital edema or **genital** edema. The classic triad of symptoms is **E**dema + **P**roteinuria + **H**ypoalbuminemia +/- Hyperlipidema +/- HYPOnatremia +/- hematuria. There can also be hypercoagulability (assume it's due to the loss of plasmin in the urine). Patients lose albumin and other proteins in the urine and the resulting low serum albumin causes 3rd spacing of fluid and thus edema. The TBW might be normal, BUT the intravascular compartment is hypovolemic and patient's can get HYPOnatremia. Urine sodium is very LOW, as is the FeNa. Etiologies include **minimal change disease**, focal segmental glomerulonephrosis (FSGS), membranous nephropathy, congenital nephrotic syndrome and membranoproliferative glomerulonephritis (MPGN).

* **MINIMAL CHANGE NEPHROTIC SYNDROME (aka MINIMAL CHANGE DISEASE)**: Look for a child with **E**dema + **P**roteinuria + **H**ypoalbuminemia +/- Hyperlipidema +/- HYPOnatremia +/- hematuria. Usually found in toddler-aged or elementary school children. This is the **most common etiology of nephrotic syndrome**.

Symptoms may include abdominal pain, diarrhea and decreased urine output with a **NORMAL CREATININE**. Biopsy will show **loss of FOOT PROCESSES**. These children generally have a good prognosis and respond well to steroids (usually prednisone) +/- an ACE inhibitor.

- **PEARL**: Any time you are giving long term steroids, consider getting a PPD first.

* **FOCAL SEGMENTAL GLOMERULONEPHROSIS (FSGS)**: Look for a child with **E**dema + **P**roteinuria + **H**ypoalbuminemia +/- Hyperlipidema +/- HYPOnatremia +/- hematuria. As the name suggests, biopsy will show focal segments of some glomeruli (not all) that have scarring (sclerosis).

* **MEMBRANOUS NEPHROPATHY**: Look for **E**dema + **P**roteinuria + **H**ypoalbuminemia +/- Hyperlipidema +/- HYPOnatremia +/- hematuria in an adolescent. Biopsy will show **diffusely thickened capillary loops** or "Thickened Membranes in Membranous Nephropathy."

* **CONGENITAL NEPHROTIC SYNDROME**: Can present with **massive edema** (aka anasarca). Treat these children with a high protein diet, albumin infusions and possibly an ACE-inhibitor. As a last resort, these children may need a bilateral nephrectomy.

* **MEMBRANOPROLIFERATIVE GLOMERULONEPHRITIS (MPGN)**: Please see above.

- **PEARL**: Of the causes listed, ONLY MPGN gives low serum complement levels (C3 and C4).

* **MNEMONICS:** n**EPH**rotic syndrome = **E**dema + **P**roteinuria + **H**ypoalbuminemia. Regarding the etiologies, try renaming nephrotic syndrome to **M**e**Ph**roteinuric sy**M**dro**M**e. The **M's** are to help you remember **M**inimal change nephrotic syndrome, **M**embranous nephropathy and **M**PGN. "**Ph**roteinuric" is to remind you that you need to look for **P**roteinuria. Remember, in th E-P-H, the H stands for **hypoalbuminemia**, NOT hematuria. Unlike glomerulonephritis, you do not always see hematuria in nephrotic syndrome. Think "–itis" = inflammation = RBCs.

## MEDULLARY SPONGE DISEASE

Medullary sponge disease can present with recurrent renal stones and hematuria. Urinalysis is otherwise normal. It's a congenital disorder of cystic dilations in the collecting system. Besides the morbidity related to the stones, it's a fairly benign condition.

**MNEMONIC**: Good prognosis = Everyone loves soft and SPONGY things!

**NAME ALERT**: Name alert is for medullary **cystic** disease.

## MEDULLARY CYSTIC DISEASE

Medullary cystic disease may present as polydipsia, polyuria, enuresis, dilute urine (they have difficulty concentrating their urine), **retinal disease** (possibly retinitis pigmentosa), FTT with short stature, anemia, **metabolic acidosis with an EXTREMELY LOW BICARBONATE LEVEL, elevated BUN and elevated creatinine**. These children have a **poor prognosis** and will eventually need hemodialysis due to a slow progression to end-stage renal failure.

**NAME ALERT**: Name alert is for medullary **sponge** disease.

## (DOUBLE TAKE) FANCONI SYNDROME

For Fanconi **syndrome**, look for polydipsia, polyuria, failure to thrive (FTT) and a **ton** of abnormal labs. It all happens due to **cystine** deposits in the kidneys causing proximal tubule damage (called CYSTINOSIS). The renal problems result in losses of amino acids, proteins, electrolytes (sodium, potassium, bicarbonate and phosphorus) and glucose. The loss of bicarb results in a **GAP** metabolic acidosis. The loss of phosphorus can present as RICKETS. Diagnose by obtaining a leukocyte cystine level.

**PEARLS:** There are MANY possible disease and points of confusion with these diseases. In Fanconi Syndrome, there is loss of sodium and the serum sodium level is low to low-normal. In diabetes insipidus, there is a loss of WATER that can result in hypERnatremia if there's no access to free water. Renal Tubular Acidosis can also result in electrolyte abnormalities and failure to thrive (FTT), but Renal Tubular Acidosis results in a NONGAP acidosis! Lastly, between Fanconi Anemia, Fanconi Syndrome, Diamond-Blackfan and Shwachman-Diamond, there is tremendous room for confusion. Use mnemonics (ours or yours) to keep them all straight!

**NAME ALERT**: The name alert is for FANCONI **ANEMIA**.

## (DOUBLE TAKE) FANCONI ANEMIA

Fanconi **anemia** is an APLASTIC anemia that is MACROCYTIC. Patients are usually > 4 years of age. Look for multiple anomalies including **short stature, café au lait spots, renal abnormalities**, microcephaly, hypogonadism and upper limb/hand anomalies. Treatment is with a bone marrow transplant.

**MNEMONIC**: Imagine you are a private investigator. You enter a mass murderers home and see **A PLASTIC ceiling FAN**. It has four ORANGE PLASTIC (**aplastic**) blades that look like **CON**es. We'll call them **FAN CONES**. Each of these ceiling **FAN CONES** has body parts hidden inside them. You look up right as **HUGE BLOOD CELLS** start to drip out of them. You find a stepladder, look inside the FAN CONES and find a hidden GONAD, a SMALL HEAD, an ARM and a HAND.... Remember, this is a CEILING fan. The Blackfan mnemonic is a standing plastic black fan.

**NAME ALERT**: The name alert is for FANCONI **SYNDROME**.

## (DOUBLE TAKE) ALPORT SYNDROME (aka ALPORTS SYNDROME)

Alport syndrome (aka Alport's syndrome) is primarily an X-linked dominant disorder. The findings are more severe in males, and include RENAL disease (starts with hematuria and progresses to end-stage renal disease), bilateral sensorineural hearing loss and eye/vision problems. Females may be noted to have hematuria that an eventually progress to ESRD. To recap, boys have hearing, vision AND kidney problems. Girls only have renal issues.

**MNEMONIC**: At **Al's PORT**, **X** marks the spot where the boats have to dock. Consider adding a story about poor AL and the KIDNEY-SHAPED boat he crashed into his PORT because he had POOR VISION and POOR HEARING. It was so bad that he couldn't see or hear everyone on the dock yelling at him to slow down!

# STATISTICS

## STATISTICS OVERVIEW

**NOTE**: Statistics questions are quick and easy "give me" points on the exam if you know where to put in the numbers and how to use the formulas in this overview. Please MEMORIZE the table below. TP & TN = True Positives & True Negatives. FP & FN = False Positives & False Negatives. PPV & NPV = Positive Predictive Value & Negative Predictive Value.

| | **DISEASE (+)** | **DISEASE (-)** | |
|---|---|---|---|
| **TEST (+)** | TP: | FP: | **P**PV = TP/(TP+FP) = **P**roportional to Prevalence |
| **TEST (-)** | FN: | TN: | **N**PV = TN/(TN+FN) = **N**ot Proportional to Prevalence (Inversely Proportional to Prevelence) |
| | SE**N**SITIVITY = TP/(TP+FN) | S**P**ECIFICITY = TN/(TN+FP) | **THIS BLOCK = TOTAL # OF PATIENTS STUDIED = ROW TOTAL = COLUMN TOTAL** |

**PEARLS/MNEMONICS:** Note in the table above that everything you need to calculate is ALWAYS within either one column or one row. Also, note that the bolded letters go in a pattern of N-P-N-P (seNsitivity, sPecificity, Npv and Ppv). This should help you remember how to label the table.

## STATISTICS TERMINOLOGY

### SENSITIVITY = TP/(TP+FN)

A high sensitivity [TP/(TP+FN)] is needed for SCREENING tests in order to rule out the presence of disease in a healthy patient. These are tests that are being done to RULE OUT a disease. So, a screening test should have a high sensitivity, which will mean that a negative result rules out the disease in that patient.

**MNEMONIC**: sp**IN** & sn**OUT**. sn**OUT** should remind you that **S**e**N**sitivity (**SN**out) is related to ruling **OUT** a disease

### SPECIFICITY = TN/(TN+FP)

A high specificity [TN/(TN+FP)] is needed for CONFIRMATORY tests. These are tests that are done to RULE IN (aka CONFIRM) a disease. So, a confirmatory test should have a high sensitivity, which will mean that a positive result rules in the disease in that patient.

**MNEMONIC**: sp**IN** & sn**OUT**. sp**IN** should remind you that **SP**ecificity (**SP**in) is related to ruling **IN** a disease.

### POSITIVE PREDICTIVE VALUE = TP/(TP+FP)

The positive predictive value [TP/(TP+FP)] looks at all of the POSITIVE results from a test and predicts how likely it is that a positive result actually means a patient has the disease. For example, what are the chances that a patient with a positive flu test actually has the flu.

## NEGATIVE PREDICTIVE VALUE = TN/(TN+FN)

The negative predictive value [TN/(TN+FN)] looks at all of the NEGATIVE results from a test and predicts how likely it is that a negative result actually means a patient does not have the disease. For example, what are the chances that a patient with a negative flu test in fact does not have the flu.

## NULL HYPOTHESIS

The null hypothesis is stated at the beginning of a research study. It assumes that the outcome being studied is the result of pure chance. In other words, it assumes that there is no variable (environmental exposure, medication, etc) that could result in an increase in this outcome. For example, if studying the association between smoking and lung cancer, the null hypothesis would state that there is no increase in lung cancer in patients who smoke.

## P VALUE

The P value represents the chance that the null hypothesis was rejected in error. In other words, what is the chance that the null hypothesis was accidentally rejected? Or, what is the chance that the difference in outcomes of a test or exposure was simply due to chance?

## SIGNIFICANT RESULTS

Results of a study are considered significant if the P value is **< 0.05**. They are considered "highly significant" if the P value is < 0.01.

## TYPE I ERROR

Type I errors basically mean that a study claimed there was a significant difference when there in fact was not one.

**<u>MNEMONIC</u>**: type "i" error. i made the error and said this is the greatest test on earth when I shouldn't have. Or, i claimed fame for discovering an association between patients born in December and a higher incidence of cancer. Turns out "i" just read the data wrong and made a type "i" error.

## TYPE II ERROR

Type II errors basically mean that a study claimed there was NO significant difference in the results when there in fact **was** one.

**<u>MNEMONIC</u>**: TWO rhymes with YOU. In a type TWO error, YOU made the mistake of claiming that my test or medicine doesn't work, when in fact it does! Think of it as a competition between two drug companies.

## COHORT STUDY

Cohort studies identify one condition (e.g., DM) and look for development of another condition (e.g., neuropathy). Cohort studies can be can be prospective or retrospective. Prospective studies are always preferred.

# NEUROLOGY

## *NEUROLOGIC TESTS, PARALYSES & PALSIES*

### SOMATOSENSORY EVOKED POTENTIALS (SEP)

ONLY order a somatosensory evoked potentials (SEP) test if you are concerned about a **demyelinating process**.

**MNEMONIC**: S.E.P. = SEParation = Testing for SEParation of the myelin from the nerve!

### NERVE CONDUCTION VELOCITIES

ONLY order nerve conduction velocities testing when concerned about a **peripheral neuropathy**.

### ELECTROMYOGRAM (EMG)

An electromyogram (EMG) is used to test for abnormal MUSCLE activity (may be due to a neuropathy).

**PEARL**: Use this in patients suspected of having a muscular dystrophy.

### MAGNETIC RESONANCE IMAGING (MRI)

Magnetic resonance imaging (MRI) is great for detailed imaging of the spinal cord and brain. Will tell you pretty much everything you want to know, but it's expensive and takes longer.

### COMPUTER TOMOGRAPHY SCAN (CT SCAN)

A computer tomography scan (CT scan) is cheap and fast (good for the ER). Great if you're looking for an acute hemorrhage, evidence of increased intracranial pressure and sometimes even a brain tumor.

### SPINAL ULTRASOUND

When looking for neural tube defects, a spinal ultrasound is a great option in newborns. It can actually be done **up until a child is 6 months of age**.

### ERBS PALSY AND KLUMPKE PALSY

Erbs Palsy and Klumpke Palsy are both associated with birth trauma, BREECH delivery, caesarian sections, clavicle fractures and LGA births.

**PEARLS: Images** of the arm/hand deformities MAY look the same. The only way to discern which palsy the child has is to evaluate the patient for specific neurologic limitations. If the baby is able to grasp, it is an ERBS PALSY. If you note a claw hand deformity in a patient able to flex at the elbow, it is a KLUMPKE PALSY.

* **ERBS PALSY (aka ERB's PALSY):** Brachial plexus injury at C5-7 resulting in paralysis of the UPPER ARM. There is a "waiter's tip" configuration and **UNILATERAL DIAPHRAGMATIC PARALYSIS**.

- **IMAGE**: http://bit.ly/o4evAy

- PEARLS: The grasp and extension of the hand are INTACT. Respiratory distress can result due to phrenic nerve injury (look for a broken clavicle) resulting in unilateral diaphragmatic paralysis. Often associated with LGA babies, breech deliveries and C-sections.
- **MNEMONIC**: **Imagine** a WAITER named "ERB" subtly asking for a tip. He CAN grasp the money you give to him.

* **KLUMPKE PALSY**: Brachial plexus injury, but lower (C7-T1). This affects the LOWER arm and hand. Worse prognosis because the nerves are typically torn. Results in a CLAW HAND deformity in which there is an INABILITY TO GRASP.

- **MNEMONIC**: Hand is stuck in a configuration in which the hand has a KLUMP OF AIR in it.

## HORNERS SYNDROME (aka HORNER'S)

In Horners syndrome, a T1 lesion results in ptosis, miosis and anhydrosis. The affected eye has a droopy lid, a small pupil (anisocoria) and is dry.

**PEARL**: Associated with KLUMPKE PALSY.

**IMAGE**: http://upload.wikimedia.org/wikipedia/commons/3/37/Miosis.jpg

## SPASTIC CEREBRAL PALSY (CP)

Spastic cerebral palsy (CP) is a **motor impairment** due to brain lesions or anomalies. The condition that does NOT progress, but the motor component can change with time. The diagnosis is usually made by 1 year of age. **Intelligence can be fully intact**. The increased incidence is due to improved survival of preterm infants.

* SPASTIC HEMIPLEGIA: Arms are affected (good cognitive prognosis).

* SPASTIC DIPLEGIA: Legs are affected (great cognitive prognosis).

* SPASTIC QUADRIPLEGIA: All extremities are affected (horrible prognosis)

* **PEARL**: Though asphyxia is often thought to be the most common etiology for cerebral palsy, it's **NOT**. It's actually only responsible for a small number of cases. **Prematurity, IUGR and intrauterine infections** have a much higher association with cases of cerebral palsy.

## ATHETOID CEREBRAL PALSY

Athetoid cerebral palsy is like spastic CP, but there is also dystonia.

**MNEMONIC**: Ever heard of athetoid chorea? If so, that could help you remember the odd movements related to the dystonia in athetoid cerebral palsy.

## WEAKNESS & PARALYSIS PEARL

**This is extremely important! In any child that has a progressive weakness or paralysis, your first step should be to evaluate him/her for respiratory compromise. Pulse oximetry is useless. Look for tachypnea, accessory muscle use, paradoxical breathing (belly breathing) and shallow respirations to help you decide whether or not you need to choose, "prepare to intubate."**

## GUILLAIN-BARRE SYNDROME (GBS, aka ACUTE INFLAMMATORY DEMYELINATING POLYNEUROPATHY or AIDP)

Patients suffering from Guillain-Barre syndrome (GBS, aka acute inflammatory demyelinating polyneuropathy or AIDP) may initially complain of back pain, **fever** and can have a facial palsy and proximal muscle weakness (trouble rising from chair or shrugging shoulders) prior to lower extremity symptoms. Classically, though, it is an ascending paralysis over several **days to weeks** in which there is ataxia and then an inability to walk. Look for diminished or **absent reflexes in the lower extremities** on exam. **Sensation is preserved** (as is bowel and bladder continence). It can progress to respiratory compromise requiring intubation. Perform a lumbar puncture to look for **albuminocytologic dissociation (increased CSF protein in the absence increased WBCs)**. FYI... they could say there is an absence of pleocytosis (pleocytosis means an increase in WBC's). For treatment, you can try IVIG or plasmapheresis.

**PEARLS:** Steroids DO NOT help. Pulse oximetry is a poor indicator of neuromuscular respiratory insufficiency. You can, however, try to obtain a negative inspiratory flow (NIF) or a Forced Vital Capacity (FVC) if the child is old enough to participate with the test (at least 5 years of age). Always keep **tick paralysis** in your differential, especially if they mention the summer time, a recent vacation, or the woods! Additionally, if someone presents with GBS a few weeks after a diarrheal illness, they might be referring to C. jejuni infection (known antecedent to GBS though mechanism is not understood). Also, when compared to any **CORD COMPRESSION SYNDROME**, GBS maintains rectal tone, bowel/bladder continence and sensation. It also has **decreased** reflexes. In cord compression syndromes, sensation, tone and continence are lost, and reflexes are **increased**.

## (DOUBLE TAKE) TICK PARALYSIS

Tick paralysis can present with a child that has a clinical picture very **similar** to that of Guillain Bare Syndrome (GBS). That means you're looking for an acute **ascending** paralysis in which there are absent reflexes. It will likely be preceded by ataxia. It's caused by a neurotoxin produced by ticks, and the solution is to **remove the tick**!

**PEARL**: If you are presented with a child that seems to have Guillain Bare Syndrome (GBS), but they mention a trip, the woods or the summer time, they are probably talking about this! Other KEY differentiating factors include the **lack of fever, normal CSF and a quicker progression of symptoms (hours – 2 days).** Other confusing choices may be Botulism (DESCENDING paralysis), or PoliomyeLITIS (+CSF findings and possible viral syndrome).

**NAME ALERT**: This is not TODD'S PARALYSIS

## (DOUBLE TAKE) TODDS PARALYSIS (aka TODD'S PARALYSIS)

If a child presents with **focal motor weakness** after a seizure, pick TODDS PARALYSIS (aka Todd's Paralysis)!

**PEARL**: There is often NO HISTORY of a witnessed seizure so have a high index of suspicion for this diagnosis in any child presenting with an acute onset of unilateral weakness of an extremity.

**NAME ALERT**: This is not TICK PARALYSIS.

## TRANSVERSE MYELITIS

Transverse myelitis is an inflammation of the spinal cord and can look very much like a cord compression syndrome (**bowel/bladder dysfunction & increased reflexes).** May or may not be related to a chronic inflammatory condition. Look for a history of fever and back pain. These patients can **die** of respiratory compromise. CSF from a lumbar puncture may show **neutrophils**. Diagnose by noting a swollen an inflamed cross section (across the entire/**transverse** section) of spinal cord on an MRI.

## EPIDURAL ABSCESS OF THE SPINE

An epidural abscess of the spine presents with a history of fever, pain, lower extremity weak, numness and tingling (paresthesias) and eventually paralysis. Exam will show **increased reflexes**. Get a STAT MRI of the spine to look for the abscess with compression of the spinal cord. Start anti-staphylococcal antibiotics and call a neurosurgeon. You should also give **dexamethasone** for this, or any other, spinal cord compression syndrome.

**PEARL**: Some of the symptoms may seem similar to Guillain Barre Syndrome (GBS), but key differences include a **lack of sensation, poor rectal tone** and a **lack of bowel and bladder control** in epidural abscesses (and any other cord compression syndrome). In GBS, continence is preserved.

## MYASTHENIA GRAVIS (MG)

Myasthenia gravis (MG) findings will include either a baby with **variable ptosis** (not constant, occurring at different times), or an older child with evidence of muscle **weakness that waxes and wanes**. On exam, reflexes are diminished but **not absent**. Myasthenia Gravis is an autoimmune disease in which antibodies to postsynaptic acetylcholine receptors **block them from functioning**. Tests include the administration of **edrophonium** (aka **Tensilon test**) or **neo**stigmine (anticholinester**ase** inhibitors) which may transiently improve weakness, in those with ptosis. If required, treatment may include **pyrido**stigmine, steroids, plasmapheresis, or worst case scenario, a thymectomy to permanently get rid of the antibodies.

**PEARLS:** The tests are listed above, but DO NOT order them for a "best next step" question. If they ask how to diagnose, the answer will likely be edrophonium/Tensilon. They're dangerous and should be left up to a neurologist. Instead, your first step should be to **look for respiratory compromise**. Evaluate the patient for possible intubation since your certification exam patient is likely going to present at a time of progressive weakness! If they are in myasthenia crisis, they can present with tachypnea and shallow respirations.

**PEARL**: There is a **strong association with THYMOMA**. Always look for it in MG patients. So another "next step" answer could be to get quantitative immunoglobulins or imaging (CT, MRI or X-ray of the chest).

**PEARL**: Congenital myasthenia gravis is a lifelong problem and patients **will have ptosis**. Babies can also have a TRANSIENT myasthenia gravis due to maternal antibodies. This **does not** include ptosis and resolves after about 6 months.

## (DOUBLE TAKE) CLOSTRIDIUM BOTULISM

Clostridium botulism is a Gram positive organism. Look for a child **less than 6 months** of age with **progressive descending** weakness. Can progress from **progressive ptosis** and a poor suck to urinary retention **within hours**. The botulism toxin **inhibits release of acetylcholine** into the neurosynaptic junction. Treatment involves either supportive care (intubate if needed) or the antitoxin if it's available.

**PEARLS:** For the exam, **progressive** ptosis in a baby probably means Botulism! Children are usually less than 6 months of age since they are too young at that age to prevent colonization in gut.

### (DOUBLE TAKE) CORYNEBACTERIUM DIPHTHERIAE

Corynebacterium diphtheriae is a Gram positive rod that can cause Diphtheria. It starts off with a low fever and URI symptoms, including a sore throat. Eventually a **pseudomembrane forms on the tonsils and pharynx**. The swelling can be so bad that intubation is needed. It can also cause motor and sensory problems along with a loss of reflexes. The D in DTaP is for Diphtheria, so it's no longer seen in the U.S. but could be seen in a child of immigrant status. Treat with metronidazole or erythromycin.

**IMAGE**: http://bit.ly/117OOkz

**IMAGE**: http://bit.ly/SoKGtL

**IMAGE**: http://www.dipnet.org/img/mem.jpg

## *INCREASED INTRACRANIAL PRESSURE & HEADACHES*

### INCREASED INTRACRANIAL PRESSURE (ICP)

Signs of increased intracranial pressure (ICP) include splitting sutures, bulging fontanelle, regression of milestones, setting-sun sign, horizontal diplopia, papilledema, nausea, vomiting, anorexia, FTT. **systemic hypertension**, **tachypnea/hyperventilation** (or **irregular** respirations) and **brady**cardia on exam. Pupils may be **dilated** and not reactive (this can be **unilateral** or bilateral). Treatment may include mannitol, furosemide, hyperventilation, a VP shunt or treat the cause!

**PEARL**: If the patient has other symptoms consistent with increased ICP but no papilledema, it's probably still increased ICP! Papilledema is a LATE finding. Also, some of the classic symptoms of increased ICP were described by Cushing. **Cushing's triad** includes increased systolic blood pressure with a wide pulse pressure, bradycardia and abnormal respirations (irregular or tachypneic). ALWAYS circle the respiratory rate because it might be the only clue for conditions like increased ICP and acid-base disorders!

**MNEMONICS:** Increased ICP is also known as Intracranial Hypertension. The systemic increase in blood pressure is the body's way of trying to match the brains BP.

**IMAGE**: (Papilledema) http://bit.ly/117WnHV

**IMAGE**: (Sun Setting Sign) http://bit.ly/TjSbzM

### LUMBAR PUNCTURE

Go ahead and order a lumbar puncture for anyone in whom you suspect meningitis, UNLESS they have a focal neurologic deficit. Lumbar punctures are contraindicated if a brain tumor/mass is suspected.

### DANDY WALKER MALFORMATION

Dandy Walker malformation can present with delayed motor development, progressive enlargement of the skull, breathing problems and signs of increased intracranial pressure (nausea, vomiting, seizures). It's a

**P**osterior **F**ossa abnormality that involves the cerebellum and the fluid spaces around it. The **cerebellar vermis** is either partially or completely missing (that's the part usually between the 2 cerebellums). There is **increased fluid in the fourth ventricle**. It's associated with **P**HACES & **Congenital Melanocytic Nevi**.

* **(DOUBLE TAKE) PHACES SYNDROME**: Diagnosis requires a large hemangioma in the face/neck area PLUS one of the following defects.

- **P**ost Fossa malformation (DANDY WALKER)
- **H**emangioma. Often in the distribution of the Facial Nerve. Look for a large **segmental** hemangioma in on the **FACE**. Segmental refers to what looks like a nerve distribution (segmented by normal skin in between). Can be associated with STROKES.
- **A**rterial cerebrovascular anomaly
- **C**ardiac anomalies: Especially COARCTATION OF THE AORTA
- **E**ye anomalies: MICROPHTHALMIA, STRABISMUS
- **S**ternal defect
- **IMAGES**: http://bit.ly/nWAflM

* **MNEMONIC**: After the exam, hopefully you'll feel like you a DANDY WALK all over the ABP question writers' **PH**ACES! Also, **P**osterior **F**ossa = **PF** = which could kind of be used to spell **PFaces**.

## (DOUBLE TAKE) PSEUDOTUMOR CEREBRI (aka IDIOPATHIC INTRACRANIAL HYPERTENSION or BENIGN INTRACRANIAL HYPERTENSION)

The symptoms associated with **increased intracranial pressure** (ICP) predominate for Pseudotumor Cerebri (aka Idiopathic Intracranial Hypertension or Benign Intracranial Hypertension). These include headache, nausea, vision changes and bulging fontanelle. Late findings are papilledema and loss of vision. The disease is **not** benign, and although CT or MRI of the head will not show herniation, a lumbar puncture will show increased intracranial pressure. It can be associated with **excessive vitamin A**, isotretinoin, tetracycline and thyroxine. Treatment may require steroids and/or shunt.

**PEARL**: PSEUDO-tumor. Don't get confused by the word "tumor." This condition is NOT associated with a mass.

**MNEMONIC**: (image)

## TENSION HEADACHES

Tension headaches tend to be frontal or band-like headaches. It there is a component of depression, children may be sleeping more and/or having trouble at school. Use acetaminophen or ibuprofen.

## MIGRAINE HEADACHES

Migraine headaches are often difficult to diagnose in children, but classically these are severe unilateral headaches associated with nausea, vomiting, photophobia and/or phonophobia. They can also be associated with auras. First try analgesics like ibuprofen or acetaminophen. Also prescribe rest in a quite and dark room. If that doesn't work, try a triptan (like nasal sumatriptan).

**PEARL**: This is a diagnosis of exclusion. If there is any focal neurologic deficit, or any question in your mind about the diagnosis, do a complete neurologic workup, starting with a CT of the head.

## OMINOUS HEADACHES

If you are given the history of an ominous headache that is worse in the mornings or associated with emesis, have a high index of suspicion for something bad, like increased intracranial pressure.

# ***MOVEMENT DISORDERS***

## (DOUBLE TAKE) DYSTONIC REACTIONS

Dystonic reactions involve sudden muscle contractions and usually some **odd postures**. Can be related to clonidine, phenothiazine, metoclopramide, promethazine and athetoid cerebral palsy. Treat **medication-induced** dystonic reactions with **diphenhydramine**.

## TICS

Tics are specific movements that are seen in a child over and over again. Onset is usually **after 3 years of age**. Location of tic **CAN CHANGE**. Children often feel a strong **need to do them** and older kids feel ashamed. **Can be suppressed** temporarily and rarely occur with purposeful voluntary movements or activities. Has no severe impact on daily living. May be exacerbated by stress or excitement. Sometimes treated with tricyclic antidepressants.

## TOURETTE SYNDROME (aka TOURETTE'S SYNDROME)

Tourette syndrome (aka Tourette's syndrome) may be diagnosed with TWO motor tics + ONE vocal tic. Will likely **change over time**. Tics may increase in stressful situation as well as when AT EASE (when AT HOME). **Clonidine or guanfacine** may be used for tics that cause significant impairment or pain (**NOT SSRI's**). Attention deficit disorder (ADD) is the most common concomitant condition

**PEARLS: ADD medications can unmask a tic. If such a scenario is presented on the exam, continue the ADD medication! Know this!**

## STEREOTYPY

A stereotypy is a pattern of movement or behavior that a child **does not feel an urge to perform**, but they do sometimes **enjoy** them (hand flapping, vocalizing portions of a video, head rocking). Usually start **before 3 years of age** and remain **UNCHANGED**. **Can be suppressed** by older children and has no severe impact on daily living.

## CHOREA

Chorea consists refers to very random dance-like movements that suddenly occur. Age of onset depends on the cause, and these may or may not resolve depending on the cause. These **CANNOT be suppressed** and often **worsen** with focused activities. They are extremely **disrupting to the child's life**.

## SYDENHAMS CHOREA (aka SYDENHAM'S CHOREA)

Sydenhams chorea (aka Sydenham's chorea) is specifically the finding of **choreiform movements after a Streptococcal infection by Group A Streptococcus (GAS).** Unlike tics, these are WORSE with purposeful movements. Can be associated with hypotonia, jerky movements and even emotional lability after a GAS infection. CSF findings are normal. Improve during sleep. Can go along with rheumatic fever (look for carditis, polyarthritis, erythema marginatum, subcutaneous nodules, etc.). Order an antistreptolysin titer (ASO titer) to look for recent streptococcal infection.

**PEARLS:** A negative ASO dose NOT rule out a Group A Streptococcal (GAS) infection. The titer may have already come down if the chorea is presenting months after an infection. For the exam, Sydenham's Chorea = **RHEUMATIC FEVER**. In reality, you can get this after a GAS infection and not have all of the other issues related to rheumatic fever. Please know the rheumatic fever section very well (see the Cardiology chapter for a very detailed discussion on RF).

## HUNTINGTONS DISEASE (aka HUNTINGTON'S DISEASE)

Huntingtons Disease (aka Huntington's Disease) is an autosomal dominant disorder due to CAG repeats that can present at any age but usually does so in adulthood. Key findings include RIGIDITY, HUNT**ING**TON'S CHOREA and EMOTIONAL LABILITY in children. Adults may have dementia and **hypo**tonia.

**MNEMONIC**: Please see the Genetics section for the autosomal dominant mnemonic. Keep in mind that HUNT**ING**TON'S inheritance pattern (autosomal dominant) is different than HUNTER'S syndrome (X-linked).

# *DYSTROPHIES*

## SPINAL MUSCULAR DYSTROPHY TYPE I (aka WERDNIG HOFFMANN DISEASE)

Spinal muscular dystrophy type I (aka Werdnig Hoffmann disease) findings include **tongue fasciculations,** hypotonia and a poor suck. It's caused by the degeneration of the anterior horn of the spinal cord, and therefore **only disrupts motor function**. There are NO sensory deficits.

**PEARLS:** There are no tongue fasciculations in kids with Duchenne Muscular Dystrophy or Botulism. Also, these children have a **NORMAL CK** because this is a NERVE problem and **NOT** a muscle protein problem.

## (DOUBLE TAKE) DUCHENNE MUSCULAR DYSTROPHY

Duchenne muscular dystrophy is an X-Linked disorder (located at Xp21 – low yield) that causes a deficiency in the muscles' DYSTROPHIN protein. Children are usually wheelchair-bound by 7 or 8 years of age and are unable to walk by their 13th birthday. They can have severe scoliosis and the major cause of morbidity (and mortality) is the **involvement of respiratory muscles**. Children have a poor cough and frequent pneumonias. The heart can also be involved (cardiomyopathy). Other symptoms include toe walking, a waddling gait, use of the **Gower maneuver** to stand, **pseudohypertrophy of calf muscles** from fatty/fibrous tissue deposition, and poor head control. Start your workup with a **CK level (like Dermatomyositis)** since it's ALWAYS elevated. If high, refer to neurologist for further workup, including and electromyogram (EMG) and **muscle biopsy**.

**PEARLS:** Though it's an X-linked disorder there is a very **HIGH rate of spontaneous mutations**, so the mom does not have to be a carrier, and there does NOT have to be a positive family history. If the mom **is** a carrier, though, she **can** have elevated CK levels even though she has no symptoms.

**MNEMONIC**: C'mon. Have you **ever** seen a picture of a female patient doing the Gower maneuver? NO! So of course this is an X-linked disease. Now imaging a child going the maneuver while wearing a tu**X**.

**MNEMONIC VIDEO:** Imagine this kid in a tu**X**: http://www.youtube.com/watch?v=bl6utCce_3g

### MYOTONIC DYSTROPHY

Myotonic dystrophy is an autosomal dominant disorder that involves **distal weakness**, especially in the **hands**. Look for a history of **slow relaxation after contractions**. Patients can have endocrine and gastrointestinal problems. The **CK is usually normal**.

### CONGENITAL LIPODYSTROPHY

Babies with congenital lipodystrophy are born very thin and long. It's due to adipose tissue resistance to insulin.

## *SEIZURES*

### FIRST-TIME SEIZURE

If presented with a **non-focal (partial)** first-time seizure in a child who is **greater than 1 year of age** and **otherwise healthy**, no further workup is indicated as long as the seizure was less than 5 minutes. Discharge from the ER with the above mentioned in this section.

**PEARLS:** After a single seizure, the **chance of future epilepsy is > 30%**. If the child has any neurologic issues (hyperreflexia, cognitive impairments, etc.), the chances are much higher. Diagnosing a seizure is better done with a good **history** rather an EEG.

### EPILEPSY & SEIZURE PRECAUTIONS & EDUCATION

Seizure precautions and education should be given to the parents of children with epilepsy. During a seizure, parents should put the child on the floor on his/her side, time the seizure, and **call 911 if it lasts more than 5 min**. When epileptic children are on wheels (bicycle), they should wear a helmet. When near water, an adult should be watching. Let parents know that recurrent seizures rarely cause death or long-term brain damage, unless a patient goes into **status epilepticus** (continuous seizure activity for > 30 minutes). For teenagers, they need to have a seizure-free interval of at least 3 months before they can return to driving (some say 6 or even 12). If a patient has regular epilepsy (not related to any other neurologic disorder), medications can be stopped after 2 years of a seizure-free interval to see how the child does off of medication.

## EMERGENCY ROOM PEDIATRIC SEIZURE MANAGEMENT

The first step in pediatric seizure management is to check the glucose! If it's normal, then give IV **lorazepam**. Lorazepam is **preferred over diazepam** because it acts fast and has a longer half-life. If you don't have IV access, give rectal diazepam as the initial therapy. After that, use an even longer acting agent for treatment **if the seizure recurs**. Long-acting anti-seizure medications include fosphenytoin, valproic acid, phenobarbital and phenytoin.

## SEIZURE TERMINOLOGY

Below are simplified definitions of common seizure terminology.

* SIMPLE SEIZURE: The patient is **conscious**.

* COMPLEX SEIZURE: The patient is **unconscious**.

* PARTIAL SEIZURE: It's **focal** in location.

- **MNEMONIC**: **PART**ial = PART of the body.

* GENERALIZED SEIZURE: It's **non-focal** (generalized) in its location.

## SIMPLE PARTIAL SEIZURES

In simple partial seizures, patients are awake and conscious. These often include motor activity of the arms or face and frequently occur as a child is going to, or coming out of, sleep. These can also be sensory in nature and might "march," or go, from one location to another part of the body. Treatment is not always offered since these often resolve in adolescence.

* **JUVENILE MYOCLONIC EPILEPSY/SEIZURES**: Patients are awake and have severe focal seizures. History will include the contraction **or** the loss of tone in an isolated muscle group. The muscle group will cause jerking movements, as though the patient is being shocked. Sometimes children are noted to be fall in a particular direction (forward or backward). These kids get lifelong valproic acid.

## COMPLEX PARTIAL SEIZURES

In complex partial seizures, patients are unconscious and have a focal seizure. These occur more often during the day.

**MNEMONIC**: It kind of sucks that the complex seizure which makes you unconscious happens to occur when you're not near your bed!

## BENIGN ROLANDIC SEIZURES (aka BENIGN EPILEPSY OF CHILDHOOD)

Benign Rolandic seizures (aka benign epilepsy of childhood) are the result of an autosomal dominant disorder and is the most common form of epilepsy in children. There are frequent facial motor symptoms with **sensory involvement of the tongue**. Seizures usually occur at night while the child is sleeping. The seizures are "simple" so patients are conscious but **unable to talk**. If treated, carbamazepine is typically used.

**MNEMONIC**: Imagine a child being UNABLE to talk because he is ROLling his FACE and TONGUE into a pillow (usually at night).

## ABSENCE SEIZURES

Absence seizures are generalized (non-focal) focal seizures that are often **provoked (or diagnosed) by hyperventilation.** Associated with an EEG showing **3 second spike and wave discharges**. Treated with **ethosuximide** or valproic Acid. It is not associated with an aura or postictal phase. Patients often look like they are staring off into space and not paying attention in class.

**MNEMONIC**: Imagine a kid having an absence seizure in class. The class bully notices him staring off into space. He takes his dirty SOCKS (ethoSUX) and puts them in the kid's face. When the seizure is over, the kid suddenly smells the SOCKS, jerks back in disgust and hits his head on the WALL (VALproic acid)!

## TONIC-CLONIC SEIZURE

Tonic-clonic seizures are a nonfocal/generalized seizure that s characterized by an initial **tonic phase** (**extension** of the extremities and back) which lasts for less than 30 seconds, and then a **clonic phase (jerking)**. Patient may have incontinence during intermittent periods of **atonia**. Afterwards, patients are **postictal** (slow, confused). Treatment options include just about everything (valproic acid, phenytoin, carbamazepine, phenobarbital and lamotrigine).

## NEONATAL SEIZURES

Neonatal seizures can present as brief inactivity, odd facial movements (like lip smacking) or staring. There's usually NO long-term issue. If it occurs within 24 hours of birth, it's probably due to asphyxia (look for mild organ dysfunction from hypoxia of in other major organs), but you MUST rule out metabolic disease. Can treat with phenobarbital (consider holding off treatment if there is only one event).

## INFANTILE SPASMS

Infantile spasms may present in children less than 1 year of age with **non-focal seizures** involving the **flexion and extension** of the whole body. Can occur in clusters and may be mistaken for a Moro reflex. Strong association with developmentally delayed children and carries a poor prognosis, especially if there was mental retardation prior to the onset. Diagnose by looking for **hypsarrhythmia** (this refers to a characteristic **EEG pattern** found **between** infantile spasms). Treat with **ACTH**.

**PEARLS:** LOOK AT THE AGE! The Moro reflex usually disappears by about 4 months of age. This is also very **strongly associated with Tuberous Sclerosis**.

**MNEMONIC**: Do you remember the mnemonic of "Tubular bazooka shooting **Ash Leafs with DANCING ticks on them?**" If not, please review the autosomal dominant mnemonic (in Genetics).

## FEBRILE SEIZURE

Febrile seizures are non-focal (generalized) seizures that last **less than 15 minutes** and are associated with a high fever. Febrile seizures are usually seen in children 6 months – 6 years of age. Workup should be aimed at the cause of the fever, and not at brain imagining. About 30% of children can have a future febrile, though most do not. The risk of future epilepsy is twice that of the general population (but still VERY low).

### BREAKTHROUGH SEIZURE

Don't always blame the mom for breakthrough seizures! Kids can outgrow their seizure medication dose, so order a drug levels for all breakthrough.

### STATUS EPILEPTICUS

Status epilepticus refers to continuous seizures, or multiple repeated seizures, lasting **greater than 30 minutes** in which the child does not return to his or her baseline afterwards. This is an **emergency** and can cause long-term brain damage.

### (DOUBLE TAKE) TODDS PARALYSIS (aka TODD'S PARALYSIS)

If a child presents with **focal motor weakness** after a seizure, pick TODDS PARALYSIS (aka Todd's Paralysis)!

**PEARL**: There is often NO HISTORY of a witnessed seizure so have a high index of suspicion for this diagnosis in any child presenting with an acute onset of unilateral weakness of an extremity.

**NAME ALERT**: This is not TICK PARALYSIS.

* **(DOUBLE TAKE) TICK PARALYSIS**: Tick paralysis can present with a child that has a clinical picture very **similar** to that of Guillain Bare Syndrome (GBS). That means you're looking for an acute **ascending** paralysis in which there are absent reflexes. It will likely be preceded by ataxia. It's caused by a neurotoxin produced by ticks, and the solution is to **remove the tick**!

**PEARL**: If you are presented with a child that seems to have Guillain Bare Syndrome (GBS), but they mention a trip, the woods or the summer time, they are probably talking about this! Other KEY differentiating factors include the **lack of fever, normal CSF and a quicker progression of symptoms (hours – 2 days).** Other confusing choices may be Botulism (DESCENDING paralysis), or PoliomyeLITIS (+CSF findings and possible viral syndrome).

**NAME ALERT**: This is not TODD'S PARALYSIS

## *ATAXIA & RELATED CONDITIONS*

### ACUTE CEREBELLAR ATAXIA

Children with acute cerebellar ataxia present with trouble coordinating movement between the neck and the hip (**truncal ataxia**) at about 1-2 weeks after a viral syndrome. Patients can also have difficulty coordinating speech (**dysarthria**), **horizontal nystagmus** and trouble walking. The viruses most commonly implicated are **Varicella (VZV), Coxsackie virus** and Echovirus (possibly EBV too). The condition usually resolves in a few weeks to a few monYuld be high on your differential for any child with an acute onset of ataxia. This is very common, usually very benign, and is very testable for the pediatrics exam.

### (DOUBLE TAKE) ATAXIA TELANGIECTASIA

In ataxia telangiectasia, patients can have recurrent pneumonias or sinusitis + High AFP (from a neural tube issue). The telangiectasias are flat, red spots of dilated capillaries that **DO NOT BLANCH**. Ataxia is from problems with the cerebellum and might be seen on the exam in a child right after he learns to walk. Patients have worsening T-cell function later in life, but in childhood the symptoms are mainly neurologic. There is an increased risk of malignancy in the 3rd decade of life.

**PEARL**: An elevated AFP often refers to an issue with broken skin, which is often found in neural tube issue. Levels are used to screen for neural tube issues in pregnant women.

**PEARL**: RECURRENT PNEUMONIAS are seen all over the exam. You will have to pick out other findings to help you narrow your differential.

**MNEMONIC**: Ataxia = Unsteady Gait = Imagine a homeless, drunk, and wheelchair-bound teen (neural tube defect) who frequently comes to your ED for **recurrent pneumonias** or **sinusitis**.

### FRIEDREICH ATAXIA (aka FRIEDREICH'S ATAXIA)

Friedreich ataxia (aka Friedreich's ataxia) may present as a child that is slow and **clumsy around the time of puberty**. Ataxia is due to cerebellar problems and a loss of proprioception. Associated findings may include congestive heart failure (CHF), diabetes mellitus and a **high plantar arch (pes cavus)**.

### BENIGN POSITIONAL VERTIGO (BPV)

Benign positional vertigo (BPV) can occur at any age, but is often seen in children under 3. Look for ataxia and possibly horizontal nystagmus. Might be associated with vomiting.

**PEARL**: Children often look **scared or pale**.

### PERILYMPHATIC FISTULA

A perilymphatic fistula can present with a child having ataxia + **hearing loss**. There may be a history of barotraumas (extreme coughing or retching).

## *MISCELLANEOUS NEUROLOGIC CONDITIONS & FINDINGS*

### JAW CLONUS & BILATERAL ANKLE CLONUS

Jaw clonus and bilateral ankle clonus is a normal findings in babies.

### UPPER MOTOR NEURON DISEASE

Findings of upper motor neuron lesions/disease include hyperactive reflexes and increased muscle tone. There are no fasciculations. Babinski reflex can be positive (meaning there is an upward going toe).

**PEARL**: The Babinski reflex can be a normal finding in children up to about age 2. After that it is considered pathologic.

### LOWER MOTOR NEURON DISEASE

Findings of lower motor neuron lesions/disease include hypoactive, or absent, reflexes. Atrophy of the muscle and fasciculations.

## HEAD TRAUMA

In cases of head trauma, if there is no focal neurological sign, no loss of consciousness and the patient vomited only 0-1 times, it is OKAY to discharge from the ER WITHOUT any imaging. If required, order a CT, not an MRI.

**PEARL**: **EPI**dural hematomas can present days later after a "lucid period" and then have rapid deterioration and death. If a patient presents with a history of head trauma + loss of consciousness several days ago, STILL get the CT scan!

## NEUROCARDIOGENIC SYNCOPE

Neurocardiogenic syncope usually occurs after prolonged standing or times of stress/anxiety. Most common etiology of syncope.

## CEREBROVASCULAR ACCIDENT (aka CVA or STROKE)

A cerebrovascular accident (aka CVA or stroke) can present with a facial droop, slurred speech or motor weakness on exam. Paresthesias are not specific. Could be seen in a chronically hypoxic patient with polycythemia, an African American or Mediterranean person with G6PD or even in a patient a carotid artery dissection from blunt force chest trauma (get carotid angiography). In general, the best diagnostic test for a stroke is an MRI. An MRI will show both a thrombotic stroke (most common) and a hemorrhagic stroke. Cerebral angiography would also be an appropriate choice. A regular CT of the head should NOT be ordered because it really only shows a hemorrhagic stroke.

**PEARL**: Regarding CT scans of the head, if you are ever specifically concerned about a head bleed (trauma, fall, history of an aneurysm or coagulopathy), order the cat scan!

## MENTAL RETARDATION

Mental retardation is defined as an IQ < 70.

## EPIDURAL HEMATOMA

A child can get an EPIdural hematoma from even mild trauma, and may present with a rapid decline in mentation. There can be a **LUCID PERIOD** during which time the patient is fully conscious before the patient suddenly becomes unconscious again. Diagnose by getting a CT scan of the head and also doing a lumbar puncture if there is any doubt. LP should show lots of RBCs even though it was a "champagne" (atraumatic) the tap.

**PEARLS:** Epidural bleeds are arterial bleed and require STAT surgical intervention (burr hole). Also, if you are presented with a patient who had trauma and lost consciousness a few days ago, GET A CT! The **lucid period** can last hours, or even **DAYS**.

## SUBDURAL HEMATOMA (SDH)

Subdural hematomas (SDH) are caused by shearing of bridging veins between the dural sinuses and are associated with high speed acceleration or deceleration. Mental status may be fast or **slow** in onset, and there is **no lucid period**. SDHs can turn into chronic SDHs. Evacuate the bleed if they show evidence of increased intracranial pressure (ICP), or if they show an increasing size on serial CT scans.

**PEARL**: In the world of pediatrics, if you see this in a baby, you should think **shaken baby syndrome**! The next step is to do a retinal exam to look for **retinal hemorrhages**!

## SUBARACHNOID HEMORRHAGE

Patients with a subarachnoid hemorrhage present with a sudden and extremely painful headache. This is the one described as being the "worst headache of my life." It can occur from head trauma, spontaneously or due to a ruptured cerebral aneurysm. Think of it as a stroke + headache. Can lead to death. Diagnose with CT head or a lumbar puncture.

**PEARL**: CT head can be negative. Look for a lumbar puncture showing RBCs in ALL tubes of the collected CSF in an equal proportion. Or a "champagne tap" with RBCs. This can also be found in epidural hematomas. Either way, both are BAD and not due to just a "traumatic tap."

## MENINGITIS PEARLS

If meningitis is suspected, do not do a pre-LP CT scan of the head **unless** there are focal deficits on exam. Do the **lumbar puncture** before the antibiotics if the patient is extremely stable or if the diagnosis is in question. **Give the antibiotics** before you do anything if there is a history of a coagulopathy, **nuchal rigidity** or any sign of **increased intracranial pressure** (ICP), including systemic hypertension, bradycardia, fast/irregular breathing or seizures.

## SPINA BIFIDA

There are many types of spina bifida (occulta, with meningocele, with meningomyelocele – all discussed below), but the term "spina bifida" basically means the vertebrae didn't form correctly. The danger is that the spinal cord and its coverings could protrude out of the protective vertebral column. Each of the related conditions may have a skin finding or cyst at the lower back. If you see any of those, or are concerned about a neural tube defect, get a **spinal ultrasound** (or MRI). If the condition is mild, patients can be asymptomatic.

**IMAGE**: http://www.neurosurgery.ufl.edu/patients/images/NEJM_neural-tube-defect.gif

* **SPINA BIFIDA OCCULTA**: Vertebrae are so minimally split that there is no opening on the back. There may be a dimple or a tuft of hair at the site. The nerves are all intact. Many people are actually walking around right now and have no idea they have it. Symptoms are absent or relatively mild (incontinence or mild sensory problems).

- **MNEMONIC**: OCCULTA means HIDDEN. The problem is hidden from the eye, and for many people it's hidden from their brain. It's so benign that they don't even know they have it.

* **SPINA BIFIDA WITH MENINGOCELE**: Vertebrae are slightly more split and open enough to let the meninges herniate out of the spinal column. A visible cyst is created which contains the meninges and CSF, but no nerves. Again, the nerves are essentially intact. Children may have mild symptoms, but the prognosis is generally good.

- **MNEMONIC**: MENINGO-CELE = MENINGES ARE SEALED in a cyst.

* **SPINA BIFIDA WITH MENINGOMYELOCELE**: Vertebrae are more widely split. Both the **meninges and the spinal cord herniate** through the vertebral column. All patients have **at least some paralysis**, but the degree varies. Patients are a setup for urologic issues (incontinence, retention, infections) and orthopedic problems. The **neurosurgery, urology, and orthopedic consultants** should all be called. The patients also require a **CT of the head to rule out a Chiari II malformation**, in which the **cerebellar tonsils are displaced downward** through the foramen magnum, and a noncommunicating **hydrocephalus** exists. The Chiari II malformation (and associated hydrocephalus) is very common in these children, and the prognosis is poor. Most patients die within 2 years.

## CHIARI MALFORMATION

There are multiple types (I – IV) of Chiari malformation, but basically this refers to a downward displacement of the cerebellar tonsils below the foramen magnum. This **can be asymptomatic**, in which case no surgeries are needed. If, however, there is evidence of brainstem dysfunction, the patient will need surgical decompression. Symptoms may include headaches, ataxia, apnea, spasticity or bladder dysfunction.

# ORTHOPEDICS & SPORTS MEDICINE

## EPIPHYSIS, PHYSIS & METAPHYSIS

Starting at either end of the bone, you have the epiphysis, then the physis (growth plate), then the metaphysis and then the midsection of the bone (diaphysis). KNOW IT.

**IMAGE**: http://bit.ly/iAsMHQ

**IMAGE**: http://bit.ly/kbaoXD

* **PHYSIS** (aka EPIPHYSEAL PLATE): Growth plate

* EPIPHYSIS: The rounded end portion of a bone that is most distal to the diaphysis (mid-section of the bone).

- **MNEMONIC**: EPI-PHYSIS = ON TOP OF – THE GROWTH PLATE. Think of this as the "distal" part of the bone. This could be farthest away from your joint (imagine using your shoulder joint and humerus for the example), or you could imagine it as being farthest away from the mid-section of a bone (diaphysis). If none of that helps, think of an ice cream cone. Slipped Capital Femoral Epiphysis occurs due to a Type I fracture when the EPIphysis slides off the TOP OF the growth plate.

* METAPHYSIS: This is also towards the end of the bone, but it is more towards the diaphysis.

- **MNEMONIC**: **M** = **M**etaphysis = closer to the **M**idsection of the bone!

## SALTER HARRIS FRACTURES

There are 5 types of Salter Harris fractures (I, II, III, IV and V) we'll discuss. As a quick summary, they basically go through the growth plate, through the growth plate + metaphysis, through the growth plate + epiphysis, through the metaphysis + growth plate + epiphysis or involve a crush injury in which the epiphysis is pushed towards the metaphysis, therefore crushing the growth plate and narrowing that space.

**IMAGE**: http://alturl.com/gtfq7

**IMAGE**: http://www.fpnotebook.com/_media/OrthoFractureSalterHarris.jpg

* TYPE I: SEPARATED or SLIPPED. The fracture is through the growth plate and the epiphysis has just slipped a little. Look for **tenderness to palpation at or near the growth plate** and a negative **X-ray**. Treat with casting for 2 weeks.

* TYPE II: Growth Plate + Metaphysis. Most common. May do a closed reduction and cast for 4 weeks.

* **TYPE III**: Growth Plate + Epiphysis. This one is BAD because the fracture goes through the epiphysis and into the growth plate. May require an OPEN reduction and often results in **poor bone growth** after the injury. This one is worse than Type IV.

* TYPE IV: THROUGH all three layers. Epiphysis + Growth Plate + Metaphysis. Definitely requires open reduction to preserve the growth plate.

* TYPE V: SMASHED. Crush injury in which the epiphysis is pushed towards the metaphysis, therefore crushing the growth plate and narrowing that space. **Can result in poor growth after the injury**. X-ray may essentially look negative, or may show a POSTERIOR FAT PAD.

- **IMAGE**: This shows the presence of an anterior and a posterior fat pad in an **unrelated** injury. http://bit.ly/lFgtOJ

- **MNEMONIC**: S-a-I-T-S can help you remember SEPARATED, THROUGH and SMASHED for Types I, IV & V. A different well known mnemonic uses SALTR. It conflicts with the mnemonic used in the terminology section, though. Feel free to look it up if you're struggling with these.

## TORUS FRACTURE (aka BUCKLE FRACTURE)

In a torus fracture (aka buckle fracture), compression of bone causes a fracture that does NOT go through the growth plate. It's not that serious and is common in children due to their soft bones.

**IMAGE**: http://bit.ly/kO8E5l

## DISTAL HUMERAL FRACTURES

Children can have **neurovascular injury** with distal humeral fractures, especially if there is a any displacement of the bone and it is **supracondylar**. These are associated with a **posterior fat pad** can result in **compartment syndrome**, pulses can be normal. Look for "**more than expected pain**."

**IMAGE**: http://bit.ly/jccKTr

## DISLOCATED SHOULDER

If a prepubertal child presents with a dislocated shoulder, **anterior dislocation is MUCH more common than posterior**. Obtain an AXILLARY view X-ray.

**PEARL**: If a **prepubertal** child presents with what looks like dislocated shoulder, it's probably NOT A DISLOCATION. It's actually a FRACTURE.

## OSTEOGENESIS IMPERFECTA

Children with osteogenesis imperfect have extremely fragile bones, lots of fractures, BLUE sclera and can have BAD TEETH.

**IMAGE**: http://bit.ly/j2njeR

## VALGUS DEFORMITY

A valgus deformity is an outward angulation of the distal part of a bone or joint.

**IMAGE**: http://bit.ly/l8F10h

**MNEMONIC**: The distal part of the bone points away from midline, "like the L" in va**L**gus.

## VARUS

A varus deformity is an inward angulation of the distal part of a bone or joint.

**IMAGE**: http://bit.ly/117Zekc

**IMAGE**: http://bit.ly/Wu7erj

## GENU VARUM (aka BOWED LEGS)

Genu varum (aka bowed legs) is not the same thing as intoeing. Babies are born this way. This is **Usually physiologic and can reassure that it will resolve by 1 yo.** **Pathologic** if either unilateral, WORSE after 1 yo, persists after age 2, presents after age 2

**IMAGE**: http://bit.ly/iDTrw5

## RICKETS

Please see the endocrine section for **extensive** discussion of rickets.

## BLOUNT DISEASE

Children with Blount disease have abnormal growth of the **medial epiphysis** of the **proximal tibia**. Much more common in African Americans.

* **PEARL**: Especially consider this diagnosis if you see a patient with **bowed legs and a fracture**. Look at the X-ray to see if the proximal, **medial** aspect of the tibia looks odd, and if **that leg** is bowed.

* **IMAGE**: http://bit.ly/kD4nEH

* **INFANTILE BLOUNT DISEASE**: Benign, no treatment needed.

* **ADOLESCENT BLOUNT DISEASE**: Look for an **overweight African American** patient. Treat with bracing or **surgery.**

## INTOEING

The causes of intoeing include tibial torsion, femoral anteversion and metatarsus adductus. All are discussed below

* **PEARL**: PICK REASSURANCE FOR A CHILD WITH INTOEING!

* **TIBIAL TORSION**: Noted in **T**oddlers around **T**wo years of age. **Most common reason** for intoeing. Usually benign and resolves by 5-6 years of age. Treat with **reassurance!**
  - **IMAGE**: http://bit.ly/WXERqy
  - **MNEMONIC**: T = TIBIAL = TORSION = TODDLERS.

* **FEMORAL ANTEVERSION**: All infants have this! This is caused by too much internal rotation of the femur and not enough external rotation at the hip. Patients can **sit in a W style**. Resolves by 8-10 yo
  - **IMAGE**: http://bit.ly/jSpmQg
  - **IMAGE**: http://bit.ly/I5Yk4Q

* **METATARSUS ADDUCTUS (aka HOOKED FOOT)**: Noted in infancy because of an **anatomic bend** in the foot. In most cases, the problem corrects without intervention. A cast or surgery is sometimes required.
  - **IMAGE**: http://bit.ly/V0LySp

## CLUB FOOT (aka TALIPES EQUINOVARUS or EQUINOVARUS DEFORMITY)

Club foot (aka Talipes equinovarus or equinovarus deformity) presents as a unilateral limp noted in infancy. Caused by an internal rotation and contraction of the **Achilles tendon**. Treat by stretching + serial casts + possible surgical release, preferably done by 1 year of age.

**IMAGE**: http://1.usa.gov/IWS5yF

**MNEMONIC**: equinova**R**us. If you can remember that va**L**gus is the outward L shape, you should be able to then recall what a vaRus deformity is and choose the correct answer.

## PES CAVUS

Pes cavus patients have high plantar arches. Associated with Friedreich Ataxia & Hurler Syndrome.

**IMAGE**: http://bit.ly/ob3rN3

**MNEMONIC**: pes cavus = high ARCHES. Imagine eating at a burger place called Gold Arches with your family. Their logo has two giant gold ARCHES, like two feet with high arches. Now say, "Eating at the Gold ARCHES my **F**amily **A**taxic and made me **HURL**."

* **KEY:** **F**amily **A**taxic = F.A. = Friedreich Ataxia. HURL = Hurler

## PES PLANUS

Pes planus patients have flat plantar arches. If incidentally found, no intervention is needed.

**IMAGE**: http://bit.ly/ob3rN3

## SLIPPED CAPITAL FEMORAL EPIPHYSIS (SCFE)

Children get slipped capital femoral epiphysis (SCFE) from a Salter Harris Type I fracture at the growth plate (physis), in which the EPIphysis slides off of the growth plate. May present with **referred KNEE pain** with a normal knee exam, and a leg that is extended and internally rotated at the hip. Obtain **AP "frog leg"** views of the hips. This is usually seen in **an obese, adolescent male**. **Treat with** immobilization, nonweight bearing, and possibly surgical pins.

**IMAGE/PEARL:** Note that this can be VERY subtle. Before and after repair - http://bit.ly/nOLGoA

**PEARL/MNEMONIC**: Imagine ICE CREAM starting to fall off of an ICE CREAM CONE. When looking at the image, be sure to look for the EPIphysis (or ice cream), to be FALLING OFF MEDIALLY!

**PEARL**: A fairly large percentage of children (> 15%) can go on to having a contralateral SCFE.

**PEARL**: Unrelated to SCFE, but if you are **not sure** what to do on an orthopedic question, and "Image contralateral side" is given as an answer choice, pick it!

## LEGG-CALVE-PERTHES DISEASE

Legg-Calve-Perthes disease is also known as IDIOPATHIC **AVASCULAR NECROSIS of the femoral head, or OSTEOCHONDROSIS of the femoral head**. As the names suggest, a lack of blood flow causes necrosis to the femoral head for unknown reasons. It occurs in children younger than 8 years of age. Diagnose by looking for an irregularly shaped femoral head on the X-ray. Treat with casting or bracing, and let the parents know that the blood supply will go back to normal with time.

**PEARLS:** This often presents with a **PAINLESS LIMP**, but **can** be associated with referred **HIP PAIN**, and knee pain (less frequent). Also, when comparing X-rays to Slipped Capital Femoral Epiphysis, you will note that the femoral head shape is **extremely irregular and jagged**.

**IMAGE**: http://www.peds-ortho.com/perthes1.JPG

**IMAGE**: http://bit.ly/WXFn7Q

## OSGOOD SCHLATTER DISEASE

Osgood Schlatter disease usually presents in active children around the time of adolescence, or a growth spurt, with pain below the knee. It occurs due to **traction at the tibial tubercle** by the patellar tendon at the point of insertion. Treat with rest for a few weeks and NSAIDS or acetaminophen. A mild prominence may be noted into adulthood.

**PEARL**: This is typically a UNILATERAL problem, though it can occur on the other side later. Diagnosis can be clinical. If asked what imaging study to get, obtain a lateral X-ray.

**IMAGE**: http://bit.ly/otS3GD

## OSTEOCHONDRITIS DISSECANS

In osteochondritis dissecans, children complain a knee that **locks up**. Look for swelling on exam. It's caused by necrosis of the articular surface of a joint, usually the KNEE. Can lead to chronic pain. Treat with immobilization. If it doesn't get better, it may require surgical removal of bone fragments.

## SCOLIOSIS

Scoliosis usually occurs around the time of a growth spurt. Children will have findings that include **asymmetric hips or scapulae**. If shown a picture, look for one shoulder to be higher than the other. When severe, it can present with neurologic symptoms, such as paresthesias or weakness of the lower extremities.

**PEARL**: Know this! Regarding treatment, if the curvature is < **25 degrees**, **observation is ok**. If the curvature is **> 25 degrees, brace the child if s/he is still growing**. If the curvature is **> 40 degrees, they will need to get surgery**.

**IMAGE**: http://bit.ly/r5TCKc

## SPONDYLOLYSIS

Spondylolysis is a true **fracture of the vertebral pars**. It can present with pain that is noted with s**tanding, extension at the hip** or the **straight leg raise test**. It can occur with vigorous physical activity.

**PEARLS:** The **straight leg raise** is positive when the patient is supine and they have pain as you passively raised their leg at 10-60 degrees. It indicates a **radiculopathy**. Note that this condition has **pain with standing**, while common BACK STRAINS are associated with relief of pain with standing.

**PEARL**: Look out for a gymnast with back pain!

**MNEMONIC**: LYSIS = BREAK! Always break down the words!

**IMAGE**: http://bit.ly/mYtkbk

## SPONDYLOLISTHESIS

In spondylolisthesis, a **slipped** vertebral body results in lower back pain. The problem and pain is usually around L5 or S1. Again, pain is **worse with standing**.

**IMAGE**: http://bit.ly/mYtkbk

**MNEMONICS:** Rename it spondylo**SLIP**thesis or spondylo**LISP**thesis.

## SUBLUXED RADIAL HEAD (aka NURSEMAIDS ELBOW)

A subluxed radial head (aka nursemaids elbow) usually occurs in young children when a child is picked up or pulled by the arm. The forearm will be **pronated** and the arm will be **flexed and close to body**. It almost looks like the patient is wearing an invisible cast, and sometimes patient's are noted to hold the affected elbow with the unaffected hand. Treat with **forced supination**.

**PEARLS:** **Know** that it involves the **annular ligament** (it slips over the radial head allowing radial head dislocation). Also, it's fine to look and feel for fractures, but there is NO NEED for imaging if the story fits.

**IMAGE**: http://bit.ly/o3sMYW

**IMAGE**: http://bit.ly/p6LkXH

## DEVELOPMENTAL DYSPLASIA OF THE HIP (DDH)

Infants with developmental dysplasia of the hip (DDH) may be noted to have a leg-length discrepancy or extra creases at the thigh. Girls and breech deliveries pose a higher risk. **Workup varies** by age and family history. Treatment requires a Pavlik harness.

* **CRITERIA FOR A CHILD LESS THAN 4 MONTHS:**
  - **ASYMPTOMATIC:** If there is a history of DDH in a **first degree relative**, get an **ultrasound**!
  - **SYMPTOMATIC:** If there are signs of developmental dysplasia of the hip (DDH) on exam, get an **ultrasound!**

* **CRITERIA FOR A CHILD GREATER THAN 4 MONTHS:**
  - **ASYMPTOMATIC**: If there is no family history, then **no workup** is needed. If there is a history of DDH in a **first degree relative** and an evaluation was never done, then since the child is now greater than 4 months of age, **plain radiographs (hip x-rays)** would be indicated. Plain x-rays are more reliable than an ultrasound at this age (femoral head ossification centers are more developed).
  - **SYMPTOMATIC:** For any child greater than 4 months of age that has signs of developmental dysplasia of the hip (DDH) on examination (regardless of family history), they need to be worked up. Therefore, **get plain radiographs** (not an ultrasound) since they're more reliable at this age.

* **PEARLS:** The age cutoff for ultrasounds is 4 months, **NOT** 6. When it comes to the Barlow and Ortolani signs, **EITHER of them being positive should prompt a workup**!

* **IMAGE**: http://bit.ly/qHTScy

* **IMAGE**: http://bit.ly/qTQrK8

## TOXIC SYNOVITIS (aka TRANSIENT SYNOVITIS OF THE HIP)

In toxic synovitis (aka transient synovitis of the hip), patients have pain with **active** range of motion (much less or absent with passive ROM). There may be a history of a low grade fever or a recent infection. A Gram stain of fluid from the joint will be negative.

**PEARLS:** This is a **diagnosis of exclusion**. Even if the story fits, you **must order joint aspiration** to a septic arthritis!

## SEPTIC ARTHRITIS

Septic arthritis will present with a patient that has a high **fever**, elevated **inflammatory markers**, possible erythema at the joint and pain with **active and passive** manipulation of the joint. Diagnose by **joint aspiration**. If the results are inconclusive, treat with an **anti-Staphylococcus antibiotic**.

**PEARLS:** If no imaging is given and you are asked for the next best step in a patient with **any** signs of a possible infection, **ASPIRATE the joint.** Do not order imaging! The most common etiology is in children is STAPHYLOCOCCUS AUREUS. If you are presented with a teen, consider Neisseria gonorrhea coverage as well

**PEARLS:** A septic arthritis is much acute than Juvenile Rheumatoid Arthritis. If you're confused on the exam, GIVE antibiotics. The differential also includes **SERUM SICKNESS**, in which case there should also be a history of urticaria and arthraLGIA (pain, but no signs of inflammation).

## OSTEOMYELITIS

Osteomyelitis is a **bone infection**. Look for pain at the metaphysis and other signs of infection (fever, leukocytosis, etc.). Use a **bone scan OR aspiration to diagnose** because it is the most specific test to rule in osteomyelitis. MRI is the most sensitive (helps rule out). Treat with IV antibiotics to cover **Staphylococcus aureus**. Treatment is usually long, 4-6 weeks.

**PEARL**: Once an organism has been identified and sensitivities are available, you may do a trial of **inpatient** oral antibiotics and then discharge if s/he does well. Once the patient is tolerating oral

## STRAINS

Strains refer to a tear in part of a MUSCLE. This is much more benign and much more common than sprains.

**MNEMONIC**: rename it **sTrUSCLE**… kind of like a toaster STRUdle.

## SPRAINS

Sprains refer to the tear of a LIGAMENT. This less common than strains and the management is more complicated.

* ANKLE SPRAINS: Most are due to **inversion** and resulting **injury to the lateral aspect of the TALOFIBULAR ligament.** Look for a mild-moderate amount of swelling, tenderness **laterally** and no joint instability on exam. Prescribe Rest, Ice, Compression with a wrap, and Elevation ("RICE"). If minimally symptomatic, RICE QID for a couple of days. If there is **difficulty** with ambulation, prescribe RICE for 2-4 weeks (especially rest). If they are **unable to ambulate**, then prescribe RICE for 6-10 weeks (especially rest).

  - **PEARLS:** For the pediatric boards, management of sprains does NOT involve casting or orthopedic referral. Those are reserved for **fractures**. Prescribe "RICE". Also, true sprains (true ligament tears) are RARE in **PRE**pubertals children because they're so flexible! If you are given a young child with what sounds like a sprain, get an X-ray because it's OFTEN A FRACTURE. Lastly, if there is any **sensory component**, consider **compartment syndrome**.

* **MNEMONIC**: Rename it sPrIGAMENT!

## ROTATOR CUFF TEARS

Rotator cuff tears will present as a slowly progressive pain in the shoulder since it usually due to **chronic** activities, such as throwing.

## ANTERIOR CRUCIATE LIGAMENT TEAR (ACL TEAR)

A child with an anterior cruciate ligament tear (ACL tear) will usually present with a **noncontact** sport related injury while pivoting or landing from a jump. They may have heard a "pop." Exam will show knee swelling, a joint effusion and an anterior drawers sign. Definitive diagnosis is by MRI. Treatment often requires surgery.

## JOINT HYPERMOBILITY

A child with joint hypermobility might present with a history of "loose joints." It sounds good, but these children are actually **more prone to getting injured with sprains**. The treatment is to encourage stretching!

**MNEMONIC/PEARL:** They'll talk about loose JOINTS, not someone being extremely flexible. The stretching stretches the ligaments and makes them less prone to injury as the joints are all loose and sliding all over the place.

## COMPARTMENT SYNDROME

Compartment syndrome occurs when trauma leads to so much swelling within a compartment of muscle that blood flow is restricted and necrosis occurs. The five classic signs are summarized by the "5 P's" of Pain, Paresthesias, Pallor, Pulseless and Paralysis. The key, however, is realizing that all 5 symptoms are rarely seen, and most of them are late findings (especially Pulseless and Paralysis). If presented with a case of trauma and any **sensory deficit** (light tough, pinprick, etc.) think COMPARTMENT SYNDROME and measure compartmental pressures.

## ACROMIOCLAVICULAR JOINT SEPARATION (AC JOINT SEPARATION)

Acromioclavicular joint separation (AC joint separation) may present with point tenderness at the distal clavicle/AC joint, and pain with **abduction of the arm** after a **fall**. Treat with a **sling** and physical therapy. If severe, call orthopedics.

## SPORTS INJURY PEARL

Once sports injury patients are pain-free, without swelling and have regained full range of motion and strength, they can be cleared to return to sports.

## CONGENITAL TORTICOLLIS

Children with congenital torticollis are born with a damaged or shortened sternocleidomastoid muscle. The head will be tilted to the ipsilateral side and the chin will point to the other side. Treat with daily stretching and physical therapy to improve the range of motion. If it doesn't improve by a year of age, consider surger.

**PEARL**: There is an odd association with developmental dysplasia of the hip (DDH). Don't get it confuse with Klippel-Feil Syndrome, in which there are fused cervical vertebrae, a webbed neck, more severe limitation in the range of motion and possibly other anomalies. And don't get that confused with Klippel-Trenaunay Syndrome, in which there is a port wine stain and overgrowth/hemihypertrophy issues.

**IMAGE**: http://bit.ly/Y2diOc

## POLYDACTYLY

Polydactyly refers to the presence of extra digits. Preaxial refers to the radial side, while postaxial refers to the ulnar side. Many of these extra digits are barely hanging on and have no bone in them. They can simply be tied off with suture. If they are well-formed, obtain a surgical consult.

* PREAXIAL SUPERNUMERARY DIGITS (aka AXIAL SUPERNUMERARY DIGITS): These are located on the 1st digit (thumb) and are more likely associated with certain syndromes (low yield).

* POSTAXIAL SUPERNUMERARY DIGITS: These are located on the 5$^{th}$ digit (pinky finger) and are less likely to be associated with a syndrome.

# RHEUMATOLOGY

## *ARTHRITIC CONDITIONS*

### ARTHROCENTESIS (JOINT ASPIRATION) PEARLS

If there is even the slightest question of a septic arthritis in a child, you **must order an arthrocentesis (joint aspiration)**. It's cheap, fairly painless and extremely helpful. Please keep in mind that the number are quite different than those used to evaluate CSF during a meningitis workup. Some pearls are listed below to help guide you on questions in which arthrocentesis has already been done.

* YELLOW FLUID: NORMAL

* POSITIVE GRAM STAIN: **Treat** for an infection, regardless of the numbers.

* WBC < 200: **NORMAL**

* WBC < 2000: Consider TRAUMA. Fluid can be clear, yellow or bloody.

* WBC > 2000 : Consider an **inflammatory OR infectious** etiology. If the Gram stain is positive, treat!

* WBC ~ 5000:

- Consider LUPUS (SLE).
- Consider RHEUMATIC FEVER, but also look for mention of **LOW or DECREASED viscosity.**

*** WBC > 50,000: Likely SEPTIC ARTHRITIS.**

* JRA: WBC around 15,000 but LOW/DECREASED viscosity

### JUVENILE RHEUMATOID ARTHRITIS (JRA)

**KNOW JUVENILE RHEUMATOID ARTHRITIS WELL!** The diagnosis requires knowing quite a few details. The child should have been under the age of 16 at the time of **symptom onset**. Symptoms must be present for at least **6 WEEKS** before the diagnosis can be made. In children, if arthritis is present it is more common in the **LARGE joints** and rheumatoid nodules are much less common when compared to adults. A positive rheumatoid factor indicates a worse **prognosis**. Do NOT order a rheumatoid factor for diagnostic purposes. It can **help with prognosis/subtyping, but a negative RF does NOT rule out RA**.

* **O**LIG**O**ARTICULAR JRA (OLIGOARTHRITIS): This refers to JRA that affects **4 OR LESS JOINTS**, and is the more common type of JRA (>50%). Positive markers (rheumatoid factor and ANA) make this subtype much more likely. It's more common in younger girls and is associated with chronic uveitis. Since **visual complaints may be absent,** patients need to have **regular eye exams**. Boys have a better prognosis.

- <u>**MNEMONIC**</u>: The **O's** for **O**lig**O** look like EYES and need regular eye exams because it is the more serious subtype.

* POLYARTICUIAR JRA (POLYARTHRITIS): This refers to JRA that affects **5 OR MORE JOINTS**. Also more common in young girls. Systemic symptoms outside of the joints is not common.

* SYSTEMIC (aka STILLS DISEASE): This is equally as common in boys and girls. There are many classic symptoms and finding to be aware of including an episodic, salmon-colored "**EVANESCENT RASH.**" Patients may also have an **extremely high LEUKOCYTOSIS** (> 30K), with **spiking fevers, lymphadenopathy** and

**possible hepatosplenomegaly**. Patients may also have **pleurisy**, **pericarditis** and the **Koebner phenomenon** (linear skin lesions appearing along a site of injury, rubbing or scratching). Serum **markers are NEGATIVE**.

- **PEARL**: **If everything else fits and the patient doesn't have an arthritis, go ahead and pick this diagnosis!** The other symptoms are commonly present well before the arthritis component kicks in.
- **PEARL**: This can be a difficult diagnosis to make and is often missed in clinical practice and on the pediatric board exam. **Please be VERY, VERY comfortable with this topic.**
- **PEARLS:** In comparison to leukemia, pain is in the AM (not at night), pain is in the joints (not the bone), mild hematologic anomalies (not severe), symptoms wax and wane (not persistent/worsening), symptoms are insidious in onset (not acute) and JRA may have a rash. **BOTH can have lymphadenopathy and hepatosplenomegaly**. In comparison to septic arthritis, remember the insidious onset of symptoms for JRA (not acute).

* TREATMENT: First line treatment = NSAIDS. Second line treatment = STEROIDS.

## SYSTEMIC LUPUS ERYTHEMATOSUS (SLE)

Signs of systemic lupus erythematosus (SLE) include arthralgias, fatigue, malar rash, discoid lesions, photosensitivity, oral/nasal ulcers (if oral, painless and at hard palate), arthritis, hematologic abnormalities, ANA antibodies (most **sensitive**), **anti-DNA** antibodies (next most **sensitive**), **anti-Smith** antibodies **(most specific/diagnostic)**, +RPR, serositis of pleura or pericardium, psychiatric or neurologic issues. Other signs include fever, weight loss, hemolysis, clots. Some other adverse effects of SLE are listed below.

* LUPUS NEPHRITIS: Associated with a high anti-DNA and low C3, **C4** and CH50.

- **PEARL/REMINDER:** PSGN only has a low C3. MPGN has a low C3 & C4.

* LUPUS CEREBRITIS: Associated with microischemia and cerebrovascular disease. Can also have seizures.

- **PEARL**: When on steroids, can have steroid psychosis. Do not confuse that with cerebritis.

* TREATMENT: NSAIDS if mild. Steroids, hydroxychloroquine, methotrexate, etc. if severe.

- **PEARL**: There's quite a bit of information above. In order to make a diagnosis of SLE, look for **AT LEAST 4 SYMPTOMS/SIGNS,** preferably from different categories (skin, heme, markers, etc.).

## NEONATAL LUPUS

Neonatal lupus occurs due to **mom's SLE antibodies** and is not very common. Look for abnormal liver enzymes, a rash, thrombocytopenia, petechiae, scaling and a **3rd degree AV heart block with bradycardia**. Babies may die if they have heart lock associated with hydrops fetalis (fluid accumulation in 2 or more fetal compartments). Diagnose with **Anti-Ro or anti-La antibodies** (aka anti-SS-A or SS-B).

## DRUG INDUCED LUPUS

Certain medications can cause an SLE type of syndrome referred to as drug induced lupus. The more common culprits include **anti-seizure medications, sulfa drugs,** lithium, hydralazine and quinidine.

**PEARL**: If you stop the medication, this is a REVERSIBLE condition.

## JUVENILE ANKYLOSING SPONDYLITIS

Juvenile ankylosing spondylitis is a chronic inflammatory condition that involves **fusion of the spine** (called "bamboo spine") and **inflammation of the hips**. Males are more often affected. Inflammation often involves the **sacroiliac joint**. **Pain is worse with rest**, so patient will have pain at night and morning stiffness. Then pain improves with activity or exercise. It is a **seronegative spondyloarthropathy** (labs are normal, ANA is negative and ESR is either negative or only slightly elevated). Patients can also have eye findings of iritis and uveitis. Treat with NSAIDS, sulfasalazine or steroids.

## JUVENILE REITER SYNDROME (aka JUVENILE REITER'S SYNDROME)

Juvenile Reiter syndrome (aka juvenile Reiter's syndrome) is a seronegative spondyloarthropathy (negative rheumatologic markers) which includes symptoms of iritis (look for pain and trouble with vision), urethritis and arthritis.

**PEARL**: This is often preceded by an infection! Look for a recent infectious picture due to enteric bugs in young children (Salmonella, Shigella, Yersinia) and Chlamydia in older/adolescent children (nongonococcal urethritis).

**MNEMONIC**: Can't SEE, can't PEE and can't CLIMB A TREE. Refers to iritis or uveitis, urethritis and arthritis.

## BEHCETS SYNDROME (aka BEHCETS DISEASE)

Behcets Syndrome (aka Behcet's or Behcets Disease) results in **aphthous ulcers**, **genital ulcers**, **uveitis**, **arthritis** and **gastrointestinal symptoms**. Labs are usually negative.

**PEARL**: This could be confused with ulcers due to HSV. If that option is presented look again for an **gastrointestinal complaints** to rule in Behcet's.

# ***NON-ARTHRITIC CONDITIONS***

## DERMATOMYOSITIS

In dermatomyositis, children can have a heliotropic rash (means pinkish-purplish) in malar area, Gottron's Papules (can be shiny and pruritic), **high CPK**, proximal weakness and telangiectasias near the nail folds. It's diagnosed with a biopsy. Can also get calcinosis cutis/calciphylaxis.

**IMAGE**: http://bit.ly/ooMHo1

**IMAGE**: (CALCIPHYLAXIS) http://bit.ly/omYj0j

## HENOCH SCHONLEIN PURPURA (HSP)

Henoch Schonlein purpura (HSP) is a vasculitis which can involve **multiple systems**, including the skin, joints, GI tract and kidneys. Classic findings include **PALPABLE** and **TENDER** purpura that **blanch.** These are most often found at the **lower extremities and buttocks**, but may be elsewhere. There may also be **peri**articular joint involvement (soft tissue only) at the knees or ankles, and a faint rash. Patients may initially present with colicky abdominal pain +/- blood in stool +/- **intussusception +/-** gallbladder hydrops. Skin findings are impressive, but the labs show a **normal platelet count**. Urinalysis will likely show hematuria +/- proteinuria which can range from mild to nephrotic range (**order a spot protein to creatinine ratio**). This

may be diagnosed on clinical findings. Complement levels can be low. A **biopsy may be obtained if there is doubt** about the diagnosis. Biopsy would show IgA, IgG and C3 deposits. For treatment, **do not give** steroid monotherapy because it will not work. The disease often resolves without intervention and the use of medications is debatable, meaning it's unlikely to be tested. For the boards, give NSAIDS for joint pain in the **absence of renal disease**, and for severe symptoms (can't eat) or for a hospitalized patient, give steroids with or without other therapies (i.e., cyclophosphamide, azathioprine, plasmapheresis, IVIG, etc.).
**IMAGE**: http://bit.ly/nsNXZo

## SARCOIDOSIS

Sarcoidosis presents with **HILAR ADENOPATHY** + multisystem complaints and findings. Patients can present with fatigue, exercise intolerance, weight loss or a **chronic cough**. Also look for any evidence of problems with the eyes, heart or kidneys. Imaging will show the classic **hilar lymphadenopathy** and possibly **peribronchial infiltrates**. If a lesion was to be biopsied, it would show the classic **noncaseating granulomas**.

## SJOGRENS SYNDROME (aka SJOGREN'S SYNDROME)

Sjogrens syndrome (aka Sjogren's syndrome) results in **dry eyes, dry mouth, parotiditis** and possibly lab findings consistent with a renal tubular acidosis (RTA). May also have arthralgias, photophobia, elevated ESR/CRP. The disease is caused by **lymphocytic infiltration of exocrine glands**. For a "first step," start with a **Schirmer's test** to look for dry eyes (a piece of paper is placed in the eye and minimal tear absorption is noted). To confirm diagnosis, biopsy the lip or a salivary gland to look for **lymphocytic infiltration**.

## RAYNAUD'S PHENOMENON (aka RAYNAUDS)

Raynaud's phenomenon (aka RAYNAUDS) will present as cyanosis and white/ulcerated finger tips due to lack of blood flow. May especially be symptomatic in the cold.
**PEARL**: While this can be associated with scleroderma or systemic lupus erythematosus (SLE), it can also be a stand alone diagnosis.
**IMAGE**: http://1.usa.gov/nWqAE2

## WEGENER'S GRANULOMATOSIS

(Rare in kids, low yield). Wegener's granulomatosis is a multisystem vasculitis that affects the sinuses, lungs and kidneys. Look for a +C-ANCA (as opposed to the p-ANCA seen in ulcerative colitis).

# PULMONOLOGY

## *CYSTIC FIBROSIS & NASAL POLYPS*

**PEARL**: You are guaranteed to see the words "cystic fibrosis" multiple times on the pediatric certification exam. The American Board of Pediatrics expects you to know this disorder well due its ability to affect multiple different systems. There is minimal bolding in this section because it's an extremely high-yield topic and you should know it ALL cold!

### CYSTIC FIBROSIS (CF)

Cystic fibrosis (CF) is an autosomal recessive disorder that **will be on your test**. Children who are carriers of the cystic fibrosis gene are asymptomatic. It result in excessive loss of chloride through sweat and is associated with significant pulmonary, electrolyte and gastrointestinal problems. The The diagnosis is made through a sweat chloride test showing a level of greater than or equal to **60 mEq/L**. A list of the many associations with this disease are listed below by category.

* **MNEMONIC**: If **sweat test >/= 60 meq = "6-tic fibrosis" =** diagnostic.

* PULMONARY SYMPTOMS: Diffuse cysts/fibrosis (hence the name), bronchiectasis, hemoptysis, pneumonias (Staphylococcus aureus, Pseudomonas, Haemophilus influenzae), chronic sinusitis, nasal polyps, pneumothorax, Burkholderia cepacia noted in sputum cultures, Pseudomonas pneumonias (and colonization).

- **PEARLS:** If Burkholderia cepacia is mentioned, go to the answers and pick cystic fibrosis!

* GASTROINTESTINAL SYMPTOMS: Pancreatitis, cholelithiasis, neonatal cholestasis, meconium ileus, rectal prolapse, chronic diarrhea, malabsorption with steatorrhea, fat soluble vitamin deficiencies (A, K, E, **D**), hypoproteinemia.

- **PEARLS**: Regarding the fat soluble vitamin deficiencies, look for problems with proprioception, ataxia, vision, coagulation, hypocalcemia,

* ACID-BASE ISSUES: Hypochloremic Metabolic Alkalosis. So a low chloride and high bicarbonate level.

* ELECTROLYTE ABNORMALITIES: Hyponatremia, Hypokalemia, Hypochloremia, Hypocalcemia. Sodium and chloride are lost in the sweat. Potassium can be lost in stool. Calcium can be low due to a vitamin D deficiency from fat malabsorption.

- **MNEMONIC**: All of the lytes are low, except for bicarbonate. This make sense given the finding of hypochloremic metabolic acidosis.

* ENDOCRINE SYMPTOMS: Diabetes mellitus, amenorrhea, failure to thrive

- **PEARLS**:
  - **Start VITAMIN E supplementation prior to age 5**
  - If a child presents with rectal prolapse or meconium illeus/plug, screen for cystic fibrosis! For meconium ileus, the boards could mention a "ground glass" on abdominal X-ray.
  - Patients can have cystic fibrosis exacerbations in which they produce quite a bit of mucus, sputum and have a difficult time breathing. Along with respiratory nebulizer treatments, treat

with an **aminoglycoside** (possibly inhaled tobramycin) **and an anti-pseudomonal –cillin** drug, such as piperacillin.

- PNEUMONIAS: **Staphylococcus infections** are more common early in life, followed by pseudomonas. Lifetime colonization is common with these two organisms as well as Haemophilus influenzae. The role of Streptococcus pneumoniae in cystic fibrosis children is debatable, but it is usually **not** a colonizer. If there is mention of a nodular pneumonia in a CF patient, choose Staph. If the boards mention lung abscesses, choose pseudomonas.
- There is a strong association between Trisomy 21 (aka Down Syndrome) and cystic fibrosis.
- **HEMOPTYSIS**: This is not a common occurrence in children. If it happens, think CYSTIC FIBROSIS. Hemoptysis can also be seen in children with congenital heart disease (CHD) and chest trauma.
- **NASAL POLYPS:** 20-70% of cystic fibrosis patients eventually developing nasal polyps. In children, and on the pediatric boards, nasal polyps are very commonly associated with cystic fibrosis.
  - **PEARL**: If you are given the history of an asthmatic patient with nasal polyps, don't choose CF. Look instead for an answer related to a classic triad of **nasal polyps + asthma + aspirin (or NSAID) hypersensitivity**
  - **PEARL**: Nasal polyps are also associated with allergic rhinitis and sinusitis.

## ***STRIDOR***

**PEARL**: Some of these topics will definitely be on your exam, and they are **very quick and easy points**.

### INSPIRATORY STRIDOR

Inspiratory stridor suggests obstruction **above the glottis**.

* LARY**N**GOMALACIA: The soft, floppy, immature cartilage of the larynx collapses and causes airway obstruction. This is the most **common cause of stridor in infants**. Look for a child with onset of stridor around **4-6 weeks** of age. Stridor is more prominent, or noted, when the child is **SUPINE** or **AGITATED**. May have retractions. Laryngoscopy may reveal rolled epiglottis. Usually resolves by 12 months of age.

- **MNEMONIC**: lar**IN**gomalacia = **IN**spiratory stridor presenting in **IN**fancy

* VOCAL CORD PARALYSIS: Look for a child with a high-pitched stridor, **but** weak cry, since **birth**. The stridor is **usually inspiratory** and is less likely to be present at rest. Children seem "uncomfortable when crying.

- **PEARL**: Patient may have a hoarse cry. This **can** present later in life with a **hoarse voice**. Also do not rule this out if the patient has a biphasic stridor (inspiratory and expiratory).
- **MNEMONIC**: If I was a baby and wanted to make a bunch of noise when I'm crying to get my parents attention, but couldn't, I would be **uncomfortable** too!

* PERTUSSIS: Look for the **single** whoop and "paroxysmal" cough.

* CROUP: Look for the **barky cough** and the **steeple sign** on X-ray (see ID).

## EXPIRATORY STRIDOR

Expiratory stridor suggests obstruction in the lower trachea.

* **TRACHEO**MALACIA: Like laryngomalacia, this usually presents around 4-6 weeks of age, and **may also be worse when supine**. The differences are that it's **usually** an **expiratory stridor, or wheeze**, and of the location (trachea). This can sometimes this CAN also have an inspiratory stridor, but expiratory is much more common.

  - **MNEMONICS:** **tracheOUTmalacia** = Stridor when breathing OUT. Also, note that both of the "–malacias" present around 4-6 weeks!

* **VASCULAR RING**: Look for a neonate having **trouble feeding and expiratory stridor since birth.**

  - **PEARL**: **Diagnose** this with a **barium swallow**, not a scope!

## BIPHASIC STRIDOR

**Biphasic stridor means there's inspiratory AND expiratory stridor.** Can be a glottic or subglottic obstruction.

**PEARL**: If you see the word **STENOSIS**, think biphasic stridor since the air struggles to get in and to get out due to this "**fixed obstruction**."

- **SUB**GLOTTIC STENOSIS: This classically presents in children as a **BIphasic stridor**, because it's a **fixed obstruction**.

  - **PEARL**: There is a STRONG association with previous **intubation**!

* **EPIGLOTTITIS**: Not all children will have stridor, if they do, look for it to be **BI**phasic. Note that this is a SUPRAglottic stenosis. Look for the drool and the thumbprint sign (see ID). Also look for immigrant status or missed immunizations (H. flu).

# *CONGENITAL PULMONARY DISEASE*

## CONGENITAL DIAPHRAGMATIC HERNIA

If a congenital diaphragmatic hernia is noted on a prenatal ultrasound, the baby should be **INTUBATED** at birth because it can result in **early respiratory distress due to pulmonary hypertension**. Children can have problems for days to weeks, and this can even be **fatal due persistent pulmonary hypertension**. Listen for decreased breath sounds on the **left side** since the bowel is in the left hemithorax. Chest X-ray will show a small, atelectatic right lung, and the heart may be pushed over causing heart sounds to be heard on the right side of the sternum. These children often have a flat or **scaphoid ABDOMEN**.

## CONGENITAL PULMONARY MALFORMATIONS

Consider one of these congenital pulmonary malformations in any child presenting with a **recurrent unilateral pneumonia**.

* CONGENITAL BRONCHOGENIC CYSTS: Most common cause of lung cysts. Usually found incidentally and patients are typically asymptomatic.

* CONGENITAL CYSTIC ADENOMATOID MALFORMATION (CCAM, aka CAM): This is a rare congenital, non-malignant mass (may have solid or semisolid components on imaging) that can be found in either side of the lungs. Presents with respiratory distress in the neonatal period.

* CONGENITAL **LOBAR** EMPHYSEMA: Most common cause of **neonatal** pulmonary cysts.

* **PULMONARY SEQUESTRATION (aka INTRAPULMONARY SEQUESTRATION, aka EXTRAPULMONARY SEQUESTRATION)**: This is cystic OR solid lung tissue that does not connect to the tracheobronchial tree. It has its own blood supply and can be associated with recurrent pneumonias, a chronic dry cough and wheezing in teenagers. The **recurrent pneumonias are always on the same side.** Surgical lobectomy is curative.

## PERSISTENT PULMONARY HYPERTENSION

Elevated blood pressure (hypertension) of the pulmonary artery causes decreased blood flow to the lungs in cases of persistent pulmonary hypertension. This **can be diagnosed at birth** with imaging showing unusually clear fields. Do an echocardiogram to look for elevated pulmonary artery pressures. If the child is in distress, do a "hyperoxia test" to see if oxygen saturations improve. If they do not, there is a persistent right to left extrapulmonary shunt.

## CHOANAL ATRESIA

In choanal atresia, children present with a history of loud/sonorous breathing SINCE BIRTH. It may sound kind of like stridor. The baby will look and sound **worse when at rest** because s/he is trying to breath through a nose that only has one open and functional airway. The child's respiratory status seems more normal/comfortable when crying and mouth breathing. Signs can include tachypnea and cyanosis.

# ***ASTHMA***

**PEARL**: Asthma is also a high-yield topic for the pediatric boards.

## EXERCISE INDUCED ASTHMA

In exercise induced asthma, children have bronchospasm and wheezing that is **only** associated with exercise. Treat with a **pre-exercise** beta agonist.

**PEARL**: If there are nighttime symptoms, they are referring to poorly controlled asthma.

**PEARL**: The differential diagnosis for a child with "exercise intolerance" include neuromuscular issues, anemia, deconditioning and cardiac disease.

## PEDIATRIC ASTHMA CLASSIFICATION

The shortcuts and mnemonics for pediatric asthma classification below may not cover every possible combination of symptoms, but it comes **extremely close** and should be plenty for the pediatric board exam.

* MILD INTERMITTENT ASTHMA: Daytime symptoms twice or less per week.

* MILD **PERSISTENT** ASTHMA: Three times or more per week.

* MODERATE **PERSISTENT** ASTHMA: Daily symptoms

* SEVERE **PERSISTENT** ASTHMA: Continuous symptoms.

* **PEARL**: There is no “moderate intermittent” asthma. This is important becau because you could easily be given symptom frequency and asked to choose treatment.

* **MNEMONICS:**
  - **DAYS PER WEEK CUTOFFS: 2-3-D-C or 23DC** provides the cutoffs for classifying asthma based on the frequency of **daytime** symptoms per **week**. 2 or less = mild intermittent. 3 or more = mild persistent. Daily = moderate persistent. Continuous = severe persistent.
  - **NIGHTS PER MONTH CUTOFFS: 2-3-W-F or 23WF or 23WtF** provides the cutoffs for classifying asthma based on the frequency of **nighttime** symptoms per **month**. 2 or less = mild intermittent. 3 or more = mild persistent. More than once per Week = moderate persistent (once per week is mild persistent). Continuous = severe persistent.

* **SPIROMETRY** - %FEV1 OF PREDICTED: Use the cutoffs below to classify:
  - MILD ASTHMA: > 80% of predicted.
  - **MODERATE ASTHMA: 60-80% of predicted.**
  - SEVERE ASTHMA: < 60% of predicted.

* **SPIROMETRY** & REVERSIBILITY

REVERSIBILITY refers to an **FEV1 that improves by 12% with bronchodilator** use. If it does, that diagnoses a reversible obstructive process (asthma). If it does not, then it **does not rule out** the possibility of asthma. The test has a **high positive predictive value** only.

* **TREATMENT SHORTCUTS/PEARLS**
  - **MILD INTERMITTENT ASTHMA:** For a patient with **mild intermittent** asthma, the use of **intermittent** inhalers as needed is appropriate.
  - **PERSISTENT ASTHMA: Inhaled corticosteroids (fluticasone, budesonide, etc.) are the mainstay of asthma treatment for anyone with PERSISTENT asthma**. This includes mold, moderate and severe **persistent** asthma. Inhaled steroids help with both inflammation **AND** hyperresponsiveness. Side effects include oral candidiasis, dysphonia (dysphonia), and cough.
  - **LONG-ACTING BETA AGONISTS:** These should be used in anyone with **MODERATE to severe asthma** who has not achieved control with inhaled steroids alone. For the purposes of the pediatric board exam, it's probably safe to assume that anyone with at least moderate asthma should be on both an inhaled steroid and a long-acting beta agonist.
  - **PEARLS:** Males, patients with low socioeconomic status, and patients who have had prior intubation have higher mortality rates. Indicators of poor control include frequent ER visits and use of an inhaler 2 or more times per month. Montelukast should be used in patients of any age if the parents don't want to give **inhaled** steroids. Do not choose to give long-term oral steroids. If a patient has sputum production, that **does not** rule out asthma.

## RHINOVIRUS

Rhinovirus is the most common etiology of a viral asthma exacerbation.

**MNEMONIC**: Imagine a fat RHINO that gets SOB and WHEEZES after just a few steps.

### RESPIRATORY SYNCYTIAL VIRUS (RSV)

Almost half of the kids who get a severe respiratory syncytial virus (RSV) bronchiolitis go on to develop asthma.

### DUST MITES

Dust mite allergies are the most common cause asthma exacerbations.

### BETA BLOCKERS & ASPIRIN

Beta blockers and aspirin can cause a fast and severe asthma exacerbation in some patients.

### ADULT ASTHMA

More than half of pediatric asthma is "outgrown" and does not lead to adult asthma.

**PEARLS:** If a child was less than 3 years old when the asthma started, or has an eosinophilia, or an elevated IgE, or a maternal history of asthma, then there is an increased risk of developing persistent asthma into adulthood.

### ASTHMA DIFFERENTIAL

A cough can also = GERD or sinusitis. Also, "all that wheezes is not asthma." So if a child presents with wheezing that does not change/improve with bronchodilators, consider a different diagnosis, such as a foreign body, vascular ring, vocal cord paralysis or tracheal stenosis.

**PEARL**: Wheezing localized to one side with an ipsilateral decrease in breath sounds should also make you think of a **FOREIGN BODY**. The same goes for a URI that won't resolve.

## *PNEUMONIA*

A pneumonia can be diagnosed based on symptoms and clinical findings alone. If a patient has a fever, chills, cough, crackles in the right lower lobe and a clear chest X-ray, s/he has a right lower lobe pneumonia. Seeing an infiltrate obviously makes the diagnosis easier. When looking for an etiology, obtain a **BLOOD CULTURE** (children are usually not very good about providing a sputum sample). This chapter will focus on recurrent and migrating pneumonias.

## *RECURRENT PNEUMONIA*

The differential is quite wide for recurrent pneumonias. This theme is frequently seen on the pediatric certification and recertification exam. Some of the conditions which are associated with recurrent pneumonia are listed in this section.

### ATAXIA TELANGIECTASIA

Ataxia telangiectasia findings include an elevated alpha-fetoprotein, possible eye findings, telangiectasias and possible ataxia.

## BRUTONS X-LINKED AGAMMAGLOBULINEMIA

For Brutons agammaglobulinemia (X-linked), look for a male child with a history of infections caused by encapsulated organisms, including H. influenzae, Streptococcus pneumoniae and Pseudomonas. Also look for an absence of lymphadenopathy and tiny tonsils.

## SEVERE COMBINED IMMUNODEFICIENCY (SCID)

Severe combined immunodeficiency (SCID) is a T and B cell deficiency. Look for persistent viral infiltrates, a history of otitis media (at 3-6 months of age) and look for a **lympho**penia or a low absolute **lympho**cyte count.

## HYPER-IGM SYNDROME (aka HYPER IGM SYNDROME)

In hyper-IGM syndrome (aka hyper IGM), there are no Ig**G**s because the switch from IgM to IgG doesn't occur (T-cells are not giving B cells the signal). Look for a **lympho**cyt**osis**, **neutro**penia.

## HYPER-IGE SYNDROME (aka HYPER IGE SYNDROME)

In hyper-IGE syndrome (aka hyper IGE syndrome), there is poor neutrophil chemotaxis which results in recurrent **Staphylococcus aureus** abscesses and pneumonias. Pneumonias may be associated with pneumatoceles or lung abscesses. Look for a pneumatocele on a chest X-ray. There may also be a history of delayed shedding of primary dentition.

## COMMON VARIABLE IMMUNE DEFICIENCY (CVID)

In cases of common variable immune deficiency (CVID), B cells do not change into plasma cell so there is a deficiency in **all of the immunoglobulin subtypes**. Look for recurrent upper and lower respiratory tract infections, as well as a history of recurrent viral infections (HSV and VZV infections).

## (DOUBLE TAKE) ASPERGILLUS

In immunocompetent patients, aspergillus may present as Acute Bronchopulmonary **Aspergillosis** (ABPA). Patients seem to have an exacerbation of asthma that WORSENS despite "appropriate" treatment with steroids. Look for increased eosinophilia and lung infiltrates. Can lead to Chronic Eosinophilic Pneumonia as well (peripheral infiltrates on X-ray). In an immunocompromised patient, this can present as pulmonary disease with **nodular infiltrates**. Treat regular Aspergillus with **fluconazole**. Treat invasive Aspergillus with **amphotericin**. Both may require long-term steroids. All Aspergillus patients should be in a negative pressure room (like MTB).

## BRONCHIOLITIS OBLITERANS WITH ORGANIZING PNEUMONIA (BOOP)

Bronchiolitis obliterans with organizing pneumonia (BOOP) is associated with recurrent pneumonias that improve a little with antibiotics. Bronchoscopy will show **grossly purulent material in the airways**, but when the material is cultured it is negative for any growth! An open lung biopsy shows thickened alveolar septa and cell hyperplasia. Usually requires steroids, but tends to eventually resolve for most patients. Etiology is unknown, but it's often found in patients with chronic inflammatory diseases.

## INTRAPULMONARY SEQUESTRATION

Intrapulmonary sequestration is a congenital pulmonary malformation that usually presents in teenagers with recurrent pneumonias on the same side. Surgical lobectomy is curative.

# ***MIGRATING PNEUMONIAS***

## (DOUBLE TAKE) TOXOCARA CANIS

Toxocara canis is a tapeworm that causes **VISCERAL** LARVA MIGRANS. Look for a child presenting with **multisystem** complaints. Usually affects the LUNGS and the GI TRACT, but can also affect the eyes. May present as abdominal pain in a child that has **hepatosplenomegaly** and wheezing on exam. Labs will show a HIGH LEUKOCYTOSIS with EOSINOPHILIA. Imaging will show lung infiltrates. The tapeworm is found in cats, dogs and dirt, so look for a kid that likes to eat dirt! It is often self-limited, but can be treated with albendazole (or mebendazole).

* **PEARL**: The words VISCERAL and MIGRANS in an answer choice should help you remember that this is a disease of the deep organs (viscera), and it migrates to multiple organs. Lung infiltrates may be noted to change, or "migrate," over time as noted on serial chest X-rays.

## (DOUBLE TAKE) ASCARIS LUMBRICOIDES

Look for a history of travel to a tropical climate or immigrant status if ascaris lumbricoides is suspected. Patients will have abdominal discomfort and pain. Can also have pancreatitis as well as INTESTINAL OBSTRUCTION. Can also have RESPIRATORY symptoms including a bronchitis that appears as SHIFTING INFILTRATES on serial chest X-rays and is associated with EOSINOPHILIA. Diagnose by examining stool for worms or barrel-shaped eggs. Treat with MEBENDAZOLE or PYRANTEL PAMOATE.

**NOTE:** Loffler Syndrome is the build up of pulmonary eosinophils in the lungs due to a parasitic infection. This is also called chronic pulmonary pneumonia. As mentioned, look for SHIFTING INFILTRATES on serial chest X-rays.

**IMAGES**: http://1.usa.gov/oQOt1s

**MNEMONIC**: **A SCAREEEE** (bunch of earthworm-like, **BENDY** creatures **CLOGGED UP** a **BARREL**.

- **Key: ascari**s, **e**osinophilia, me**benda**zole, intestinal obstruction & barrel-shaped eggs

**MNEMONIC**: Imagine **A SCAREEEE PYRATE** (**pyr**antel pamo**ate**) playing a **A NEW PIANO** (pneumonia) on his ship. As the waves get rough, the NEW PIANO starts to **SHIFT** all over the deck (shifting infiltrates).

# ***MISCELLANEOUS PULMONARY DEFINITIONS & CONDITIONS***

## VOCAL FREMITUS

Increased vocal fremitus (VF) indicates the presence of pneumonia or a consolidation. It's noted when you are able to feel vibrations at a particular area on a patient's chest. Decreased VF indicates the presence of a pleural effusion or pneumothorax. Breath sounds are said to be "bronchial" over a consolidation, and decreased over an effusion.

**MNEMONIC**: Try to think of it this way, your lungs are like a sponge. If part of it is filled with a little bit of gunk (pneumonia/consolidation), there are still airspaces that allow sound to get in. Sound **cannot** get into a sponge that is in a glass of water, so vocal fremitus is absent. Also sound cannot exist in a vacuum, like a pneumothorax.

## COR PULMONALE

Cor pulmonale is also known as right heart failure and is caused by pulmonary hypertension, which is often due to an obstructive airway process. The right-sided heart failure can only be reversed (at least a little) if the etiology of the obstructive pulmonary process is diagnosed and treated. Signs of cor pulmonale may include lower extremity edema, hepatomegaly, clubbing or a gallop on the cardiovascular exam.

**PEARL**: If a child is noted to have large tonsils or adenoids with a lung or heart issue, think about possible obstructive sleep apnea (OSA) since it can lead to pulmonary hypertension and right-sided heart failure.

## TACHYPNEA

Tachypnea may be the only indication on exam that a patient has a metabolic acidosis.

## HYPERCAPNIA (aka HYPERCAPNEA)

Hypercapnia (aka hypercapnea) refers to an **elevated CO2**, not a fast respiratory rate. Can result in agitation, flushing of the skin and even headaches (cerebral vasodilation).

**PEARL**: Opioids cause a central apnea resulting in hypercapnia (elevated CO2), but since oxygenation is ok, the O2 sats or PO2 may be normal.

## ACUTE LIFE THREATENING EVENT (ALTE)

An acute life threatening event (ALTE) includes apnea, unresponsiveness and cyanosis! **Hospitalize** any patient after having an ALTE. Do a thorough workup for infectious, neurologic, respiratory, metabolic and social issues.

## ALPHA-1-ANTITRYPSIN DEFICIENCY

Alpha-1-antitrypsin deficiency results in lung and liver disease which may not present until adolescence. Emphysema is noted on imaging with **bullae at the bases of the lungs**. Liver disease can result in ascites, cirrhosis, GI tract varices, portal vein hypertension, splenomegaly and coagulopathy. A biopsy of the liver will show **globules**. Labs can show a high GGT, and possibly an elevated alkaline phosphatase.

**PEARL**: Patients may not have any lung symptoms.

**MNEMONIC**: Bullae are at the Bases, close to the other problematic site, the LIVER!

## RESPIRATORY DISTRESS SYNDROME (RDS)

Respiratory distress syndrome (RDS) is caused by a relative surfactant deficiency. Look for a premature baby that did not receive steroids prior to birth. Chest X-ray will show bilateral ground glass findings and air bronchograms. The risk of RDS is significantly HIGHER in an infant of a diabetic mother (IDM) or a baby born by c-section. The risk is LOWER if the child had IUGR, a mother on drugs, prolonged rupture of membranes

(PROM) or an **LS ratio > 2.0.** Intubate the baby if the pH is < 7.2 or PCO2 is > 60. Otherwise, give O2 and shoot for a goal PO2 of 50-70.

**PEARL**: Consider a patent ductus arteriosus (PDA) if the baby is not getting any better after 2-3 days.

## NASAL FOREIGN BODY

Look for a nasal foreign body if a child presents with blood-tinged nasal mucus, unilateral purulence or "generalized body odor."

## FOREIGN BODY ASPIRATION

A child with a foreign body aspiration will usually presents within 24 hours and is associated with decreased or abnormal breath sounds on one side. The object typically goes down the right bronchus. Can also present as a cough that doesn't seem to want to go away in a patient is mobile. Can eventually lead to a **postobstructive pneumonia.**

## VOCAL CORD NODULES

Vocal cord nodules may present with a hoarse voice or cry that is BETTER in the morning and worse with use.

**PEARL**: Vocal cord paralysis would not change in character.

**MNEMONIC**: It's like an overuse disorder. Have you ever had a lecturer that has voice fatigue towards the end of a lecture? Imagine him or her having vocal cord nodules.

## CHRONIC COUGH

Workup for a chronic cough should include a **chest X-ray, a sweat chloride test and a PPD**. If all are negative and the patient is > 6 years of age, the next step should be to get spirometry. Other considerations include pertussis, a foreign body, asthma and a psychogenic cough (especially if it's a honk-like and **disappears at night**).

## PNEUMOTHORAX

Look ofr a pneumothorax in a tall, thin adolescents who uses marijuana and presents with chest pain. Treat with 100% OXYGEN if it is small. If it's large (> 15%), try a needle aspiration and see if that helps to get rid of the air between the lungs and the pleura. If not, place a chest tube.

**PEARL**: Pain DOES NOT dictate management. A small pneumothorax is very benign.

## FLAIL CHEST

A flail chest occurs when ribs break and become detached from the chest wall. It is life-threatening since the flailing broken ribs can lead to a pneumothorax. Place a chest tube, give O2 and possibly consider surgical repair.

## BRONCHIECTASIS

In bronchiectasis, there is destruction of the bronchial tree leading to dilation that is often visible on chest X-rays. It can cause a cough and even hemoptysis. Cough is often worse with changes in position. It's associated with cystic fibrosis, alpha-1 antitrypsin deficiency and emphysema.

## ***HIGH YIELD CHEST X-RAY FINDINGS & PEARLS***

### PULMONARY VASCULAR CONGESTION

4 [ TGA, Truncus, TAPVR, LV outflow tract obstruction

Pulmonary vascular congestion can be noted on chest x-rays of children with transposition of the great arteries/vessels (TOGA), truncus arteriosus, total anomalous pulmonary venous return (TAPVR) and left ventricular outflow tract obstruction (LVOT).

### PATCHY AREAS OF DIFFUSE ATELECTASIS

Meconium asp

If patchy areas of diffuse atelectasis are noted in a newborn with focal air trapping and increased lung volumes, pick **meconium aspiration**. Meconium is rarely passed in utero before 34 wks. If the baby is vigorous, there is **no need to intubate**. If floppy, DON'T stimulate… **suction below cords**, and **then** intubate! The most common complication is persistent pulmonary hypertension.

### FLUID IN HORIZONTAL FISSURE

TTN

If fluid in the horizontal fissure is noted in a newborn with tachypnea, pick **transient tachypnea of newborn**. The chest X-ray should also have appropriate expansion/inflation. If it shows under-inflation, then either the X-ray needs to be repeated, or this is not the diagnosis.

### UNDERINFLATED CHEST X-RAY

RDS or Bands:Neutrophils >0.2 GBS pneumonia

A chest X-ray showing underinflation, ground glass findings and air bronchograms is either **respiratory distress syndrome** or **congenital Group B Streptococcus (GBS) pneumonia**.

**PEARL**: Do **NOT** look at the WBC to help you differentiate between RDS and GBS pneumonia. It is very unreliable in the newborn. Look instead at the **ratio of bands to neutrophils.** If it's **> 0.2**, that is more suggestive of GBS.

### DIFFUSE OPACITIES WITH CYSTIC AREAS

BPD

If diffuse opacities with cystic areas are noted in a premie or infant, consider bronchopulmonary dysplasia (BPD) from prolonged oxygenation or barotrauma. Treat with diuretics and watch out for electrolyte abnormalities (especially **hypo**calcemia). Some of these kids can end up having poor neurologic development.

# PSYCHIATRY & SOME SOCIAL ISSUES

## ATTENTION DEFICIT DISORDER (aka ADD, ADHD, and ATTENTION DEFICIT HYPERACTIVE DISORDER)

Impulsive + Inattentive + Disorganized thoughts in a child = Attention deficit disorder (aka ADD and ADHD). Must be **DIAGNOSED AFTER 6 years of age**, but **symptoms must be PRESENT prior to 7 years of age**. Boys > girls. 50% persist into adulthood, especially the inattentive/disorganized part. Sugar doesn't exacerbate it. Best treatment is **pharmacologic** + "other" interventions (psych referral).

**PEARLS:** If a child was started on a stimulant medication and now pays more attention, that does not diagnose ADD. Before making the diagnosis, make sure lead poisoning, iron deficiency, vision problems, hearing problems, thyroid disorders, sedating medications (antihistamines, seizure medications), CNS infection and CNS trauma **have bene ruled out**. Other conditions to consider include hay fever, Klinefelter Syndrome, absence seizures, maternal drug/alcohol abuse, patient drug/alcohol abuse and depression.

## LEARNING DISABILITIES

Learning disabilities are usually discovered in older children **when school gets harder**. Patients with learning disabilities have problems in **specific areas** (math, reading, writing, speaking, listening or reasoning). Learning disabilities can be overcome with training, but they **are not outgrown.**

**PEARLS:** A difference in the **Verbal IQ** and the **Performance IQ** (spatial/visual skills) means the child is **at risk** for a learning disability. **Letter reversal** is normal until **7 years of age**. After that, it is a sign of dyslexia. **Stuttering** is ok until ~ 4 years of age. If the stutter persists when at school age, do a workup.

**MNEMONIC**: 100% of language should be intelligible by 4 years of age. So a stutter should be gone by then too!

## SCHOOL PHOBIA

School phobias tends to occur in single-parent households. Treat by having the parent escort the child to school, have the child spend more time with peers, and decrease the amount of time being spend with the parent.

## DEPRESSION

Depression is a very nebulous diagnosis in kids and on the exam. For the boards, consider choosing this if a child's behavior starts to change and interfere with routine life activities (like school). Pick this over drug abuse unless there is something specific to go by.

## DEATH RESPONSE IN CHILDREN GRIEF 6

* CLASSIC STAGES OF THE DEATH RESPONSE: Shock/paralysis → Denial/avoidance → Anger/Resentment/Outpouring of emotions → Bargaining → Depression/Realization → Testing/looking for realistic solutions → Acceptance. Having said that, in children the response often **depends on the age** groups they are in.

Shock
Denial
Anger
Bargain
Depression
Acceptance

- PRESCHOOL: Regression +/- tantrums.
- SCHOOL AGE: Trouble at school, trouble sleeping, possible physical complaints/somatization (headaches, abdominal pain).
- TEENAGERS: Acting out.

* **PEARL**: If you get a question about death and what to do as a pediatrician, choosing reassurance will be wrong! Choose something to do that is benign (not medications).

## DIVORCE

Children respond differently to divorce depending on how old they are at the time (like death).

* 5 YEARS OLD OR LESS: Regresses to the most recent milestone.
* 6-9 YEARS OLD: **Crying**, guilt, fantasies about reunion, fear of rejection.
* 9-12 YEARS OLD: Mourns the loss of a safe family structure. May be angry at one, or both, parents.
* TEENS: Act out. Depression. May isolate themselves and act indifferent. May have suicidal ideation.

## PARENTAL ADJUSTMENT TO A CHILD WITH MALFORMATIONS

Parents need to adjust to a child with malformations. They may initially not want to hold, see or feed the baby. Pediatricians are responsible for encouraging bonding. The classic stages are as follows: Shock/Fear → Denial → Sadness/Anger → Acceptance.

## CHRONICALLY ILL FAMILY MEMBER

Children may have trouble coping with the idea of having a chronically ill member of the family. Protective factors include having financial security in the home, being male (girls try to act like mothers), younger age and having a large, complete (no divorce) and joyful home.

## CONVERSION DISORDER

Conversion disorder is an **unintentional NEUROLOGIC** complaint often after a stressful or traumatic event (unilateral extremity paralysis, loss of vision, etc). Usually resolves within 1-2 days.

**PEARLS:** This is **not** a fake complaint. It's usually **ACUTE**.

## SOMATIZATION

Somatization is an **unintentional** complaint (not necessarily neurologic). In children this usually presents in the form of headaches and abdominal pain. This is sometimes considered to be a child's way of "taking control" of uncontrollable and stressful situations.

**PEARL**: This can be a chronic complaint or problem. Again, it's not fake.

## PSYCHOSOMATIC

Psychosomatic exacerbation is a **true exacerbation** of a chronic medical illness due to stress. The symptoms subside fairly quickly, even if the stress does not. An example of this would be a **true exacerbation** of eczema due to a stressful life event.

## BREATH HOLDING SPELLS

Breath holding spells are most common between **6 to 18** months of age. Children have **apneic episodes in response to pain, fear, anger, frustration or injury**. Children can become **unconscious** or even have a **seizure** with a postictal period. EEG is always normal. No specific treatment is required except making sure that children don't hurt themselves during a seizure.

**PEARL**: Patients **do not have to** pass out for it to be called a breath-holding spell. Children stop having them by about 5 years of age (usually much earlier).

* **CYANOTIC BREATH-HOLDING SPELLS**: These are the most common. Anger or frustration seem to be the primary triggers, though the trigger can vary. Usually children will cry so hard that at the end of expiration they **turn blue**, lose muscle tone and then become unconscious. They usually wake up 1-2 minutes later.

* **PALLID BREATH HOLDING SPELLS**: More commonly due to pain, fear or an injury. Children become **pale** and then lose consciousness.

- **MNEMONIC**: "You look like you saw a ghost!" That's a phrase commonly used to describe people when they suddenly become afraid and become pale. Think of pallid breath-holding spells the same way. Uncontrollable factors cause spell, whereas in cyanotic breath-holding spells it seems to have more to do with difficulty controlling emotions.

## NIGHT TERRORS

Night terrors usually occur during the **first 1/3rd of a child's sleep,** and they typically occur at about 90 minutes into a child's sleep. It occurs in **boys** more than girls and there is sometimes a **family history** of it. Children typically **do not recall** the event and may even be a little **confused** if they are woken during the episode. Children can actually walk during these episodes, and are thus at increased risk of injury. Occurs in stage 4 of the sleep cycle.

## NIGHTMARES

Nightmares occur during the **last 1/3rd of a child's sleep** and occur during **REM sleep**. Children are easily woken and **will recall** the episode. Children **do not walk** during nightmares and remain immobilized in bed, so there is no risk of injury.

## CHILD DISCIPLINE

The use of appropriate child discipline is considered EXTREMELY important. It provides children with emotional stability and structure.

**PEARL**: The AAP and the ABP frown on corporal punishment. Don't pick it.

## THUMB SUCKING

Most kids outgrow thumb sucking. Recommendation should be to provide positive reinforcement at times when the child is not sucking his/her thumb. Starting at age 5, a more aggressive approach is ok.

## CHILD ABUSE

**Risk factors** that increase the likelihood of child abuse include unwanted/unexpected pregnancies, prematurity, a greater number of young children in the home, and having children born with malformations/anomalies. Factors that **do not increase the risk** of abuse include gender of the child, overall size of the family and parents' education level.

**<u>PEARLS</u>: Child neglect is said to be the most common form of child abuse**. If a mother is noted to go against usual recommendations because it's just "easier," work on the psychosocial issues rather than the specific act. For example, not meeting a child's emotional or cleanliness needs.

* **ABUSE RELATED FRACTURES**: Fracture of the scapula, sternum, spinal processes, multiple ribs, skull and any spiral fracture.

- **<u>IMAGE</u>**: http://bit.ly/oaVFZi - Healing rib fractures

* **BUCKET HANDLE FRACTURE**: This fracture and **CORNER FRACTURES** are **highly specific for abuse**. They are fractures of the metaphysis caused by sudden pulling which causes avulsions.

- **<u>IMAGE</u>**: http://bit.ly/nOtXFn - Bucket handle fracture at the red line in the drawing.
- **<u>IMAGE</u>**: http://bit.ly/q11blX - **Corner fracture**
- **<u>NAME ALERT/PEARL:</u> Buckle** fractures are NOT associated with child abuse. Other fractures/conditions that more likely due to an **accident** include linear skull fractures, supracondylar fractures of the elbow and clavicle fractures.

## IMPACT OF MEDIA ON CHILDREN

Watching too much TV increases aggressive behavior and has a true influence on children's diets due to advertising. TV also seems to obscure reality, as well as increase a sedentary lifestyle. Many children watch > 25 hours of TV per week. The official recommendation is to allow no more than 2 hours of TV per day, and preferably with adult supervision.

# <u>*MISCELLANEOUS TID BITS & PEARLS*</u>

## PREVENTATIVE MEDICINE TERMINOLOGY

Preventative medicine includes primary, secondary and tertiary prevention. **Primary prevention** refers to preventing a disease from every happening (vaccinations). **Secondary prevention** involves catching an new disease **early** so that it can be treated (screen tools, pap smears). **Tertiary prevention** aim to reduce the negative impact of an active disease in order to get a patient back to their previous state, or to reduce the complications of it (like giving insulin to a diabetic).

## MOUTH GUARDS

For a child with braces, a mouth guard should be worn during play of **all sports.** For everyone else, one should be worn for collision sports. Shot put and discus somehow qualifies for one too.

## CONTACT SPORTS PARTICIPATION

Contact sports participation should be avoided in children with only one organ left of what should be a pair of organs (testicles, kidneys, ovaries, eyes), and children with hepatosplenomegaly. Also, the same goes for children who have had repeated concussions.

## BICYCLE SAFETY

Bicycle safety recommendations are simple, but the vast majority of children do not follow proper precautions. All children riding a bicycle should wear a **helmet**, and they should have **front and rear reflectors** on their bikes. Most bicycle related injuries are preventable. Head injuries are responsible for almost all bicycle related deaths.

## FIRST DENTAL EXAM

The first dental exam should sought at 3 years of age

## HOT WATER HEATER

Temperature of a hot water heater should be set to **no more than 120 degrees**.

**MNEMONIC**: Imagine boiling a dozen/12 EGGS on top of a HOT WATER HEATER. 12 should remind you of 120.

## BOAT SAFETY

For proper boating safety and death prevention, all children should wear a life jacket at all times on all boats.

## TONGUE TIED (aka TONGUE TIE)

Being "tongue tied" (aka tongue tie) means that a child is **unable to** get the tongue out past the incisors. This is due to a short frenulum. If it results in a problem with feeding, surgery is required.

**PEARL**: An inability to say "th" does not mean a patient is tongue tied, and it warrants no interventions. Reassure the family that it will resolve when the child goes to school since that's when the ability to say "th" finally comes.

## ENURESIS AND ENCOPRESIS

Enuresis and encopresis occur when children involuntarily wet or soil themselves. Most kids have bowel/bladder "control" by 3 years of age. Parents should provide positive feedback when a child shares his/her need to urinate. This helps with faster toilet training. **The term "enuresis" should not be used as a diagnosis until a child is at least 3 years of age**.

* **PRIMARY ENURESIS**: Means a child never had a time period of continence and continues to wet his/her bed. A parental history of primary enuresis increases the likelihood of it in a child significantly. Use bed alarms for long-term use, and consider using intranasal desmopressin (DDAVP) for children going on a sleep over.

* **SECONDARY ENURESIS**: Means a child begins bedwetting again after a dry period in which s/he was potty trained and continent of urine.

* **ENCOPRESIS**: This is when a child voluntarily or involuntarily soils himself/herself after having previously been potty trained. This can be related to stool withholding. Treat by putting the child back into diapers for a while and providing positive feedback to him/her. May use laxatives temporarily for older children.

# PEDIATRIC LAB VALUES

**NOTES**: This section on pediatric lab values has some GREAT TIPS & TRICKS to getting some easy points from the American Board of Pediatrics' board exams. This is not 100% accurate. It has been simplified and many of the values have been rounded up or down in order to make the memorization easier. For example, instead of giving a range of 96 – 111 for newborns' serum chloride, and then 102 – 112 for older children, an overall range of 100 – 110 has been given.

## COMPLETE BLOOD COUNT (CBC)

* **LEUKO**CYTES (aka WBC): For newborns, the WBC can as high as 30,000 on DOL 1.

* **LYMPHO**CYTES: Absolute **lympho**cyte counts (ALC) are much higher in young children when compared to teens and adults. The ALC is usually HIGHER than ANC until about 8 years of age. ALC should NOT be < 40 in preadolescents. If it is, consider the patient to have **lympho**penia.

* HEMOGLOBIN
  - NEWBORN: Anemia in a newborn is indicated by any hemoglobin value **LESS THAN 13**.
  - NADIR: Occurs at approximately 6 weeks of age. The value should be about 9.

* PLATELETS
  - NEWBORN: Consider thrombocytopenia in a newborn to be any platelet count LESS THAN 150 for males and 200 females. Look for clues in the history that point towards maternal ITP as etiology (this can last weeks).
  - OLDER KIDS: For children 2 months and older, set the lower limit of normal at approximately 200-250 in your mind.

## COAGULATION STUDIES

* PARTIAL THROMBOPLASTIN TIME (PTT): Consider any value > 45 in a coagulation study to be prolonged.

* PROTHROMBIN TIME (PT): Consider any value > 15 to be prolonged.
  - NEWBORNS: May have a prolonged PT up to 9 months of age.

## NORMAL PEDIATRIC ELECTROLYTE VALUES

* SODIUM: 135 - 145

* URINE SODIUM: A value of < 20 should indicate a low urinary sodium. The kidneys are either holding on to sodium tightly, or they are not holding on to water tight enough. If you get confused, it's probably in your best interest to assume that a low urine sodium reflects low/decreased sodium excretion (i.e., a low FeNA).

* POTASSIUM: 3.5 – 5

* CHLORIDE: 100 - 110

* URINE CHLORIDE: Use 15 as your cutoff. > 15 is high, < 15 is low.

* BICARBONATE LEVEL:

- NEWBORNS: May have a normal bicarbonate level **as low as 17**. Sources vary on this, but use 22 as the upper limit of normal for a full-term newborn. Use 20 as the upper limit of normal for a premature newborn.
- CHILDREN: 20 – 24. For ABG's, use 24 as the standard.

* BLOOD UREA NITROGEN (BUN): Use 18 at the upper limit of normal

* CREATININE: From birth to ~ DOL 10, the level reflects mom's creatinine so up to 1.3 is probably normal. After that, use the following as an approximate guide to the upper limit of normal:
  - 1 months = ~0.1
  - 1 yo = ~ 0.3
  - 5 yo = ~ 0.5
  - 9 yo = ~ 0.6
  - 13 yo = ~0.8

* CALCIUM
  - TOTAL CALCIUM: 8-11
  - IONIZED CALCIUM: Upper limit = 4.5

* MAGNESIUM: 1.2 - 2

* PHOSPHORUS: 3-6

* ALBUMIN: 3.5 – 5

* TOTAL PROTEIN: ~ 5-8

* AST: < 40
  - LOW YIELD FACTS: Newborns may have a normal AST value of up to 120. Children aged 10 days to 2 years can have a normal value up to 80. After 2 years of age, though, it should definitely be < 40.

* ALT: < 45

## ALKALINE PHOSPHATASE

An alkaline phosphatase reading of 450 is definitely high for any age group. Infants and toddlers can have a value of ~400 that may still be considered normal.

## GAMMA-GLUTAMYL TRANSPEPTIDASE (GGT)

Gamma-Glutamyl Transpeptidase (GGT) < 50

## DIRECT BILIRUBIN

Direct Bilirubin > 0.6 is abnormal, **even in a newborn**. The newborn jaundice issues are mostly indirect hyperbilirubinemia.

# PEDIATRIC VITAL SIGNS

**NOTES**: This section on pediatric vital signs also has some GREAT TIPS & TRICKS to getting some easy points from the American Board of Pediatrics' board exams. Again, the numbers have been rounded in order to make them easier to memorize.

## PEDIATRIC RESPIRATORY RATES

We're generally taught that a normal respiratory rate (RR) in a newborn may be up to 60 breaths per minute. What about a 3 year old? I suggest using the mnemonic below to guide you.

**MNEMONIC**: Rates of 60, 50, 40, 30, 20 should be considered high (tachypnea) in children at ages 1, 2, 3, 4 and 5. Keep in mind though, that many consider a RR of 40 to be the upper limit of normal for infants aged 2 – 12 months.

## PEDIATRIC HEART RATE OR PULSE

* NEONATES: Use a heart rate / pulse of 100 as your cutoff for sinus bradycardia, and 160 as your cutoff for sinus tachycardia in newborns.

* OLDER CHILDREN: Use 50 as your cutoff for sinus bradycardia, and 160 as your cutoff for sinus tachycardia.

## PEDIATRIC BLOOD PRESSURE

The normal pediatric blood pressure range for a child is based on his or her age percentile for height. Keep the following values in mind for children who are at the 50th percentile for height:

* For a 1 YEAR OLD, a systolic BP of > 100 is at > 90$^{th}$ percentile
* For a 5 YEAR OLD, a systolic BP of > 110 is at > 90$^{th}$ percentile
* For a 10 YEAR OLD, a systolic BP of > 115 is at > 90$^{th}$ percentile
* LINK FOR BLOOD PRESSURE NORMS BY AGE & HEIGHT PERCENTILE:

http://www.nhlbi.nih.gov/guidelines/hypertension/child_tbl.pdf